Ernst Mach's Vienna 1895–1930

BOSTON STUDIES IN THE PHILOSOPHY OF SCIENCE

VOLUME 218

ERNST MACH'S VIENNA 1895–1930

Or Phenomenalism as Philosophy of Science

Edited by

J. BLACKMORE

R. ITAGAKI

and

S. TANAKA

History of Science Society

KLUWER ACADEMIC PUBLISHERS

DORDRECHT / BOSTON / LONDON

A C.I.P. Catalogue record for this book is available from the Library of Congress.

ISBN 0-7923-7122-4

Published by Kluwer Academic Publishers,
P.O. Box 17, 3300 AA Dordrecht, The Netherlands.

Sold and distributed in North, Central and South America
by Kluwer Academic Publishers,
101 Philip Drive, Norwell, MA 02061, U.S.A.

In all other countries, sold and distributed
by Kluwer Academic Publishers,
P.O. Box 322, 3300 AH Dordrecht, The Netherlands.

Printed on acid-free paper

Printed in the Netherlands.

THIS BOOK IS DEDICATED TO THE MEMORY
OF ARTHUR ONCKEN LOVEJOY 1873-1962,
AMERICA'S GREATEST HISTORIAN-EPISTEMOLOGIST.

Ernst Mach (Courtesy of the Ernst Mach Institute)

TABLE OF CONTENTS

LETTERS IN ENGLISH TRANSLATION

PUBLISHED SOURCES IN GERMAN	LOCATION OF LETTERS
(1) J. Thiele. *Wissenschaftliche Kommunikation.* Kastellaun: A. Henn Verlag, 1978, pp. 153-156	Deutsches Museum, Munich. Husserl Archiv, Louvain/Leuven.
(2) R.G. Ardelt, *Friedrich Adler.* Wien: Bundes-Verlag, 1984, pp. 163-164.	Verein für Geschichte der Arbeiterbewegung, Vienna.
(3) J. Blackmore & K. Hentschel. *Ernst Mach als Aussenseiter.* Wien: Braumüller, 1985, pp. 57-79.	Verein für Geschichte der Arbeiterbewegung, Vienna
(4) J. Blackmore & K. Hentschel. *Ernst Mach als Aussenseiter.* Wien: Braumüller, 1985, pp. 82-83.	Deutsches Museum, Munich.
(5) J. Blackmore & K. Hentschel. *Ernst Mach als Aussenseiter.* Wien: Braumüller, 1985, pp. 118-160.	Wilhelm Jerusalem Archive, Jerusalem, Israel. Deutsches Museum, Munich.
(6) J. Blackmore & K. Hentschel. *Ernst Mach als Aussenseiter.* Wien: Braumüller, 1985, pp. 150-152.	Deutsches Museum, Munich.

(7) Location: Forschungsstelle und Dokumentationszentrum für Österreichische Philosophie, Graz.

(8) N. Bohr. *Collected Works.* Amsterdam: North-Holland, 1976, p. 490. W. Moore. *Schrödinger - Life and Thought.* Cambridge: Cambridge University Press, 1989, pp. 225 & 226. E. Schrödinger. *The Interpretation of Quantum Mechanics: Dublin Seminars 1949-1955 and other Unpublished Essays.* edited by M. Bitbol. Woodbridge: Ox Bow Press, 1995, pp. 121-122.

(9) F. Stadler (ed.). *Scientific Philosophy: Origins and Developments.* Letter translated by W. Heald & G. Zoeller. Dordrecht: Kluwer, 1993, pp.194-208.

Chapter One

INTRODUCTION

Section Guide
1. Prolegomena
2. Biographical Sketch
3. Epistemology
4. Textbook Ontology

1. PROLEGOMENA

While both philosophers and historians almost always love truth and the search for truth, and both often carry out extensive research, there can be noticeable differences when historians write about the history of philosophy and when philosophers write about it. Philosophers often look at the past with categories and interests taken from the present or at the least from the recent past, but many historians, especially those who love research for its own sake, will try to look at the past from a perspective either from that period or from even earlier. Both camps look for roots, but view them with different lenses and presuppositions.

This prolegomena has been added to prepare some philosophers for what will hopefully only be the mildest of shocks, for seeing the history of philosophy in a way which does not treat what is recent or latest as best, but which loves the context of ideas for its own sake, a context which can be very foreign to contemporary likes and dislikes. To be sure, we historians can deceive ourselves as easily as philosophers, but we tend to do so about different things.

During the longest period of human history it seems probable that most human beings identified the physical world in direct or presentist terms. The sun *seems* to go around the earth, so it must actually do so. Appearances "out there" *seem* to be the external physical world so they must actually be that world. And the smallest things we *seem* to notice must be the smallest things in fact.

It was Galileo Galilei, however, who argued in the strongest and most persuasive way that direct description of the sensory appearances could be very misleading and was often misunderstood. The sun did *not* go around the earth, but the earth and the other planets went around the sun (as Aristarchus in Hellenistic times could have argued given the size and presumably greater gravitational attraction of the sun). The appearances were *not* the external physical world but imperfectly represented it by means of sensory perception (and the resulting conscious images of external objects). And the smallest noticeable sensory appearance was *neither* matter *nor* the smallest part of matter. (All appearances were mental, and the smallest parts of matter were too small ever to have their size veridically represented by appearances.)

Above all, Galileo and most of his major successors insisted that the primary goal of physical science was not to "save the appearances" but "to understand the real physical world" including material powers and forces, which on the basis of heliocentrism, representism, and later atomism were believed to exist beyond all appearances. Indirect or

J. Blackmore, R. Itagaki, and S. Tanaka (eds.). Ernst Mach's Vienna 1895-1930. 1-27.

representist physical science replaced direct or presentist physical science, though careful study of the appearances could still have great empirical value as evidence.

The great advantage of this indirect world view is that it allows much more practical causal explanation, especially if to exist *means* to have causal capacity. Most human beings want to use wise means to help further wise ends, and to be effective the means chosen must be concrete and appropriate. Matter consisting of mass and energy is concrete, while mere appearances seem mental, subjective, and virtually impotent. Sir Isaac Newton's laws of force have long been part of common sense as they still are today.

But there have always been minority movements since the Scientific Revolution which have refused to accept Galileo's idea about the primary goal of science. This opposition view has been especially prominent among philosophers and intellectuals living in German-speaking lands. The primary reasons for this are clear. Martin Luther opposed heliocentrism, great Goethe opposed indirect realism about nature, and "the Germanies" as they were often called before their unification, which long had more universities than other countries, preserved Scholastic and pre-Galilean ways of thinking longer than in most other countries in Central and Western Europe. The Scientific Revolution largely occurred during what was called "The Thirty years War " (1618-1648) which destroyed much of Germany and slowed its scientific development. In short, the older direct or presentist position about nature seemed to last longer in Germany than in France, England, Holland, and Locke-influenced America. One sign of this has been that while sensory appearances are commonly viewed as subjective outside of Germany, though useful as private evidence, inside German-speaking countries they are often admired as *anschaulich* as if intuitive, good, and a main goal of scientific procedure is to "absorb their immediacy".

The leading 18th century philosophers who opposed representism were Berkeley, Hume, and Kant, but while the first two clever thinkers had little influence on English science, Kant would exercize a tremendous influence on much German thinking about science. Kant alleged that identifying the external physical world with what was beyond sensory and conscious limits was "metaphysical", subject to antinomies, and inconsistent with science, thereby rejecting the views of most physical scientists who had as their goal understanding the real external physical world. This rejection was of course a minority view among scientists even in Germany, but it had significant influence on most philosophers, even those thinkers who called themselves "philosophers of science".

Thus, it came to pass during the late nineteenth century that while most scientists continued to support heliocentrism, representism, and atomism and the accompanying indirect and representist assumptions about the external physical world that several well-known philosophers of science did not. Nor are we just thinking of Germany and Switzerland where Richard Avenarius carried a phenomenalist torch. John Stuart Mill in Eingland, Pierre Duhem in France, and Ernst Mach in the Habsburg Empire all advocated a return to "save the appearances" science as if science should restrict its concern to appearances and ways to relate them. And with time these philosophers of science would gradually influence more and more scientists, at least for a while.

To be sure, most physical scientists would continue to respect the great discoveries made by the Scientific Revolution, namely, that the planets go around the sun, that the appearances merely represent the external physical world, and that much that is too small to be represented by our senses is real, especially atoms and molecules, but philosophical opponents would attempt to re-define what acceptance of these theories meant as if one could "accept" great discoveries without believing that what was discovered was real or claims about them true. One side thinks such "acceptance" is honest and the other does not. Indeed, if the goal of philosophy has become to use reason to defend the irrational or to accept arguments on other grounds than probable truth, then perhaps it is no more than apologetics and may well deserve the scorn which many practical people direct at it.

This book is a history of an important minority movement which rejected the representist views about matter which Galileo, Descartes, Boyle, Locke, Newton, Darwin, and most scientists have seemed to accept since the 17th century. In particular, it is a rather detailed history of the spread of Ernst Mach's phenomenalism and his version of "save the appearances" physical science. Mach's easily understandable prose attracted many readers from about 1895 to 1910, but was strongly opposed first by other presentist movements in Austria and Germany such as Husserl's phenomenology, and second, after first sympathizing with it, it was eventually opposed by the two main founders of "modern physics", Max Planck and Albert Einstein. It was even criticized by Vladimir Lenin, a Bolshevik leader who became dictator of Russia, who would use that opposition as part of a litmus-test to help determine who was a "true communist".

But many aspects of "Machism" continued to be influential even in the "Red Vienna" believed to exist after World War I and as is well-known were largely incorporated during the 1920's into what became called "The Vienna Circle" and later "Logical Positivism". Part of this resuscitation was made possible by followers of Mach such as Philipp Frank, Otto Neurath, Hans Hahn, Rudolf Carnap, and others who later became members of The Circle and Logical Empiricism, thereby adding a type of continuity to a minority movement which has probably had more overt methodological influence in recent times than any other organized school, at least among philosophers of science and to some extent even among scientists.

Let us agree with Machists and Logical Positivists that truth is not a popularity contest and that majority votes about philosophy should not be decisive. Majority opinion does not and cannot *determine* truth, though it may constitute a type of evidence, reflect weight of evidence, and may symbolize what is true. Nor can this editor accept the notion that philosophy in general and epistemology and ontology in particular are not subject to proof or refutation by scientific evidence. The discoveries by Galileo and later scientists about heliocentrism, perceptual representism, and atomism *do* constitute evidence . . .

There are many opinions about Mach and his supporters, and all deserve a hearing, hence we have gathered together and presented an anthology with many different perspectives and approaches. It is not just a monograph. Nor should we overlook the many valid contributions which Ernst Mach made to psychology and sense-physiology from Mach Bands and his discovery of Gestalt phenomena to his explanation of the equilibrium function of the inner ear. Indeed, even some of his contributions to methodology of physics have value such as his study of analogy and similarity and how they have led to many important discoveries (See the chapter by Michael Stöltzner on Otto Neurath.)

If we seem obsessed by epistemological and ontological classification it is because our primary concern in this book is with Mach the philosopher and his influence, and we are determined to understand that philosophical influence in as fair, reliable, informative, and useful a way as possible, but admittedly there were other arrows in his quiver besides phenomenalism and a philosophy of science based largely on it, even if none of them were quite as controversial or historically important as his epistemological effort to roll back important assumptions underlying the Scientific Revolution (though in fairness he thought he was advancing science rather than retarding it by advocating a return to "save the appearances" science).

2. BIOGRAPHICAL SKETCH

Ernst Mach (1838-1916) was born in what was called the village of Chirlitz/Chrlice in the Habsburg crownland of Moravia which was close to Brünn/Brno, the capital city of the province. Both his parents were German-speaking in an area where Czech was increasingly spoken and which is now part of the Czech Republic. While Mach in later years would strongly regret the spread of nationalism within Central Europe and would oppose it as an irrational factor, nevertheless until his departure from Prague for Vienna in 1895 he was himself strongly associated with the German-speaking minority in that troubled city even to the extent of not openly admitting that he had been taught and presumably learned the Czech language in school. Nor in his autobiographical documents or apparently elsewhere did he admit that many of his ancestors had decidedly Czech names and presumably also spoke that language. (The name "Mach" was common to many German *and* Czech-speaking families.) But a *Stammbaum* written by his sister Marie and an article by Josef Sajner,[1] which has been translated into English by us, have made the Czech connection somewhat more clear.

Mach was largely Bohemian and Moravian in ancestry which like many other inhabitants of those regions meant that his background was a mixed one with both German and Czech-speaking forbears.[2] This means among other things that both Austria and the Czech Republic can and have claimed him as their own. Furthermore, since Ernst Mach lived in Vaterstetten near Munich during the last three years of his life, there is a nearby *Gymnasium* named after him, and much of his *Nachlass* has recently been transferred from Freiburg im Breisgau to the Deutsches Museum in Munich, it is also true that Bavaria and Germany could claim him. Furthermore, he was buried in a cemetary near Munich.

But most of Mach's childhood, that is, from 1840 until 1852 was spent in Untersiebenbrunn, which was a small, German-speaking village about half-way between Vienna and Bratislava, the current capital of Slovakia. His father was a tutor during the period, first of Baron Bretton's two sons and later until the Revolution of 1848 the children of Graf Stadion in Vienna, a major political figure in Austria (Minister of the Interior). But after the defeat of the liberal forces in Vienna he began teaching his own children more intensively while devoting additional time to his farm. Johann Mach was highly educated himself and taught Ernst Mach Greek, Latin, much history, and seems to have encouraged his son to instruct himself in basic math, physics, and music.

Ernst Mach often mentioned that the decisive philosophical influence on his life came at the age of fifteen[3] when he read a book in his father's library by Immanuel Kant, his most polemical book, namely, *Prolegomena zu einer jeden künftigen Metaphysik, die als Wissenschaft wird auftreten können* (1783) (Prolegomena to every Future Metaphysics which would Aspire to be a Science). Ernst Mach seems to have learned and subsequently taken for granted that science and metaphysics were incompatible, metaphysics was bad, and that metaphysics attempted to understand what allegedly existed beyond sensory or conscious bounds. This naturally led to the view that science was concerned with what could be immediately sensed or noticed and not with what Galileo, Descartes, and Newton thought existed beyond conscious bounds such as what they regarded as the real physical world.

But Mach did not know about the epistemological change during the Scientific Revolution from "save the appearances" to "understand reality science", and years later when he himself became a historian of science he seemed to view Galileo Galilei as an "empiricist" and Newton as a "lapsed empiricist", that is, he failed to grasp the indirect and representist character of their epistemology, nor did Kant or subsequent German and Austrian sources enlighten him about that.

Whereas most "realists" would try to strengthen Kant's "thing-in-itself" by adding spatial and temporal parameters to what they regarded as physical, most "idealists" (and many Germans) would do the opposite, that is, eliminate the *Ding an sich* as an unnecessary metaphysical appurtenance. Mach would do this when he was 18, which left him with epistemological and ontological phenomenalism.

In 1852 Mach returned to Moravia in order to finish a last three years of Gymnasium in the city of Kremsier, but his dislike of grammar and political history and problems with his physics teacher seem to have kept him from receiving high marks. Nevertheless, he later thought that he had been very fortunate to have a teacher in biology who taught him a theory of evolution before Darwin's famous book came out in 1859, but the version was based on Lamarck's ideas, and in later life Ernst Mach would still argue that Lamarck's approach including his doctrine of the inheritance of acquired characteristics was more profound than that of Darwin, though like many people of his time he did not always carefully distinguish between biological evolution by natural selection and by inheriting acquired characteristics. Spencer's Lamarkism was also widespread.

Mach entered the University of Vienna in 1855 and graduated four years later. His interests were very largely in experimental physics and in sense physiology. He became a *Privatdozent* in physics but his own experimental work whether in physics, physiology, or psychology was focused on sense perception. His private reading in philosophy seems to have been largely in the works of George Berkeley and Gustav Fechner. He adopted much from Fechner, whose book titled *Elemente der Psychophysik* had come out in 1860. Subsequently, Mach also called sensations "elements" while believing that appearances apart from how they were related were the same in physics and psychology, but it would sound better to call them "elements" when related in a "physical" way and "sensations" only when related in a "mental" manner. He often said that he had earlier been troubled by the fact that psychology and physics seemed to be grounded on different philosophical assumptions and he felt grateful to the Leipzig philosopher for suggesting a way such that one could employ the same epistemological and ontological assumptions in both fields, i.e. phenomenalism.

Gustav Fechner is also sometimes thought to be the founder of scientific psychology because he introduced the first widely accepted psychophysical law, namely, the Weber-Fechner logarithmic relation between "physical" input and a "mental" output. Mach, however, gradually began to distrust the reliability of the law in many cases, nevertheless, during the early 1860's as a *Privatdozent* in Vienna he continued to teach Fechner's ideas, published an account of Fechner's theory[4] and even began to apply a variation of Gustav Fechner's understanding of atoms in some of his work in physics and psycho-physics. But Ernst Mach's comprehension of atoms as influenced by the presentist epistemology of Fechner and the anti-atomism of Berkeley and Kant was never ever very clear, fully accepted by Mach, or successfully applied in his work, and from 1864 on - apparently with a sigh of relief - Ernst Mach's phenomenalism came fully to the fore even in his scientific work and he gave up physical atomism altogether, though his theory of sensory elements would later be called "atomistic" by many psychologists, but that is another story.

From 1864 to 1867 Ernst Mach became, first, professor of mathematics, and then, professor of physics in Graz. He also became engaged to his future wife there and continued his interest in music at the home of the city mayor. In fact, it was Mach who would later encourage the mayor's son to become an opera composer. The son, Wilhelm Kienzl, also composed the national anthem for the first Austrian Republic after World War I.

In 1867 Mach moved to Prague where he married his fiancée and would remain for twenty-eight years. They would have five children. In 1872 Mach published his first book in philosophy *Erhaltung der Arbeit* (Conservation of Work). Even though it was relatively short (under a hundred pages), in abbreviated form it anticipated almost all of his later ideas in epistemology and ontology including his elementism and theory of *Denkökonomie* or economy of thought. But Mach's teaching was almost entirely in aspects of experimental physics and he would gain few philosophical followers in the city of Prague. His main lecture and laboratory focus was on acoustics and optics, and it was there that much of his well-known work on shock waves was performed. His book on geometrical optics which was started in the 1860's was published posthumously in 1921 with its notorious preface which seemed to reject Einstein's theory of relativity.[5]

Though Mach's interest in politics at the time appears to have been slight, he was twice elected Rector of the University of Prague; the first occasion was for the whole university (1879-80) and the second period for the German university (1883-84), that is, after the Czech students and teachers had set up their own university. But he resigned his position before the second rectorate was finished partly in order to pursue his writing and partly because he was tired of nationalistic conflicts.[6] His most influential book among physicists, *Die Mechanik,* was published in 1883 and his major work on epistemology appeared in 1886 and was later given the title *Analyse der Empfindungen* (Analysis of Sensations). Mach also prepared several physics textbooks during the late 1880's and early 1890's for pupils from the lowest to the highest grades.[7]

But in all honesty, he was very little-known to the Viennese public until he was offered a position at their university in 1895 as a Professor in the "History and Philosophy of the Inductive Sciences". Republications of most of his books soon followed and he rapidly acquired a large number of philosophical supporters and admirers, and not only in Vienna but from 1900 to 1910 in many European countries and even in America. Indeed, the German owner of an American company set up a publishing house (1888) with a German-American editor largely to help publish Mach's articles and books in the United States. In fact, Mach's book *Popular Scientific Lectures* (1895) was published in English before it appeared in German.

Though Mach generally had a good reputation as a lecturer, most of his influence as a philosopher would not come through his students as was commonly the case in much Germany and Austria and particularly with Franz Brentano but through his books, which on the whole seemed comparatively easy to read. This fact had the unexpected consequence that many students from Central and Eastern Europe for whom German was a second language preferred to read Mach's books since they seemed to be relatively easier to understand. Furthermore, they seemed to presage the future. Even students for whom German was a first language but whose major field was outside of philosophy often read Mach's works, which for the most part were almost completely free of jargon and laborious technical distinctions. In fact, his writing style often seemed more French than German, something not totally inexplicable given the fact that Mach's father had frequently read novels to his children at night in the original languages when they were young.

After teaching only three years of history and philosophy of science in Vienna Mach suffered a stroke in 1898 which paralysed the right side of his body, paralysis which lasted until his death 18 years later. Nevertheless, with the help of his wife and eldest son Mach became able to type with his left hand and continued to publish articles and books. For example, his lectures in philosophy and methodology of science from 1895-98 were published in 1905 as *Erkenntnis und Irrtum*, a book which was subsequently translated into English as *Knowledge and Error*.

Mach's eldest son who had become wealthy before World War I through his patents on inventions, completed a laboratory in 1913 for his father in Vaterstetten, Bavaria, which encouraged Mach to leave Vienna for Germany in that year and allowed Mach with the help of his son to conduct further experiments, of which some seem to have been intended to test the light-constancy aspects of Albert Einstein's special theory of relativity employing Ludwig Mach's own interferometer, an optical device he had co-invented some time earlier. But some local electrical disturbances and eventually the German government putting high tension wires near the testing area along with Ludwig Mach's financial problems after the great German inflation of 1920-23 and possible weaknesses in physics seem to have frustrated the project.[8]

Ernst Mach died during the middle of World War I in 1916. He had been like Einstein at that time a pacifist on principle which may have influenced Albert Einstein's very positive obituary of him which was published in the same year. Mach's eldest son, Ludwig, knowing his father's views, even tried unsuccessfully to keep his father from learning about the outbreak of the conflict.

3. ONTO-EPISTEMOLOGY

Ernst Mach's philosophy is still honored in the moderate words of current historians of philosophy such as Rudolf Haller and Friedrich Stadler (1988). Haller writes about the possibility of Ernst Mach's influence on Ludwig Wittgenstein::

> Ludwig Wittgenstein would formulate it later in his *Tractatus logico-philosophicus* following Mach: 'The world is the totality of facts' and the task of scientific investi-gation is to form an image of those facts in thought.[9]

And Stadler adds with even more optimism:

> Even at first glance one sees the two-fold importance of current Mach scholarship: First, there is the *historical relevance* of Ernst Mach as the central figure of intellectual Vienna at the turn of the century as is reflected in Mach's correspondence, and second, there is also the current *scientific and philosophical relevance* of Ernst Mach as an experimental and theoretical physicist, sense-physiologist and psycho-logist, historian of science, and not least as educator and textbook writer.[10]

But starting during the 1960's with Gerald Holton[11] and continuing later with John Blackmore,[12] Ryoichi Itagaki,[13] and other critics, they have found it difficult to overlook that Ernst Mach's phenomenalism along with Logical Positivism can be hard to reconcile with some very important developments in science, while physicists have found it increasingly awkward to ignore Mach's opposition to the reality of atoms and reluctance to accept Einstein's theory of relativity. Perhaps understandably, this change in mood has upset some of the stronger remaining friends of Ernst Mach and his contributions to science and philosophy. Gereon Wolters (1988) and Paul Feyerabend (1987) have complained bitterly. Wolters writes in a footnote:

> Ernst Mach's controversy with Planck [1908-1911] cannot be analysed here. But one should expressly point out that Mach in his writings scarcely names Planck and has never criticized him personally, even indirectly. One should keep in mind that Mach had already basically completed his life's work when in 1900-01 Planck originated quantum theory. Nor is there evidence that Mach ever paid close attention to the theory. Planck's outburst against 70 year-old Mach who was in very poor health is indeed best explained as the emotional reaction of a renegade. As Jost (1979)[14] has shown, Planck had been during his earliest period as a professor, as he himself confessed, a follower of Mach's philosophy and 'concerning the existence of atoms had himself been a sceptic until his conversion in Fall 1900'.[15]

And Feyerabend planned a clever counterattack:

> In the following essay I shall try to separate Mach's physical arguments from his 'epistemology'. It will turn out that a separation is not difficult to achieve. Mach's physical arguments taken together, constitute a philosophy of science which differs from positivism, is in line with Einstein's research practice (and some of Albert Einstein's more general observations on research) and entails entirely reasonable objections against nineteenth century representations of atoms and the special theory of relativity. We shall also see that where Mach and Einstein differed it was Einstein who talked positivism while Mach gave a much more complex account of scientific and commonsense knowledge. Mach's 'epistemology', however turns out to be no epistemology at all. It is a general theory (or theory - sketch) comparable in form (though not in content) to atomism, and quite different from any positivistic ontology.[16]

The primary purpose of *our book* is to realize the first purpose which Friedrich Stadler mentioned above, that is, to detail "the *historical relevance* of Ernst Mach as the central figure of intellectual Vienna at the turn of the century". This relevance includes Mach's influence on Viennese thinkers who in turn have significantly impacted later philosophy of science. As for the primary purpose of *this introduction to this book*, apart from disagreeing with Paul Feyerabend about the importance and relevance of Ernst Mach's epistemology for his work in science and philosophy of science, it is partly intended to give some historical background but mostly to put forward a relatively fair and inclusive system of epistemological and ontological classification. After over a century of philosophical controversy, we want to *understand* Mach's epistemology and philosophy of science and that of his followers and successors and in a way which *transcends* the arguments of both supporters and opponents. And as a historian there is a tendency to look to the past at least as much as to the present to gain that insight.

Many movements seem to thrive until their roots are discovered. This means that philosophers and historians often have contrary aims. For example, most philosophers of science seem to want to advertise their position as new, fresh, and "the wave of the future", as if they have successfully kept up with recent developments. But many historians like to trace back contemporary movements to their apparent sources, even if those have become discredited or refuted. One result is that more and more philosophers try to write the history of their own discipline so that uncomfortable roots and branches can be successfully reinterpreted, bifurcated, or buried forever. But there are still many honest people in the world, both philosophers and historians, such that in spite of rhetoric and advertising truth will prevail.

On the other hand, by looking for rotten roots we may encourage what philosophers call "the genetic fallacy", that is, judging the truth of an idea by its origin instead of by means of relevant evidence. The philosophers are surely right in this respect, but it is not only still legitimate to look for roots but required if we are to *understand* current thought more accurately and completely. There is some difference between understanding and judgment. Indeed, the more we understand, the more likely we should be to be able to form sound judgment. Furthermore, if a current idea is very similar to one which was refuted in the past, then pointing this out is very relevant to the determination of truth now, indeed so relevant that concealing the similarity by alleging that someone is guilty of using the genetic fallacy is clearly reprehensible, shameful, and wrong. Indeed, to discourage, distort, or block historical research through fear of the genetic fallacy would not only help encourage uncritical acceptance of fake novelty, but could well set back the search for truth by obscuring the fact that similar ideas may already have been refuted. In short, hampering historiography far from being in the interests of philosophy hurts sound progress and understanding in all fields.

Mach's philosophy and its wide dissemination circa 1895 to 1930 have become roots of recent philosophy of science as Stadler has mentioned above, but which under Mach's own name or that of Logical Positivism have also been largely discredited. Especially wounding to Mach's supporters as already suggested in part were criticisms from Planck, Einstein, and in the political realm from Lenin and by the impression that Mach rejected both the reality of atoms and truth of relativity theory. But if we check contemporary philosophy of science we will detect numerous Machian traces especially in epistemology and ontology. Many young thinkers still contrast science with "metaphysics" as if they were incompatible. In short, Many people still reject a trans-conscious physical world as if "metaphysical" or "meaningless". But let us reject such pejorative classification and invidious distinctions. Instead of:

(1) "realism versus metaphysics"

with "realism" meaning the epistemological position called direct or empirical realism, let us distinguish between:

(2) "direct realism and indirect realism"

Similarly, instead of:

(3) "science versus metaphysics"

let us distinguish science from its epistemological and ontological assumptions while keping in mind that they may be inseperable and that epistemological positions and assumptions commonly change the meaning of the basic terms we use. For example, "phenomena" does not mean the same thing for people who identify them with sensations (phenomenalists), with sensed "objects" (direct realists), and with trans-conscious objects (indirect realists). The struggle to replace slanted and prejudicial classification with fair and inclusive categories has not yet been won, but we will do our best in this book to avoid the worst examples from philosophy, such as those quoted above, i.e. (1) and (3).

In addition to pejorative language there are also misleading "equivalents" such as when Atoms and molecules are called "theoretical entities" or when believing in the utility of the "atomic theory" is treated as being the same as believing in the reality of atoms (though it is hard to believe theories could blow up Hiroshima and Nagasaki). The rejection of "mechanistic science" is still thought by some to have refuted the representist "metaphysics" of Galileo, Descartes, & Newton. Einstein's theory of relativity is imagined by many to have confirmed that space, time, motion, and sometimes even truth are "merely relative". And quantum theory has been interpreted to mean that particles are unreal or "meaningless" when not being observed or measured. Mach and later positivists are of course not the only sources of unfair classification but when traced back many do have "Machian roots", but for every criticism in this book we also feel duty-bound to look for constructive aspects of influrence as well.

But in order to trace roots and branches in philosophy it is necessary for historians to introduce an adequate system of relatively neutral classification, since without valid similarities and distinctions nothing can be understood, and this can be hard on many readers who have grown up with one-sided categories which have been developed to favor a particular point of view while remaining ignorant or unreasonably opposed to alternatives. Spelling out classification can also require careful study, and no system is either fully neutral or fully inclusive, but we can avoid most slanted labels and invidious distinctions. Also, even historians sometimes allow that while something rarely comes from nothing that on some occasions a few ideas even in philosophy do seem genuinely original or "emergent" and hard to classify. On the other hand, absence of an apparent source does not prove originality. Further research will almost always locate the real roots of ideas and perspectives. The history of ideas is important.

By describing the spread of Mach's philosophical ideas and how his epistemological and ontological assumptions have survived through many changes of designation and classification and even when so-called "linguistic philosophy" has allegedly made such considerations moot we should also be able to put contemporary philosophy in a much more informative light. If a movement has seemed to acquire a bad connotation, then those supporters who still think it is sound will often feel tempted to give it a new, fresh name, and as long as no one is deceived by this practice it is moral and legitimate. The problem is that many people *are* deceived, but when this happens we historians can help play an important role by pointing out the continuing features of movements which have retitled their calling cards, but we also need to adopt more neutral and inclusive classification ourselves in order to avoid being buffaloed by label-shifting from advocates of those particular philosophies which want to put their movement in more attractive focus.

Almost all particular philosophies are minority movements and like commercial products need persuasive rhetoric, advertising, or even sophistry if they are to gain large numbers of supporters. A few can be gained through evidence and valid argumentation, but big numbers commonly require big deception. We serious historians are obliged to expose such behavior but without being *unduly* critical or "casting the first stone". Does being neutral toward immorality encourage it? To some extent, it may, hence restrained, moderate criticism may be appropriate, perhaps casting the second or third pebble.

But in carrying out our historical responsibility to reconnect past with present movements especially when the latter try to advertise themselves as new, one must not expect praise. No one likes his dirty linen exposed, even if we are largely talking about labels and unfair classification and not content. Furthermore, most current good monikers for positions and philosophies through repeated criticisms will gradually sour and in the view of supporters "need an even newer designation" to replace the discredited verbal handle of the moment. Nevertheless, supporters of genuinely independent perspectives should feel a measure of gratitude for the added enlightenment about the connections between Mach's point of view and several more recent and contemporary positions and schools of thought, many of which claim in a false or quite misleading way to "oppose Machism".

Persuasion can be particularly difficult for minority movements which are thought to have been refuted. Badgered supporters are often tempted not merely to dress themselves up in new garments and deny any connection with anything discredited in the past but to use strongly pejorative comments about their opponents. This one-sidedness rarely works with people familiar with relatively neutral classification, but if beginners can be caught before they have learned the latter, then persuasion generally becomes much easier. Not all minority movements or minority philosophers stoop to such tactics (and of course we are not talking about race, religion, or ethnicity here, but only philosophy), but alas, many of the more influential thinkers do lower themselves to support "what works" in too great a hurry and without enough reflection on the likely unethical, false, or genuinely unscientific consequences.

After Mach's ideas "were refuted" from 1910 to 1920 by opposing the reality of atoms and seeming to doubt relativity theory, within a few short years he became a father-figure of Logical Positivism which proved to be a master of persuasive propaganda, especially with students and those young scientists insufficiently familiar with basic epistemology, ontology, and neutral classification. There was of course more to Logical Positivism than skillful rhetoric, but it was that feature which stuck in the craw of most opponents, including a probable majority of well-educated scientists and philosophers.

It has often been said that no historian who is deeply interested in his subject can be fully neutral, but he can be fair, reasonable, and informative, and by revealing the connection of contemporary thought to the spread of Ernst Mach's phenomenalism throughout Central Europe and eventually beyond we will attempt to be all three, even if this initial volume only covers his impact in Vienna in detail. Ideas have homes too, since they may well literally exist only in particular human minds or brains which only exist in particular locations.

If we were inclined to prejudice our readers *against* Mach we would call his phenomenalism "idealism" or even "subjective idealism" which are what the similar views of Berkeley and Hume about matter were often called in the 19th century and even as late as Lenin's philosophy book of 1909, but the first term is almost always pejorative and the second has largely become so.

Similarly, if we wanted to prejudice our readers *in favor* of his epistemology and ontology we would label his position either "empiricism" or "neutral monism", even though the first term is properly methodological and can be adopted by people holding a wide range of epistemological positions and the second term is misleading because phenomenalism is not neutral between direct and indirect epistemologies and ontologies but is direct or presentist. Thus fairness also drives us toward introducing an even more inclusive and as yet uncontaminated system of philosophical classification.

But let us carefully distinguish language from meaning. We can mean the same thing using many different languages. Meaning is primary not words, and while it would be nice if meaning were "objective" or "intersubjective" in fact it is private and subjective in that meaning depends on the intent of the speaker or writer, as it exists in terms of his best understanding fairly and reasonably interpreted, and failure to understand that intent *defines miscommunication.*

One may not agree with Ernst Mach's phenomenalism, but he was an honest man. He realized that both epistemological and ontological phenomenalism were philosophical positions which he held in terms of belief and normmally of assumption, at least in philosophy of science. Phenomenalism was not a mere language, nor could expressing himself in "physical object tlangauge" which he normally did change that belief, that is, his basic or underlying phenomenalism. One should not confuse a philosophical position with expression of it. Words do not determine meaning, nor are they meaning. Rather people use language to help them express what they intend, assume, or believe. This is common sense, and it is a shame that many Logical Positivists and linguistic philosophers have abandoned this fundamental aspect of common sense. Let us now turn to a relevant discussion of philosophical *positions.*

Let us start by contrasting consciously direct with consciously indirect epistemology about the external world, and second, let us gradually relate other direct or presentist philosophical movements to "Machism", both those prior in time, those which were contemporaneous, and those which represent continuations of his own, though often identifying themselves with very different labels and even claiming to reject Mach's phenomenalism or even all epistemology and ontology. Nor should we ignore important differences. If two opinions were identical in all respects, then they would be one.

As for those aspects of his philosophy such as *Denkökonomie* or theory of economy which have one meaning if subordinated to presentism and another if presentism is subordinated to it, we will not be able to pay full attention to them, but some mention will be paid in passing.

How is presentism including that of Mach generally contrasted with representism in philosophy? All human beings when they *clearly* refer to something must assume that the target or object of reference must either be conscious or unconscious (to the person making the reference). If it is conscious, then it is presented, but if it is referred to but not conscious, then it is represented. It is normal for sensory colors and feelings to be treated as if consciously presented to us, but some philosophers also think that chairs, tables, and other physical objects can also be consciously noticed or at least some aspects of them. Such people are presentists in epistemology. They may also define what is *real* in terms of what can be sensed or be made conscious in which case they are also ontological presentists.

But there are other philosophers who think that we never directly notice physical objects at all in any respect or at least never consciously experience external physical objects, that is, which exist outside of our heads. These philosophers have adopted an indirect or representist epistemology. Most representists think that the difference between a real physical object and a mere geometrical phantom is that the real objects possess mass, weight, or energy in addition to mere mathematical or geometrical qualities such as size, shape, and number . This is a difference in ontology, namely between matter-representists and form- or Platonic-representists. The ontology of matter-representism is normally establshed or confirmed by defining reality in terms of causal force or power as if only what is *concrete or potent* were real, while the ontology of form-representism is established or confirmed by defining reality in terms of form, logic, laws, mathematics, or universals as if only what was *abstract or impotent* were real.

The most decisive reason why the question of what is real in the sense of existence (as opposed to real as genuine as in "that is a real diamond") is so important, especially for representists, is that for them only what is real in the sense of potent can have causal influence. Abstractions and fictions lack that influence, even though our understanding of them may have mental importance. Theoretical science tries to understand idealized types of causes by means of idealized laws, but in order to understand particular causes, that is, something non-idealized and historical, a more concrete approach which emphasizes particular historical powers or forces is normally employed. This becomes especially important when human beings try to influence the world around them. Means are used to help further ends, but effective means must possess real energy, matter, or force, not mere constant conjunctions among appearances or something which can be deduced from abstract laws. To repeat, it is this focus on the importance of reality as a potential or actual cause or force which naturally leads many people to prefer representism to presentism about the external physical world. It seems more practical.

On the other hand, people who think that science should focus more on certainty and that nothing is more certain than direct sensory experience, just as naturally prefer presentism. Furthermore, many religious people who would like to protect religion from scientific criticism believe that by restricting the scope of scientific understanding to direct description of appearances and to relations between appearances that this will somehow help protect religion from scientific criticism by helping religion to uncritically suggest what exists *beyond* sensory bounds. Furthermore, many religious people attempt to literally believe that God is omnipotent, that is, all powerful, which if interpreted literally would suggest that both nature and mankind are causally impotent. David Hume's effort to reduce causes to presentist terms, that is, to mere constant conjunctions among appearances - contrary to both his historical writings and to his own seeming indifference or opposition to religion - could be easily interpreted to fit that apparent religious need.

In short, it is often thought that presentism is more focused on sensory description and representism on action and historical causes, but both epistemological positions, being very basic, are probably for the most part not deliberately adopted but are unwittingly accepted from an early age before one even thinks of trying to defend or justify one or the other. A temperamental preference for description of appearances may help push one toward presentism just as a temperamental preference for action and causal explanation in terms of forces may push one toward representism.

However, what we would mean by *mature* understanding must include conceptual mastery of *all* major alternatives, and at least partly this means fair understanding from each alternative's own point of view. But it also means an attempt to understand from outside all perspectives. It is in this sense that mature understanding of epistemology would encompass both presentism and representism from inside each and from outside of both as well as from how each tends to understand the other.

While epistemology is traditionally understood as theory of knowledge, what we will call *mature* epistemology is basically theory of reference plus theory of human understanding and includes at least three main features in addition to what we have mentioned, first, it is deeply concerned with consciousness, since different conceptions of its scope largely determine a key difference between presentism and representism, second, it is deeply concerned with reference since unlike presentists representists almost always use or assume with only a few exceptions that what we intend to refer to is also what we actually refer to such that we can refer to what cannot be made conscious This could also be called liberal or volitional reference. And third, mature epistemology recognizes that changes in one's epistemological position or assumptions almost always changes the associations we make with words (connotation) and what we use the words to help suggest (denotation). For example, key philosophical expressions such as true, real, fact, and cause have one set of associations or meanings for presentists and another for representists. The same is true for physicists. A scientist holding presentist assumptions will not mean the same thing by atoms, matter, space, time, or science as a scientist with indirect or representist assumptions.

If presentists accept what is generally called a correspondence theory of truth (which they normally do not), then the correspondence is usually with sensations, sensed objects, or something which is or can be made conscious. For representists what *they* mean by the correspondence theory of truth is normally with what they regard as the external physical world which they believe exists completely beyond the limits of consciousness. Similarly, presentists tend to mean something empirical or conscious by "facts" while representists normally mean something real or true whether we can sense or experience it or not. In general, what presentists mean by matter is criticized as "mental", "subjective", or even "solipsistic" by representists, while what representists mean by matter is criticized as "mechanistic", "metaphysical" or even "meaningless" by many or most presentists.

While representists almost always think that the existence of something is prior to understanding of it, that is, that ontology logically, temporally, and causally precedes epistemology, presentists often think that reality is what is experienced as if epistemology is at least logically prior. But our point here is that presentists *and* representists commonly use the "same" words but with *different* ontological meanings attributed to them such that we must understand epistemological differences and each side's use of words *before* we can properly understand genuine ontological differences. It is in this sense that for both presentists and representists that epistemology is prior to ontology.

The concern of ontologists for what is irreducible or ultimate leads to monistic expressions like materialism and immaterialism, phenomenalism and relationism, and to dualistic labels like mind-matter dualism, content-form dualism, and natural-supernatural dualism. There are also pluralistic expressions which are used to refer to what are more than two "irreducible" or "ultimate" types of reality. But each of these expressions often changes meaning, that is, different associations are made with the terms or they are used to suggest different things, when employed by people with different epistemological preferences or assumptions.

For example, while the term "phenomenalism" is often thought to label a type of presentist epistemological or ontological position, it seems many positivists and linguistic philosophers treat it as merely "linguistically different" from "realist language". They want to treat epistemological and ontological distinctions as "outside of science" even if everybody must refer to what can be amde conscious or what cannot be made conscious, that is, must assume an epistemological position *which is not merely linguistic*. As for denying the reality of consciousness, this might be fine for comatose or dead people, but hardly for anyone else. Even the denyer is presumably conscious of denying.

But differences in definition go even further, even within the presentist camp. Phenomenalists and direct "realists" will disagree about what physical phenomena are.

Just as a presentist's physical phenomena are not a representist's physical phenomena, so even a phenomenalist's physical phenomena are different from a direct "realist's" physical phenomena.

To repeat, epistemology commonly changes the meaning of ontological terms as well as the associations made with ontological expressions. A presentist's materialism is not materialism to a representist but phenomenalism or even solipsism, while a representist's materialism is not a presentist's materialism, but in pejorative terms are only "mere metaphysics, dogmatism, idealism, or nonsense". We use the term "representism" to contrast with "presentism" below even though the term indirect is more inclusive than "representism"

Table #1

Position	*fair classification*	*old classification*
What presentists mean by "matter"	matter presentism	phenomenalism or direct realism
What representists mean by "matter"	matter representism	mind-matter dualism or indirect realism
What presentists mean by "form"	form presentism	phenomenology
What representists mean by "form"	form representism	Platonism

The fact that table #1 includes onto-epistemological classification such as "matter-presentism" and "matter-representism" clearly suggests that we are dealing with different definitions of "matter". Most representists think that sensations, feelings, and other seemingly mental phenomena do not fit what is normally meant by the expression "matter" or even that Democritean favorite "refined matter", and that even if in some way there is a connection or even a historical evolution from one into the other that the conscious world just doesn't fit into the same category as the primary and energetic qualities normally attributed to matter. Presentists on the other hand with their tendency to identify atoms, molecules, electrons, and matter generally with "pictures" and other non-veridical substitutes taken from conscious phenomena are apparently not troubled by alleged major differences between matter and conscious appearances.

Some representists like George Santayana identified the non-physical world with what he called "essences" rather than with mental phenomena, but the vast majority who support indirect ontology about the physical world causally distinguish between appearance and reality while conceding that sensory appearances are also real in some sense, since they too seem to have a measure of causal influence, but that basically they are just not the same things, and even appear to be "incommensurable" in type.

The next table includes traditional onto-epistemological positions which would combine epistemology and *causal or explanatory ontology* . Our understanding of them can benefit from being compared and contrasted with those listed in table #1 which combine epistemology and either *matter-ontology* or *form-ontology*, even if they can be difficult to distinguish in practice. By "indirect explanation" we mean in terms of forces or powers, particularly as concrete means to help further particular human ends. By the phrase "direct explanation" we mean either constant conjunctions among appearances as favored by Hume and Mach or Kant's necessary conjunctions among appearances. As for an appeal to so-called deterministic laws among forces or powers, this could be regarded as a tool to help provide indirect explanation but is perhaps closer to what in table #1 we have termed form-representism than to matter-representism..

Table #2

Traditional Epistemological Classification

Direct Realism	Indirect Realism	Phenomenalism
direct description indirect explanation	indirect description indirect explanation	direct description direct explanation

The important point is that if direct realists replace indirect with direct explanation, then their position dissolves into a type of phenomenalism, *even if they retain a "physical object language"*. While table #2 looks simpler than table #1, if as mentioned above it combines ontology with epistemology in a different way, then one must consider if that difference is factual in practice or merely linguistic.. Furthermore, table #2 does not show as clearly as table #1 how changes in epistemology can alter what we *mean* by particular ontological terms and positions. Nevertheless, both tables should help those readers who want to understand epistemology and ontology in reasonably fair and inclusive ways and not just to fit a particular point of view.

In terms of the three criteria (far) above for *mature* understanding of epistemology, it seems that Mach's understanding of epistemology was relatively mature, regardless of whether one agrees with his views or not, but this conclusion does not necessarily hold for all of his followers. Indeed, most so-called Logical Positivists would experience serious problems in trying to reach epistemological maturity. First, they rarely focused on the nature and scope of consciousness, second, they rarely used a sufficiently liberal theory of reference to allow them to accurately distinguish presentism from representism, and if they referred to the latter at all they labeled it as "metaphysics", a term with many arbitrarily linked meanings and a lot of pejorative baggage. And third, instead of recognizing that presentists and representists almost never associate the same ideas or meanings with key philosophical terms, Wittgenstein, Carnap, and many Logical Positivists who accepted the so-called "linguistic turn" in philosophy falsely imagined that the *opposite* was true, namely, that language could determine epistemology as if the differences between presentism and representism were "merely linguistic". This seemed plausible to them partly because they thought that meaning, reference, and language were primarily matters of convention as opposed to being actually determined by what particular people intend, believe, and assume.

In general, Logical Positivists tried to understand representism solely with criteria taken from presentism, rather than from a perspective capable of fairly understanding both sides. Many representists also fail to gain clarity about the three main criteria for mature epistemology, but if presentists restrict the scope not merely of understanding and reality to what can be made conscious or "empirical" but even reference, then there is no way that they can accurately understand representism at all, and alas, many Logical Positivists *did* try to restrict reference or "meaningful reference" narrowly.

Ernst Mach seems to have been a presentist in both epistemology and ontology of science. He claimed in one passage that only sensations were real and in another that only sensations and relations were real. Sometimes his position was immanentist as if only what was being sensed now were real, but normally he did not add that temporal condition, though whether he took it for granted or not remains a question. But he had no objection to theories, hypotheses, and analogies provided that it was understood that in the final stage of physics or science that they would disappear and only direct description of conscious sensations by mathematical functions would remain. By denying that he was a philosopher or had a philosophy he apparently meant that the phenomenalism was part of his methodology of science, but that outside of science he was free to adopt something more generally practical. But what he regarded as his best as opposed to other operative understandings was clearly presentist.

Mach also held an epistemological position in the sense of theory of knowledge in that he thought that only one ontological type of referent could be "known" with absolute certainty, namely conscious sensations. Most matter representists affirm the general understandability of matter but reject knowledge both in the sense of direct conscious experience and in the sense of absolute certainty. They think that most conscious experience involves unconscious editing such that recognition and identification of, and relation to what we notice can be misrecognition, misidentification, and misrelation. As for what exists beyond conscious bounds, representists commonly think it can often be understood by means of inference or analogy and confirmed by properly applied measuring tapes, rods, scales, cameras, and electronic and other machines.

Furthermore, many presentists *and* representists doubt that anything is certain or absolutely certain in the sense of being *unconditionally* certain. This position would seem to imply, however, that nothing true, real, or right can be demonstrated or proved to be more than highly probable at best. Also, most rational thinkers, regardless of their epistemological preference, use weight of evidence to determine degree of probability as relative or conditional certainty. Even Rudolf Carnap, a youthful Machist and later perhaps the most influential Logical Positivist, who unfortunately misled himself into thinking that language and form could replace epistemology, eventually adopted a probabilistic approach to understanding.

Ernst Mach was the father of more than one type of phenomenalism and a sort of uncle of phenomenology, though his influence occurred before those movements became clearly distinguishable from each other, something which has not yet completely taken place in some disciplines outside of philosophy such as physics and psychology. Both philosophical schools were presentist in epistemology and ontology, but as suggested above in our table phenomenalism emphasized the changing features of sensation, while phenomenology as developed by Edmund Husserl, who went through an early Machist phase himself, placed increasing emphasis on intuition of the formal or ideal sides and aspects of sensation and consciousness. Edmund Husserl attempted to discredit Mach's phenomenalism by alleging that it was guilty of "psychologism", namely by treating logic in a psychological rather than in a formal or idealized way. Mach, however, denied that Husserl's approach was itself other than a type of psychology.[16.1]

Phenomenology has also developed a more inclusive conception of consciousness which differs from "normal" belief about it by absorbing more aspects of human intent and understanding. Indeed, many phenomenologists and their existentialist followers treat what they mean by consciousness as virtually all-inclusive as if Aristotle were right that by intuiting the "essence" of an object we are also conscious of its "real" nature. Both phenomenal types of presentism above have claimed at times to be fully "presupposition-free", something which most representists find hard to accept as true.

From 1890 to about 1910 Viennese philosophy was largely caught in a crossfire between Mach and his phenomenalist followers on one side and former students of Franz Brentano on the other, especially by followers of Alexius Meinong, but after the latter group largely died off (around 1920) there was a further split with one group of thinkers emphasizing a revival of traditional German philosophy as led by Neo-Kantians such as Reininger, especially in the University of Vienna Philosophical Society, with other orientations joining small outside groups such as "Schlick's Circle". Mach's admirers largely clustered here and led by Schlick and Carnap with many ideas from Wittgenstein, they gave greater emphasis to mathematical logic and to a physical object linguistic facade covering Machist epistemological and ontological assumptions. Most of Franz Brentano's influence had come through students like Husserl and Meinong whereas almost all of Mach's philosophical influence came through his writings, and in those days books seemed to last longer than people.

Nevertheless, outside of Vienna, because of Husserl and his influence in Freiburg, Göttingen, and Munich with disciples also attracted by his books, it seems that the *word* phenomenology is currently more fashionable than the *word* phenomenalism among professional philosophers, but both tendencies are presentist and Ernst Mach's basic epistemology continues to flourish under some interpretations of recent movements or labels such as "scientific realism" and "internal realism" where subtle differences and verbal disguises have tended to obscure the presentist base.

We decided to present articles on Mach's *phenomenalist* followers in this book *before* treating his *phenomenological* friends and opponents, even though to some extent the latter came first in time. We have taken this step because phenomenalism is easier to understand than phenomenology, which in the hands of Brentano, Husserl, and Meinong include almost innumerable distinctions which can be very difficult to fully unravel and interrelate.

In general, phenomenalists have been more influenced by Ernst Mach's attempted reduction of knowledge and reality to sensations and relations between them, while phenomenologists have been impressed by the formalistic aspects of Mach's theory of elements and in some cases by his contributions to gestalt psychology. Mach denied he was a solipsist on the ground that there was no ego, self, or subject. Representists, however, insist that in addition to rejecting the existence of the real physical world he even denied his own existence, which was a kind of super-solipsism or even nihilism. But some introspective philosophers may even welcome solipsism, "since nature and other people are only nuisances anyhow".

4. Textbook Ontology

Let us end this introduction by turning to the philosophical content of a few physics textbooks as that is reflected in the attitude of various textbook writers on the reality of atoms and molecules, since this seems to be little known and distinctly bears on the German and Austrian situation in which Ernst Mach as a philosopher and textbook writer was closely involved. During the middle of the 19th century an obscure French physics teacher called Pierre Benjamin Adolphe Ganot (1804-1887) completed two physics textbooks, a very long one (660 pages) in 1851 with some additions and a title change in 1854 (708 pages) and in 1858 or 1859 a short textbook (only 538 pages). He published the books himself.

Two important points must be emphasized: first, his two textbooks seem to have become by far the most popular physics textbooks not only in France but in the world and versions of them were published well into the 20th century . And second, on the first page of text in both the *Traité Elémentaire de Physique Expérimentale et Appliquée* and in the *Cours de Physique Expérimentale*, that is, both textbooks as well as in translations into different languages, the reality of atoms and molecules is clearly expressed and taken for granted. In the process, he manifestly supports what we have called matter-representism, that is, mind-matter dualism.

Professor Seiji Takata in Japan has made a list of the extant editions of Ganot's two textbooks which he could find and from as many countries from which he could locate bibliographical data.[17] Naturally, he may have missed some editions or countries since many libraries do not keep old textbooks and many bibliographies are not complete. Nor are even the bibliographical lists from the Bibliothèque Nationale, The British Museum, The Library of Congress , and other key sources exhaustive. But he has discovered some remarkable facts about publication and use.

Adolphe Ganot's big physics textbook (*Traité*) went through at least 30 French editions from then year 1851 to 1931, 20 Spanish editions, 19 English-American editions. There were also Dutch and Russian translations as well as English translations published in India and Australia. In Japan, Dutch, French, and English translations were used.[18] The "small" textbook (*Cours*) seems to have gone through fewer editions, 11 in France from 1858-59 to 1908, 8 editions in the United States, and a translation into Dutch.

But while many sources have agreed on the popularity and predominance of Ganot physics textbooks and translations around the world, we do not as yet know the actual number of the books sold or exactly who they were sold to, even if it is clear they were employed in introductory physics for both university and pre-university students. Nor would it be reasonable to infer that most professors of physics necessarily held the same view on atomism as the books they made their students read. Still, it was almost certainly a disposing or predisposing factor on some teachers and many students who enrolled in introductory physics courses

The first textbook we will quote from was a 1880 reprint of "The Old Edition" circa 1860 which was called the Peck-Ganot *Introductory Course of Natural Philosophy*, and which seems to have been very popular in the United States. The opening words run:

> "Section I. - Definitions and General Properties of Matter.
>
> 1. **Physics - Physial Agents.** - NATURAL PHILOSOPHY, or PHYSICS, treats of the general properties of bodies, and of the causes that modify these properties without altering their constitution.
> The principal causes that modify the properties of bodies are: *heat, light, gravitation,* and *Electricity*. These causes are called *Physical Agents or Forces.*
>
> 2. **A body** is a collection of material particles; as a stone, or a block of wood. A body which is exceedingly small is called a *Material Point.*

> 3. **Molecules and Atoms.** - Bodies are made up of small particles called *Molecules* and those again are composed of still smaller elements, called *Atoms*."

As mentioned, almost all English-language physics textbooks seemed to follow Ganot's favorable attitude toward atoms and molecules rather closely for the forty years between 1860 and 1900, though Mach's publications in America would eventually affect or change some minds. Other textbooks repeated the above lines almost word for word, even including reference to atoms and molecules on the first page as if it were accepted that atoms and molecules were real. But even in those American textbooks where atoms and molecules were not mentioned on the first page, their existence seems to have been taken for granted. In general, the textbooks emphasized historical discoveries and experimental physics with little or no mention given to ether or other speculative theories. The books aimed to create interest in physics.

Another popular French physics textbook had been written by A. Privat Deschanel and was translated into English as *Elementary Treatise on Natural Philosophy*. It would go through six editions by 1883 and also refers on the first page of text to atoms as if their reality were assured.

Some readers might well suppose that these textbooks for high schools, institutes, and introductory college and university classes were naturally simplistic and materialistic and need not have represented the sophisticated views of most physicists, but much physics aims at simplicity, and the above positive view of atoms and molecules was much more compatible with mind-matter ontology than it was with reducing all reality to sensations. The reason for this is that for most people it is implausible to identify sensations and feelings directly with matter as atoms and molecules. In short, a rather simplified view of a sophisticated physical position is not always a serious distortion of it. Some textbooks even allowed for the possibility of divisible atoms.

Were there any German or Austrian editions of either Ganot textbook? Yes, Professor Takata has found *one*, an edition from Ganot's *Traité* of 1851 which was dated 1858 which appeared under the title *Lehrbuch der Physik* as edited by Heinrich Adolphe Weiske and which was published by L. Wotz in Leipzig,[19] but he has not yet found a copy of the actual book. In short, if there was an impact it was probably small. As for the less popular textbooks by Deschanel no apparent study of them has yet been made.

The natural question is: Does the relative absence of German editions of extremely popular textbooks under Ganot's name indicate a less than enthusiastic attitude by German physicists and students toward the reality of atoms and molecules? The question is difficult but not impossible to answer. We do have access both to a recent history of German textbooks and to a number of German textbooks from the period, that is, from the late 19th and early 20th centuries.

The recent history of German physics textbooks by Gunter Lind is titled *Physik im Lehrbuch 1700-1850*. It is a very detailed work and was published in 1992. But even though the work actually mentions some textbook writers after 1850 there is no mention of Ganot or Deschanel. But the author does mention a number of early physics textbooks which take a stand on atoms and molecules including Ernst Gottfried Fischer (1805), Georg Friedrich Parrot (1809), Georg Gottlieb Schmidt (1813), Andreas Baumgartner (1824), Georg Wilhelm Muncke (1819), and Jakob Friedrich Fries (1826), but only Muncke among the textbook writers mentioned by Lind for that period seems to have supported atomism, and he qualified that view considerably.[20] According to Lind most of the writers alleged "We know nothing of atoms, we know nothing of powers, we know only effects."[21] To reinforce this bare impression Lind devotes a whole section to "Die Erkenntnistheoretische Begründung der Physik durch Kant" (Kant's Epistemological Foundation of Physics).[22] In short, to the extent that Kantian and other philosophy may have influenced physical views on the reality of atoms, it seems not to have been positive.

But while Lind does not mention Ganot, Deschanel, or their books he does discuss the extraordinary impact which French writers would generally exercise starting in the first half of the nineteenth century even on German textbooks, mentioning the names of Réne Juste Haüy (1804-05), Jean Babtists Biot (1819), Gabriel Lamé 1838/41, and Claude Servais Mathias Pouillet 1839-43 as leading examples of authors of textbooks translated into German.[23]

A distinguishing feature of many of these textbooks was apparently their mathematical content. Whereas the popular Ganot textbooks advertised how free they were from mathematics, German editors would apologize for not including more mathematics! Of these translations, the one by Pouillet would become most important, and under the title of *Müller-Pouillets Lehrbuch der Physik und Meteorologie*, Vieweg: Braunschweig, which was published from 1844 to at least 1931. The book was intended for beginning students in universities but gradually became massive including four and five thick volumes for each edition and eventually becoming more reference books than textbooks. The editor of the 8th through 10th editions (1876 to 1906 and later) was an Austrian Leopold Pfaundler, a professor of Physics first at Innsbruck and later in Graz. The works eventually included separate books on mechanics, optics, acoustics, and thermodynamics with Pfaundler usually writing at least one book himself. (One unfortunate result of increase in size was a confusion about editions, since each book with its own topic also had its own history and editions). While the books were published in Germany and appear to have become widely used throughout German-speaking Europe, we have not yet found translations of these textbooks into other languages.

Pfaundler seems to have been a good friend of Ernst Mach. There are nine letters from him to Mach from 1878 to 1896 in the Deutsches Museum, and Pfaundler not only mentioned several of Ernst Mach's contributions to experimental physics and some of Mach's laboratory instruments, but at the very end of the foreword to the ninth edition (1886) he makes his only admission of assistance to any person: "I am obliged to give special thanks to my honored colleague Prof. E, Mach in Prague, whose friendly advice and whose publications have been especially helpful."[24]

But the story is not so simple. Before having been influenced by Mach it seems that Pfaundler accepted Du Bois-Reymond's indirect realism about the external world even writing the following in an early Introduction (1876?): "Mankind is never in a position to perceive bodies in their essence and know this". Pfaundler even cites Emil Du Bois-Reymond's 1872 work: "Über die Grenzen des Naturerkennens". But after encountering Mach and his ideas, (and perhaps becoming confused in the process) Pfaundler, first, eliminates much reference to atoms and molecules from his Müller-Pouillet textbooks and then argues in the 1906 edition that physicists should avoid philosophy. (See the footnote at the bottom of page one in the 1906 edition of his volume in the series on mechanics).[25] Later editions of this multi-volume enterprise at least in the volumes which he himself wrote carry this attitude forward plus scepticism of atoms and molecules, and the attitude persists even in the hands of other editors into the 1920's. Much like earlier German writers he seems to have come to understand the reality of atoms and molecules not as a scientific issue but as a philosophical one. It would be Professor Max Planck in his long controversy with Ernst Mach (1908-1911) who would finally persuade many German and Austrian physicists and other scientists to take a more realistic attitude toward atoms and molecules, but the Vienna Circle itself is proof of how strongly the earlier attitude could persist, at least among many young mathematicians and philosophers of science.

We have also studied other German and Austrian physics textbooks from the 19th century which had been published during the 1880's and 1890's, and first of all, they only referred to atoms and molecules in the short section on chemistry, and second, again only as something "hypothetical". Ernst Mach's first textbook in physics published in 1863, *Compendium der Physik für Medicner*, did make a number of partly positive, partly cryptic references to something strange which he called "atoms" but when he returned to writing textbooks in the 1880's his attitude had become almost completely negative, even if a little bit more understandable.

The Viennese authors Ernst Mach and Alois Höfler clearly understood physics in presentist terms at that time, though Höfler did treat the atomic theory distinctly more seriously than Mach. (Joachim Thiele in his very useful bibliography lists Mach as the original author of 2 physics textbooks for university students and 23 editions of textbooks for Gymnasium and middle school pupils from 1887 to at least 1917.)[26] About half were published in Germany and half in Austria, but all were in the German language. (i.e. Thiele lists no translations.)

Mach and Höfler corresponded with each other extensively about physics education from 1887 to 1891 (thirty-five letters from Mach alone)[27] and after Ernst Mach became half-paralyzed in 1898 he offered the co-editorship of a journal in science education to him (1901) which Höfler accepted in 1903 [*Zeitschrift für den physikalischen und chemischen Unterricht* (Berlin)]. Höfler's first textbook in physics in 1881 was actually a revision of an earlier book which had originally come out in 1854.. Both men wrote for a wide range of pupils and students from the youngest to the oldest. Textbooks by both men were re-published well into the 20th century, Mach's until almost the end of the First World War and Höfler's well after his own death until 1930.[28]

Since we have not examined all of the later editions which tended in Mach's case to be partly amended or re-written by other authors, it is hard to say whether either Mach or Höfler abandoned their earlier designation of atoms and molecules as "hypothetical". But all in all, it does seem that the German and Austrian physics textbooks which we have examined take a very different attitude from the Ganot and Deschanel books. Ganot's works with their strong support of the reality of atoms and molecules on the very first page of text were translated and widely used around the globe, while the comparable textbooks by Mach and Höfler and the much larger works edited by Mach's other major educational friend Leopold Pfaundler do not yet show any evidence of having been used or published outside of German-speaking countries, and the German-language copies seen were all rather sceptical about the existence of atoms and molecules. This strongly suggests a major difference in epistemology and ontology. Ganot and his Western friends took representism about the external physical world more or less for granted while the German-language textbooks we have seen did not.

As a historian in this context, we are not allowed to say whether that was good or bad, but it clearly suggests that there was a split about epistemology, especially a division between German-speaking and non-German-speaking scientists, and that Mach was on the German-speaking, presentist, and anti-atomistic side. It also suggests that Ernst Mach as a textbook writer along with Pfaundler and Höfler may have helped to dispose or predispose many German and Austrian students toward taking a sympathetic attitude toward presentism and an unsympathetic one to representism than otherwise might have been the case.

On the other side of the coin, many presentists from about 1890 to 1910 would interpret some of the more recent theories and discoveries in physics as refuting "mechanistic science" (including the kinetic theory of gases and statistical mechanics) in favor of what was being generously called "pure thermodynamics", "the electro-magnetic world view", Wilhelm Ostwald"s "energeticism", or Ernst Mach's "elementism" and "theory of economy". And even though these anti-atomistic theories would be cut down to size by the work of Max Planck and Albert Einstein in putting the reality of atoms and molecules even further beyond reasonable doubt (Josef Loschmidt had even calculated the size and density of gas molecules as early as 1865 and new elements were being discovered almost every year during the 1890's), many presentists would keep trying to interpret quantum and relativity theory in ways compatible with presentist epistemology and Mach's philosophy. The first two "Machists" we will discuss in this book, Friedrich Adler and Philipp Frank, would be especially active in trying to advertise Mach as the spirit behind the new physics as if in addition to his relativism about space, time, and motion he showed the way by rejecting "metaphysics" (i.e. representism) in favor of "consistent empiricism" (i.e. sensory presentism) in physics.

Let us turn to Ernst Mach's initial words in his two most sophisticated textbooks, one for teachers and one for students. The textbook which was prepared for physics teachers in Germany and Austria included two chapters by his assistant Gustav Jaumann. Both books were published in Leipzig and Vienna in 1891. The titles of both books were the same *Leitfaden der Physik für Studierende* (Textbook for Students Studying Physics). The only major differences were that the book for students was missing the two chapters by Gustav Jaumann, which Ernst Mach thought were too difficult for students.

Jaumann who would be an assistant for about ten years apparently lost hope of receiving a position elsewhere and would neglect many of his duties, which Mach's son Ludwig, who seems to have been gifted at constructing laboratory equipment, would take over from Gustav Jaumann in the German University of Prague. Jaumann, who unlike most of Ernst Mach's students was gifted in mathematics, seems to have disliked Mach's son intensely. He would finally receive a position in Brünn shortly before Mach left Bohemia in 1895. But while there was no love lost at the time between Mach and Jaumann, Mach would later appreciate the fact that Jaumann was a strong supporter of Mach's presentism and would also oppose the reality of very small particles such as atoms and electrons. Indeed, It is said that physicists in Brünn/Brno may have only begun to teach atomic and relativity theories after World War II.[29] The first page of the first chapter in Ernst Mach's two 1891 textbooks began as follows:

"I. Introduction

1. The contents of that which we perceive through the senses (colors, tones, smells) we call the *Sense-World or Nature* . . .

2. Also in complete darkness, in absolute stillness, where the senses are totally ineffective, we are still able to picture what is perceivable by means of imaging or thinking.

3. What remains perceivable as a rule can be noticed by several senses, provided that the senses are in good working order. We call this matter. Spatially bound matter we call bodies."

In his two textbooks of 1891 Mach does eventually mention atoms, but only as a mere hypothesis and rather belatedly beginning on page 317 and as if it were only relevant for chemistry. He starts:

"10. The Atomistic Hypothesis

490. One often uses instead of the expression "combination number of elements" the term the *atomic weight* of elements . . . That one designates these unnamed numbers as if they were weights has its source in a hypothesis which Dalton introduced and which has drawn much attention."

Given the fact that belief in the reality of atoms was widespread from the middle of the 19th century on or even a bit earlier it may seem difficult to consider those philosophers who opposed the reality of atoms and molecules to be progressive or genuine friends of science. Nor is it even relevant here whether atoms were thought of as tiny billiard balls, vortices, power points, or divisible particles. If like the rest of the external physical world they were trans-conscious in principle and fact, then presentists who wanted to remain loyal to their epistemological or onto-epistemological assumptions had to finesse them away in some seemingly plausible manner, often by allowing use of the atomic theory while rejecting the reality of atoms; by treating the latter as mere "helpful concepts"; by reducing them to something merely linguistic; or by arguing that epistemology and ontology were "external to science". (Two of Mach's most influential followers, Philipp Frank and young Rudolf Carnap, would pass through many stages in their rather clever efforts to seem to make their ideas fully compatible with physical science while actually rejecting the existence of atoms.)

Some presentists even questioned heliocentrism as Mach did and even rejected the representative theory of perception in its currently modified form at least as a theory with physical implications about the real world. Apparently, as most representists would see the problem, presentists wanted to restrict the scope of science back to "saving the appearances", a pre-Galilean position. But why is it so important to describe presentism and its followers? From a representist perspective, precisely because one can still discover presentism in much contemporary philosophy of science, and much of it still has Machian roots, which need to be exposed

Furthermore, from a presentist perspective one ought to benefit from recognizing a key source of earlier presentism in contemporary philosophical discussion. For example, when one hears "scientific realism" contrasted with "metaphysical realism" if a presentist makes the distinction, then he is likely to think that it basically distinguishes science from metaphysics. But if a representist should makes the contrast, however, and he is fair-minded, then it is between presentism and representism. or in onto-epistemological terms it is largely between matter-presentism and matter-representism.or at least between direct realism and indirect realism about the external world.

NOTES

[1] Jozef Sajner, "Ernst Machs Beziehungen zu seinen Heimatort Chirlitz (Chrlice) und zu Mähren", *Bohemia*, **24** (1982) 358-368.

[2] Karl Kuhn, "Ernst Machs Herkunft und Abstammung", *Heimatbildung*, **19** (1938) 269.

[3] Ernst Mach, *Analysis of Sensations*, New York, 1959, p. 30.

[4] Ernst Mach, (Vorträge über Psychophysik), *Oesterreichische Zeitschrift für praktische Heilkunde*, **9** (1863) 146-366. The pagination is not consecutive but is broken into twelve parts.

[5] Ernst Mach, *Die Prinzipien der physikalischen Optik*, J.A. Barth: Leipzig, 1921.

[6] Ernst Mach to Josef Popper-Lynkeus, Prague, December 30, 1883. This part of the letter is included in: John Blackmore, *Ernst Mach*, University of California Press: Berkeley & Los Angeles, 1972, p. 79. See also Dieter Hoffmann, ""Ernst Mach in Prag" in Ernst Mach - *Studien und Dokumente zu Leben und Werk*, edited by D. Hoffmann & H. Laitko, Deutscher Verlag der Wissenschaften: Berlin, 1991, pp. 141-179..

[7] E. Mach, *Grundriss der Naturlehre für die oberen Classen der Mittelschulen - Ausgabe für Gymnasien*, F. Tempsky: Prag & Wien, 1891. See also: Ernst Mach & Johannes Odstrcil, *Grundriss für die unteren Classen der Mittelschulen - Ausgabe für Gymnasien*, Tempsky: Prague, 1887.

[8] See Blackmore 1972, op. cit., pp. 282-285. For criticism of Blackmore see Gereon Wolters, *Mach I, Mach II, Einstein und die Relativitätstheorie*, de Gruyter: Berlin & New York, 1987.

[9] Rudolf Haller, "Grundzüge der Machschen Philosophie", *Ernst Mach - Werk und Wirkung*, R. Haller and F, Stadler (Hrsg.), Verlag-Hölder-Pichler-Tempsky: Wien, 1988, p. 72.

[10] Friedrich Stadler, "Ernst Mach - Zu Leben, Werk und Wirkung", Ibid., p. 12.

[11] Gerald Holton, "Mach, Einstein, and the Search for Reality", *Daedalus*, Spring, 1968, pp. 636-673. See also his articles "Ernst Mach and the Fortunes of Positivism" & "More on Mach and Einstein" in *Science and Anti-Science*, Harvard University Press: Cambridge. 1993, pp. 1-73.

[12] Blackmore 1972, op. cit. See also his books: *Ernst Mach als Aussenseiter*, edited with Klaus Hentschel, Braumüller: 1985 & *Ernst Mach - A Deeper Look*, Kluwer: Dordrecht, 1992.

[13] Ryoichi Itagaki, "Why Did Mach Reject Einstein's Theory of Relativity?", *Historia Scientiarum*, Nr. 22, 1982, pp. 81-95. See also "Three Batches of Reasons for Mach's Rejection of Einstein's Theory of Relativity", Blackmore 1992, Ibid., pp. 277-295.

[14] Res Jost, "Boltzmann und Planck: Die Krise des Atomismus um die Jahrhundertwende und ihre Überwindung durch Einstein", in H. Nelkowski u. a. (Hrsg.) *Einstein Symposium Berlin aus Anlaß der 100. Wiederkehr seines Geburtstages 25. bis 30. März 1979*. Berlin, 1979, 128-145.

[15] Gereon Wolters, "Atome und Relativität - Was meinte Mach?" in Haller & Stadler 1988, op. cit., p. 486, footnote 4.

[16] Paul K. Feyerabend, "Machs Theorie der Forschung und ihre Beziehung zu Einstein", Ibid., pp. 435-436.

[16.1] See Ernst Mach to Wilhelm Jerusalem, München-Vaterstetten, June 8, 1913.

[17] Seiji Takata, "Ganot's Textbooks of Physics Introduced into Japan", *Historia Scien-tiarum*, **33** (1987) 25-41. Takata also wrote two other articles on Ganot's textbooks in Japanese. See *Kagakusi Kenkyu*, Series II, No. 147, **22** (1983), 130-136; and Series II, **23** (1984) 176-180.

[18] Ibid., pp. 30-37.

[19] Ibid., p. 28.

[20] Gunter Lind, *Physik im Lehrbuch 1700-1850*, Springer Verlag: Berlin, etc., 1992, pp. 233-246.

[21] Ibid., p. 246.

[22] Ibid., pp. 266-277.

23 Ibid., p. 235.

24 Leopold Pfaundler, "Vorrede zur neunten Auflage des ersten Bandes", *Müller-Pouillet's Lehrbuch der Physik und Meteorologie -in drei Bände*, Druck und Verlag von Friedrich Vieweg und Sohn: Braunschweig, 1886, p. xiv.

25 Leopold Pfaundler, "Einleitung", *Mechanik und Akustik*, Band **I**, (Müller-Pouillets Lehrbuch der Physik und Meteorologie, zehnte umgearbeitete und vermehrte Auflage in vier Bänden herausgegeben von Leopold Pfaundler). Friedrich Vieweg und Sohn: Braunschweig, 1906, p. 1.

26 Joachim Thiele, "Ernst Mach-Bibliographie:, *Centaurus*, **8** (1963) 224-225.

27 Adolf Hohenester, "Ernst Mach als Didaktiker, Lehrbuch- und Lehrplanverfasser", *Ernst Mach - Werk und Wirkung*, Hrsg. R. Haller & F. Stadler, Verlag Hölder-Pichler-Tempsky: Wien, 1988, p. 162.

28 Ulf Höfer, *Verzeichnis des Nachlasses von Alois Höfler (1853-1922) - Teil I.*, Forschungsstelle und Dokumantationszentrum für österreichische Philosophie, Graz, 1994, p. 34.

29 This was told to the author on a visit to Prague in 1968, but is partly confirmed by the following article: Ivan Úlehla, "A Remark on Mach's Influence on Czech Universities", *Ernst Mach and the Development of Physics - Conference Papers*, edited by V. Prosser & F. Folta, Universitas Carolina Pragensis: Prague, 1991, p. 478.

REFERENCES

Blackmore, John. *Ernst Mach - His Life, Work, and Influence*. Berkeley & Los Angeles: University of California Press, 1972.

Blackmore, John, and Klaus Hentschel. *Ernst Mach als Aussenseiter*. Wien: Braumüller, 1985.

Blackmore, John (ed.). *Ernst Mach - A Deeper Look*. Dordrecht: Kluwer Academic Publishers, 1992.

Feyerabend, Paul . "Machs Theorie der Forschung und ihre Beziehung zu Einstein". *Ernst Mach - Werk und Wirkung*,. Edited by R. Haller & F. Stadler. Wien: Hölder-Pichler-Tempsky, 1988, pp. 435-462.

Folta, Jaroslav. See Prosser, Václav.

Haller, Rudolf, and Friedrich Stadler, (eds.). *Ernst Mach - Werk und Wirkung*. Wien: Hölder-Pichler-Tempsky, 1988.

Hentschel, Klaus. See Blackmore 1985.

Höfer, Ulf. *Verzeichniss des Nachlasses von Alois Höfler (1853-1922)* - Teil I. Graz: Forschungsstelle und Dokumentationszentrum für österreichische Philosophie, 1994.

Hoffmann, Dieter, and Hubert Laitko (hrsg.). *Ernst Mach - Studien und Dokumente zu Leben und Werk*. Berlin: Deutscher Verlag der Wissenschaften, 1991.

Hoffmann, Dieter. "Ernst Mach in Prag". Ibid., pp. 141-179.

Hohenester, Adolf. "Ernst Mach als Didaktiker, Lehrbuch- und Lehrplanverfasser". *Ernst Mach - Werk und Wirkung*,. Edited by R. Haller & F. Stadler. Wien: Hölder-Pichler-Tempsky, 1988, pp. 138-163.

Holton, Gerald. "Mach, Einstein, and the Search for Reality". *Daedalus*. Spring 1968, pp. 636-672.

Holton, Gerald. *Science and Anti-Science*. Cambridge: Harvard University Press, 1993.

Itagaki, Ryoichi. "Why Did Einstein Reject Einstein's Theory of Relativity?". *Historia Scientiarum.*, 22 (1982) 81-95.

Itagaki, Ryoichi. "Three Batches of Reasons for Mach's Rejection of Einstein's Theory of Relativity". *Ernst Mach - A Deeper Look*. edited by J. Blackmore. Dordrecht: Kluwer Academic Publishers, 1992, pp. 277-295.

Jost, Res. "Boltzmann und Planck. Die Krise des Atomismus um die Jahrhundertwende und ihre Überwindung durch Einstein". *Einstein Symposium Berlin aus Anlass der 100. Wiederkehr seines Geburtstages 25. bis 30. März 1979*. H. Nelkowski u. a. (Hrsg.). Berlin: 1979, pp. 128-145.

Kuhn, Karl. "Ernst Machs Herkunft und Abstammung". *Heimatbildung*. 19 (1938) 268-269.

Laitko, Hubert. See Dieter Hoffmann...

Lind, Gunter. *Physik im Lehrbuch 1700-1850*. Berlin: Springer Verlag, 1992.

Mach, Ernst. "Vorträge über Psychophysik". *Österreichische Zeitschrift für praktische Heilkunde* 9 (1863) 146 discontinuously to 366.

Mach, Ernst, and Johannes Odstrcil. *Grundriss für die unteren Classen der Mittelschulen - Ausgabe für Gymnasien*. Prag: F. Tempsky, 1887.

Mach, Ernst. *Grundriss der Naturlehre für die oberen Classen der Mittelschulen - Ausgabe für Gymnasien*. Prag & Wien: F. Tempsky, 1891.

Mach, Ernst. *Die Prinzipien der physikalischen Optik*. Leipzig: J.A. Barth, 1921.

Mach, Ernst. *Analysis of Sensations*. New York: Dover, 1959.

Nelkowski, H. (ed.). *Einstein Symposium Berlin aus Anlass der 100. Wiederkehr seines Geburtstages 25. bis 30. März 1979*. H. Nelkowski u. a. (Hrsg.). Berlin: 1979.

Odstrcil, Johannes. See Ernst Mach 1887.

Pfaundler, Leopold. "Vorrede zur neunten Auflage des ersten Bandes", *Müller-Pouillet's Lehrbuch der Physik und Meteorologie - in drei Bände*, Braunschweig: Druck und Verlag von Friedrich Vieweg und Sohn:, 1886.

Pfaundler, Leopold. "Einleitung". *Mechanik und Akustik. Band I.* (Müller-Poillets Lehrbuch der Physik und Meteorologie. zehnte umgearbeitete und vermehrte Auflage in vier Bänden). Herausgegeben von Leopold Pfaundler: Braunschweig: Friedrich Vieweg und Sohn, 1906, p. 1.

Prosser, Václav, and Jaroslav Folta (eds.). *Ernst Mach and the Development of Physics - Conference Papers*. Prague: Universitas Carolina Pragensis, 1991.

Sajner, Josef. "Ernst Machs Beziehungen zu seinen Heimatort Chirlitz (Chrlice) und zu Mähren". Bohemia. 24 (1982) 358-368.

Stadler, Friedrich. "Ernst Mach - Zu Leben, Werk und Wirkung". *Ernst Mach - Werk und Wirkung*,. Edited by R. Haller & F. Stadler. Wien: Hölder-Pichler-Tempsky, 1988, pp. 11-63.

Takata, Seiji. "Ganot's Textbooks of Physics Introduced into Japan", *Historia Scientiarum*. 33 (1987) 25-41.

Takata, Seiji. *Kagakusi Kenkyu*. Series II. No. 147. 22 (1983) 130-136. And 23 (1984) 176-180.

Thiele, Joachim. "Ernst Mach-Bibliographie". *Centaurus*. 8 (1963) 224-225.

Wolters, Gereon. "Atome und Relativität - Was meinte Mach?". *Ernst Mach - Werk und Wirkung*. Edited by R. Haller & F. Stadler. Wien: Hölder-Pichler-Tempsky, 1988, pp. 484-507.

Úlehla, Ivan. "A Remark on Mach's Influence on Czech Universities". *Ernst Mach and the Development of Physics - Conference Papers*. edited by V. Prosser and F. Folta. Prague: Universitas Carolina Pragensis, 1991, pp. 477-478.

Chapter Two

FRIEDRICH ADLER

partly by himself

Section Guide

1. Biographical Remarks by the Editors
2. "Adler as Physicist" by Joël / Einstein
3. "A Machist Forword" by F. Adler
4. Selected Mach-Adler Correspondence

1. BIOGRAPHICAL REMARKS

Many of Ernst Mach's early admirers and supporters would become much more famous than Friedrich Adler, who is little known outside of Austria, but none would have more influence on historical events. He had a knack for being at the right place at the right time. Indeed, no understanding of Mach's influence in Vienna or Zürich is possible without focusing on this remarkable young man who combined interests in physics, socialism, and Mach's world view. Furthermore, the brief nudges he gave to history would in their under-recognized way help bring down an empire, save a nation from communism, open Albert Einstein's academic path, and help give a Machist and pacifist flavor to Austrian socialism, or what was then called *Austro-Marxismus*.

Friedrich Adler (1879-1960) was born in Vienna. He was the son of Victor Adler (1852-1918), a medical doctor who would become the longtime leader of the Austrian Social Democratic Party and a major figure in the establishment of the first Austrian republic in 1918. Fritz Adler wanted to follow in his father's footsteps as a top socialist, but his father discouraged him and persuaded the young man to study science instead. He entered the Technical University of Zürich in the year 1897 where he studied chemistry, mathematics, and physics. He received his doctorate with a dissertation which was on the dependence of the specific heat of chromium on temperature. He would publish scientific articles in 1904 which were connected with his work in thermodynamics. From 1905 to 1907 he continued his scientific work in the Deutsches Museum in Munich.[1]

Adler became a *Privatdozent* at the ETH in Zürich from 1907 to 1911 which is where he also taught classes and attempted to support a family. He had married one of the many Russian students in Zürich as early as 1903 and would have three children, all girls. He would also live upstairs in the same building as a physicist of the same age called Albert Einstein and became friends with him. They were both pacifists, left-oriented, and both became great admirers of Ernst Mach.'s independent spirit.

Adler would allege that Ernst Mach's philosophy helped him resolve an apparent contradiction between Marxist philosophy and what he came to call "mechanistic materialism" (See Section III). Adler would even maintain that Mach's phenomenalism allowed him to identify matter with experience and Marxist dialectic with development,[2]

J. Blackmore, R. Itagaki, and S. Tanaka (eds.). Ernst Mach's Vienna 1895-1930. 29-59.

thereby reconciling mind and matter.

Non-Machists might find identifying matter with experience and Marxist dialectic with development to be simplistic or even bizarre, but Adler seems to have honestly thought that a serious problem had been solved and that he could treat as identical what most people would regard as very different from each other. Rudolf Carnap on a more sophisticated level would also appeal to linguistic reduction after having apparently gone through a Machian phase, but even he would stop short of identifying matter with experience. But whereas Adler's more famous contemporary would much later allege seemingly opposite things, namely, that he had both taken Ernst Mach's philosophy seriously and never taken his epistemological and ontological position seriously,[3] Adler seems to have remained almost completely loyal to Mach's theory of elements at least well into the 1920's, except for some self-questioning circa 1910 (See Section 4). But while Rudolf Carnap's late autobiographical comments are all over the map with respect to Mach thus making it difficult to be sure what his real position was, it was an absence of late publications by Adler on Ernst Mach which make it difficult to be sure what *his* epistemological and ontological positions were after the mid-twenties, as if one man published too much and the other too little about this matter for the period in question.

Indeed, Adler's attraction toward Mach's philosophy beginning around 1903 and his overwhelming interest in socialist politics would increasingly take precedence over his work in physics. One can see this happen in his publications of this early period where there is clearly an attempt to combine physics either with Mach's phenomenalism or with the physical ideas of socialist heroes like Friedrich Engels.

(in English translation)

Physics and Machism:	Physics and Marxism:
Metaphysics in Ostwald's Energeticism (1905)[4]	"A Materialistic Conception of History & Math" (1905-06)[5]
Physics as Phenomenological Science (1907)[6]	"Friedrich Engels and Science" (1906-07)[7]
"The Discovery of the World Elements" (1907-08)[8]	"What is the Use of Theory?" (1908-09)[9]
A translation into German of Pierre Duhem's book: *The Goal and Structure of Physical Theories* (1908)[11]	"The Unity of the Scientific World Picture" (1909)[10]
	"'Machism' and the Materialistic Conception of History" (1909-10)[12]

Young Adler was by no means the only phenomenalist in Zürich. For example, Richard Avenarius had been teaching a philosophy very similar to that of Mach in the city for twenty years until his death in 1896,[13] teaching which helped prepare the way for Mach's major influence in the Swiss city before the First World War. Avenarius, whose father had been a book publisher in Paris and who had influential relatives or namesakes in Moscow, seems to have had a significant influence on the large numbers of Russian medical students in Zürich, perhaps not directly since his ideas were expressed in terms of his own special jargon and neologisms, but partly through students and translations.

Avenarius also edited the journal *Vierteljahrschrift für wissenschaftliche Philosophie* for many years which by linking science with phenomenalism seems to have influenced many scientists who opponents like Max Planck would soon argue should have known better. Shortly after the death of Richard Avenarius in 1896, the young philosopher, Edmund Husserl, who had himself been earlier influenced by Mach, wrote many book reviews over a period of several years in that same journal specifically targeting what he called *Psychologismus,*[14] or treating logic in psychological terms, a supposed "offence" which both Mach and Avenarius had practiced. But in the opinion of many "realists" it seemed uncertain whether Husserl's *Phänomenologie* which he would soon be advocating was any closer to most science than the *Phänomenalismus* of Mach and Avenarius.

Avenarius's called his philosophy *Empirio-Kritizismus* and for several years many Russian Marxists used the name and were attracted to his ideas. The best-known Russian Machist was Dr. Alexandr Bogdanov who had apparently learned it from V.V. Lesevich who had studied directly with Avenarius in Zürich. Adler and his wife seem to have been responsible for translating and publishing a paper by Bogdanov on Mach in German.[15] Bogdanov, who had been trained as a medical doctor, became a leading Bolshevik during the first decade of the 20th century, second only to Lenin himself. The two men seem to have parted ways largely over policy matters concerning whether Bolsheviks should attend the Russian Duma and also about philosophy. Vladimir Lenin did not believe that phenomenalism was compatible with either science or political activism. It seemed unable to allow for causes as particular physical forces.

Indeed, Vladimir Lenin, like young Adler, was also in Zürich circa 1908. Nor was the Russian leader very happy that so many Zürich socialists had been "infected" by the phenomenalism and "idealist" philosophy of Mach and Avenarius. Nevertheless, young Adler soon became an admired socialist leader in the city which itself had a socialist government. Even Lenin seems to have liked him personally.

Indeed, when a position in physics became open at the city university in the same year, that is, 1908, Adler became a leading candidate, but a number of surprises altered matters. First, his mentor Professor Alfred Kleiner instead of supporting Friedrich Adler feared that he was too concerned with philosophy,[16] and second, Adler's Machist rejection of the reality of atoms and electrons threatened to make his appointment unpopular in the physics community as a whole. But on the "positive" side it turned out that Kleiner was even more opposed to the leading candidate Albert Einstein,[17] and *Regierungsrat* Heinrich Ernst, a high government official, preferred Adler. But since as mentioned above, Fritz Adler's heart was primarily in politics anyhow, he praised Einstein as the better candidate "objectively speaking" and tried to withdraw from competition.[18] He urged the socialist city fathers, university authorities, and Einstein himself that the latter was the right man.

Einstein gave up his job in the Swiss Patent Office in Bern and accepted a position as *extraordinarius professor* of theoretical physics at the old University of Zürich (not the ETH across the street). This did not mean that Adler accepted Einstein's special theory of relativity, in fact like many other Machists when he had time to examine it carefully as well as the general theory in 1917 he would reject major aspects of both, much to

Einstein's distress.[19] But in any case Einstein was obviously so talented and had been so discriminated against by university authorities in the past that Adler, who felt that his own mission in life was to help the downtrodden, did all he could to further his career. Adler wrote to his father about the young genius on June 19, 1908:

> I have forgotten to say who will probably receive the chair in physics. He is a man who from the standpoint of the [right[people is in principle much more deserving than myself. And when he receives it, apart from wistful regret, it will make me happy. The man's name is Einstein, who studied at the same time as I did, and we attended some of the same lectures together.
>
> Our development was rather parallel. Like myself we both married fellow students and had children, but no one supported him. For a long time he didn't have enough to eat and especially during his time as a student the professors at the Polytechnic [ETH] were disrespectful toward him and even forbade him use of the library, etc. He had no social understanding of how to get around people. Finally, he found a place at the Patent Office in Bern, but in spite of all of this misery he continued to advance his understanding of physical theory, and today he is among the most successful and best acknowledged of those who work in the tradition of Boltzmann. And Boltzmann's orientation [in physics] is in fashion, and not that of Mach.[20]
>
> Naturally, the people who treated him so badly in the past now suffer from a bad conscience. Furthermore, it is a scandal not only here but in Germany too that such a man should still be sitting in a patent office . . . . This very semester he has received his Habilitation after great difficulties which legally allows him to come to Zürich [from Bern] which he is doing now, though probably only as a transition station to a better position somewhere else. For he will almost certainly receive a call from Germany relatively soon. I accept the objective facts about what I assume will be his future course. Indeed, it is wonderful that this man who has been so much put upon will succeed. It just goes to demonstrate that *he who can, will* . . . . [21]

The following year, saw the publication of Vladimir Lenin's attack on the philosophy of Mach and Avenarius: *Materialism and Empirio-Criticism - Critical Comments on a Reactionary Philosophy* (1st Russian edition, 1909). Adler had defended Mach's position in print the previous year and Lenin used these references to criticize "Fritz" for being simpleminded. Adler, however, believed that in spite of Lenin that actually Machism and Marxism were compatible epistemologically, a point of view which other physicist-philosophers like Philipp Frank and some Logical Positivists would also accept.

The main reason Adler rejected Lenin's position seems to have been that the Bolshevik leader identified matter with substance, a position which Adler following Mach considered to be "metaphysical" as if incompatible with science. Adler also thought that Mach's philosophy had helped physics rise above "the mechanistic world view" but that Lenin as a non-scientist had failed to keep step with recent developments in the physical sciences (See Section IV).

On the other hand, Lenin rejected young Adler's position and his "compatibility hypothesis" between Machism and Marxism, apparently because it ignored the facts of consciousness, namely, that Machian "elements" or Adler's "experience" were mental or conscious while matter in the external world for Lenin and most people was physical and trans-conscious, hence they could not be legitimately identified with each other.[22] Lenin probably also thought that it would be hard to develop a practical theory of causes as political or historical forces from Mach's reduction of matter to mere groups of sensory elements or to Friedrich Adler's conscious "experience". Ernst Mach had proposed his phenonomenalism as a foundation for science, but Lenin saw, first, that most scientists rejected Mach's philosophy, and second, that a different foundation was needed in the practical world, especially if it were to be concrete and palpable enough to facilitate a socialist revolution.

Friedrich Adler, however, who only belatedly considered consciousness in an epistemological context, remained Machist, though as we shall see there would be some slight discrepancies which would come out in their correspondence. Speaking about letters, Adler, whose wife was Russian and thus could read Lenin's work, also wrote to Mach about Lenin's attack but shrugged it off as the work of an amateur in philosophy and that therefore Mach need not worry about it (See Section IV, letter #38). But as is well-known, even after Lenin came to power in Russia in 1917 he continued to attack and persecute Bolsheviks who sympathized with Mach's philosophy, a campaign which lasted for decades and not just in Russia but in communist movements around the world. By the 1930's and 1940's numerous Logical Positivists had also come under the ban since they too were regarded as "Machists".[23]

Since both Lenin and Adler were in Zürich at the same time and Adler was perhaps the most influential socialist who supported Mach's philosophy outside of Russia it is natural to suppose that Lenin's bitter attack on Machism had been partly inspired by Adler's publications. But strangely, Lenin was not as severely critical of "Fritz" as he was of Bolshevik Machists like Bogdanov. Both Lenin and Adler considered the other to have only a very "naive" understanding of philosophy, but Lenin apparently thought he saw something promising in the politically-minded young physicist for the future, and from a communist perspective he would be right. Adler's deeds would have more influence than his words.

In 1910, Fritz Adler became editor of the local Zürich socialist newspaper and he temporarily dropped his interest in both physics and Mach's philosophy, even becoming too busy as he said to continue corresponding with Mach.[24] In the following year, he was called back to Vienna largely to take over organizational matters and to help the Austrian Social Democrats discourage Czech socialists from forming their own "national" party. But the price was high. He had given up his *Dozentur* in physics in Zürich and now since his wife did not want to leave Zürich, they separated as they would often do during their rather stormy marriage of over fifty years. Their three children would also in time take sides. Indeed, one might maintain that as a firm rationalist that he patiently chose to endure a lot of hen-pecking from family members, but without letting it influence his normal behavior. Nor could he prevent the nationalistically-minded Czechs from going their own way. Indeed, his moderate stand in the Party which opposed many strikes and demonstrations and all violence lost him the support of numerous socialists in Zürich, Vienna, and Prague. Here is a list of some of his publications of the period.

"The Tactics of Austrian Social Democracy" (1909-10)[25]

"The Value of Parliamentarianism" (1910-11)[26]

"Inferiority in Internationalism" (1910-11)[27]

Perhaps understandably, when Adler became Party Secretary in Vienna (1911-1916) he threw himself into office work and short-term problems to such an extent that academic and philosophical publications all virtually ceased.[28] He was now doing what he wanted to do, but a closely related love was editing journals on Party matters. His main editorial outlets of the time and later were:

1910-1911 Editor of *Volksrecht. Sozialdemokratisches Tagblatt* (Zürich)

1911- Editorial writer for *Der Kampf* and from 1913- Co-Editor (Vienna).

1911-1914 Editor of *Das Volk* (Vienna)

1918- Editor of *Der Kampf* (Vienna) and contributor until 1934.

1938- Editor of *Der Sozialistische Kampf* (Brussels and Paris)

1942-1944 Editor of *Austrian Labor Information* (New York)

But Friedrich Adler's historically most significant moment would come during World War I and would be wildly out of character. Contrary to the expectations of many socialist leaders, when the war broke out instead of party members joining a general strike against the conflict in line with pacifist dogma, almost all supported their own country and eagerly marched in line with their respective national brothers "to fight the enemy". This happened all over the Continent on both the side of the Triple Entente of Britain, France, and Russia and on the side of Germany and Austria. But this did nott mean that opponents of the War had fully vanished. Young Adler and many others remained strongly opposed to the conflict.

Meanwhile, Ernst Mach in his half-crippled physical condition in the small village of Vaterstetten, Bavaria continued to publish, though after the polemics with Max Planck (1908-1911) where he threatened to leave "the church of physics if belief in atoms were required"[29] most of his publications were in psychology and anthropology.[30] Indeed, because of his eldest son's wish not to unduly disturb him he was only belatedly told about the War, but like Adler and Einstein he was a pacifist on principle.

Indeed, when Adler gave up his *Dozentur* in physics and returned to Vienna in 1911, he was doing the same thing which Mach had threatened to do (June 1910), except that Adler actually did "leave the church of physics", though it might be argued, first, that Ernst Mach's publications in other fields and resignation from the Austrian Academy of Sciences (April 28, 1913) constituted de facto departure, and second, that Adler retained an interest in physics as his anti-Einstein book (1920) would demonstrate.

By late 1916, a few months after Mach's death, Adler decided that he had to do something dramatic in order to help stop "The Great War" as it would be called in many circles then and later. The Austrian parliament which included many socialists as led by his father had been closed since 1914 which meant that rational political debate was almost impossible. Whatever Fritz Adler did it would have to shake the whole political system in his native country.

It seems that the Austrian Prime Minister, Count Carl Stürgkh, a strong-minded patriot who had closed the Austrian parliament even before the assassination of the Archduke in Sarajevo and refused to re-open it, was in the habit of eating his big meal of the day at Hotel Meissl & Schadn starting at about 12:30.[31] The hotel overlooked the famous Kärntner Strasse which ran from the *Staatsoper* at one end to the *Graben* at the other. The day was October 21, 1916, a Saturday. Adler sat two tables a way. At about 2:30. Adler paid his bill and began to pass by the Prime Minister who was lighting a cigar.[32] He then stopped, pulled a pistol from his coat and fired three times at the head of Count Stürgkh which was only inches away. He then half-shouted: "Down with tyranny! We want peace!"[33] Stürgkh died instantly and Adler was apprehended and thrown into jail.

Emperor Franz Josef was stunned like most of his countrymen, became depressed, and died exactly one month later. The Austrian parliament was re-opened. The new Emperor, Kaiser Karl, wanted peace as did many of the members of the reconstituted legislative body. Adler's action would add an important element to a long and decisive chain of events.

Precisely at the time that civilian leadership in Germany and Austria would seek a way out of the War causing a split with Ludendorff, Conrad, and the military leaders who still wanted to win, in England, France, and America new, aggressive civilian leaders came to power, namely, Lloyd "Hang the Kaiser" George; Georges "Tiger" Clemenceau; and the handy re-election of Woodrow Wilson for a second term allowed him to think more seriously of entering the conflict on the side of the Allies, which The United States did on April 6, 1917. Lenin would help Russia leave the War after the Bolshevik coup d'état, but the entry of America more than made up for the loss. In short, as home morale sank in Germany and Austria acerbated by a shortage of food and raw materials, it rose markedly in the West.

In spite of home front difficulties and loss of spirit, the German military continued to be successful in the field until the summer of 1918, that is, until grossly surprised by the collapse of the Balkan front where the brave but exhausted twelve divisions of Bulgarians plus an insufficient number of Germans had been holding back 29 divisions of Serb, English, Italian, and French soldiers. In mid-September 1918 a Serbian attack by six divisions aided by other nationalities suddenly routed the Bulgarians and threatened three capitals, Istanbul to the east and both Sofia and Budapest to the north.[34] As soon as the Hungarians and some Croats on the Italian front heard about the Serbian breakthrough they began to desert in order to go home, some to protect their native lands and others to help form with Serbia what would be called "Yugoslavia" or the land of the South Slavs. The resulting collapse of the Austrian front in Italy was then aggravated by young Kaiser Karl who announced that henceforth the Austro-Hungarian Empire would become a federal state, which implied that the Slavic groups in the Empire would become at least autonomous and perhaps independent. Since Slavs made up half of the Austrian army, the result by October 1918 became a race by all minorities to get home as quickly as possible. Bulgaria, Turkey, and Austria-Hungary were now virtually defenseless and made peace with the victorious Allies. This left Germany as the sole remaining member of "The Central Powers", and even the front lines in France were weakening, though not in terms of the panic which had swept all of Germany's few allies. The German request for an armistice was accepted and went into effect on November 11, 1918, though the Allies continued to blockade Germany afterwards until she fully accepted the Versailles *Diktat* half-a-year later.

Friedrich Adler of course did not cause all of this, but he had helped the ball roll toward Allied victory by acting to weaken Austrian resolve on the home front. He naturally thought his violent act would help bring peace rather than defeat and more misery for his homeland, but pacifist murderers are not the only ones guilty of wishful thinking, especially during wartime. Many Germans and Austrians would believe Hitler when he later said that "socialists, Jews, and pacifists" behind the lines "had betrayed the soldiers at the front". Did he have his fellow Austrian Friedrich Adler in mind who was all three?

Had Fritz Adler been an ordinary assassin, he probably would have been quickly executed, but Einstein served as a character witness for him and his father was leader of the Socialist faction in the newly re-opened Austrian parliament. Nevertheless, he was condemned to death in May 1917, though as a result of a brilliant speech his sentence was reduced to 18 years in prison.[35] In fact, he only served two years because the end of the War in October and November 1918 brought a government to power where the socialists were the largest party with the result that he was quickly released. In the confusion of the moment, however, just as the First Austrian Republic was being formed, his father died, but not before seeing his party eise to power.

But returning to his time in jail, Friedrich Adler as a privileged political prisoner had lots of free time which encouraged him to renew his interest in both physics and Mach's philosophy. Indeed, Mach's son Ludwig asked Friedrich Adler to proofread both the latest edition of Mach's 1896 book on thermodynamics and Mach's long-delayed book on optics.[36] Adler did both proofreading jobs carefully and conscientiously and seems never to have given an opinion on the controversial preface to the book on optics which came out in 1921,[37] a preface which announced Ernst Mach's opposition to Einstein's special theory of relativity.

Since Adler was also critical of Albert Einstein's theory at the time, it is perhaps understandable that he saw nothing suspicious about the preface. Friedrich Adler was known as a very honest person and if he had thought for a moment that Ludwig Mach had falsified his father's views he would surely have mentioned it somewhere, especially since having proofread the book he was in the best position to comment on the preface had he thought anything might be wrong with it.. Indeed, several years later he and Einstein even worked to help Mach's eldest son financially, which neither man would have done if the preface had been suspect.[38]

Still, even today there are some people who believe that Ludwig both "organized" the preface himself by combining together some of his father's previous remarks (which he may have done as editor) and distorted his father's views (which seems most unlikely).[39] The unpleasant fact which these critics want to avoid is that Mach and many of his closest philosophical followers including Adler, Jaumann, and Petzoldt strongly supported his phenomenalism in science and hence opposed the reality of atoms and by doubting Einstein's light constancy principle[40] also tended to be sceptical about his relativity theory as well.

Concerning the debate with Einstein, Adler was impressed by Einstein's work like everyone else, but when he had time to study the special and general theories of relativity closely in prison he developed his own theory of space and time (See Section II) and became very critical toward Einstein's theories. The two physicists then engaged in a long correspondence where Einstein severely criticized Adler's ideas. But gradually as the two men began to understand each other's assumptions more clearly (and Adler began to catch up with the literature from 1909 to 1917), Einstein became more tolerant,[41] since he realized that Adler's position followed from a few assumptions which were not totally unreasonable but which he himself could not accept and that Adler kept correcting his position in order to adjust to many of Einstein's criticisms. Nevertheless, when Adler finally published his much revised work in 1920 the reception by other physicists was not very favorable. Adler seems to have abandoned physics once again.

In order to understand Friedrich Adler's theory which primarily concerns foundation priorities rather than physics per se one has to keep in mind the widespread resistance to the light constancy principle. Einstein seemed to be introducing two incompatible axioms, one alleging that all motion was only relative and the other claiming that the speed of light in a vacuum was not relative. Furthermore, he seemed to be trying to arbitrarily force the two axioms to be compatible with each other, thereby apparently creating a vast number of otherwise unnecessarily paradoxical results. Above all, by his emphasizing measurement which seemed to require the application of light to measure light or as if the light constancy principle were presupposed, he appeared to be creating measurement artifacts like "relativistic mass" which would not take place if measurement did not use light or did not employ assumptions about its character and constancy. In short, numerous people thought there was something circular in his treatment of light constancy such that his theory used a flawed theory of measurement to determine reality. Thus, Adler like many other people, especially as represented circa 1920-1922 in the book *Hundert Autoren gegen Einstein* (1931),[42] attempted to develop a theory of relativity which did not rest on measurement in terms of light constancy or which attempted to "relativize" or delete the latter axiom. Indeed, even Philipp Frank, another friend of both Mach and Einstein, is said to have derived with the assistance of Hermann Rothe the Lorentz transformation without the axiom of light constancy.[43] On the other hand, one of the major reasons Max Planck seemed to adopt Albert Einstein's special theory so quickly - and his support was surely a major factor in its acceptance - was precisely his conviction that light constancy as Einstein understood it added a new absolute to physics.

Let us back up for a moment. First, while Mach's criticisms of Newton's theory of absolute space, time, and motion may have influenced Einstein, nevertheless, Ernst Mach was not as opposed to Newton's views as is commonly thought, rather, as he frequently announced he wanted to strengthen classical mechanics by interpreting it in terms of Mach's phenomenalism,[44] which if done would modify so-called "mechanistic science", a type of science which Adler violently rejected. And second, Adler's theory was not as close to Mach's view as is often imagined. Actually, he went back to Newton and like a few other people tried to translate Newton's distinction between absolute and relative space, time, motion, rest, and place in a more "reasonable" manner. Mach had simply considered all motion relative, but if we admit that there is a difference between caused or accelerated relative motion and the merely relative motion of a train which is standing still but which we in a moving train seem to see moving, then one might call the first *relative motion* and the second *merely relative motion*. Then we might want to carry the distinction over into a difference between *context-absolute* and *context-relative* . For example, to use an example which Adler could not have used, if there is a satellite fixed in a position in the sky so that it moves with the Earth, say a communication satellite, then at least hypothetically if it had a camera it could photograph a train moving which is accelerating or caused to move, but it would not detect movement of the sitting train (which we had falsely thought was moving), that is it could measure relative motion but not *merely* relative motion or in terms of our subsequent terminology it could notice context-absolute motion but not context-relative motion.

While the above approach has largely only terminological and kinematic differences from that of Newton, Adler went well-beyond this by introducing a context-conception of "Copernican systems", that is, of planets circling around any sun or celestial body and by making the heart of each Copernican system dynamic by identifying its causal foundation with a particular center of gravitation. He also alleged that there are probably many other systems in the universe where planets move around suns and that those are also kinds of Copernican systems, including our Moon-Earth-context.

There were also groups of celestial systems or what would soon be called galaxies. Of key importance, however, is that there are many subordinate gravitational centers in different systems, or in his eventual terminology in different "places, systems, and zones". As for the universe as a whole, since we are ignorant of whether it is finite or infinite it is not certain whether his context differentiation applies or if it does, then how. The point, however, is that there could well be more inclusive or "higher" systems with more fundamental or far-reaching gravitational centers such that some are more privileged than others. A problem would naturally arise where many different gravitational centers seem to overlap or be at work such that different contexts could be very difficult to fully distinguish from each other.

Concerning Albert Einstein's general theory of relativity with its equivalence of gravitational and inertial mass and its alleged "curved space", while by our referring to energy it is not a completely kinematic theory, it surely seemed to many people to be insufficiently dynamic to be the final word. It seemed incapable of being applied to local acceleration in a satisfactory manner, even if one accepts the inclusion of the so-called "Mach Principle". On the local level, most gravitational motion seems distinguishable from and very different from forced or caused acceleration, and in everyday life they are certainly explained differently. Gravitational acceleration on Earth at least in an ideal case seems to follow Galileo's law of falling bodies, but most forced or resisted acceleration on the surface of the Earth is uneven and varies with pressure and resistance in a way which seems very different, especially if influenced by human intervention.

Adler's approach places physical priority on epistemology and ontology whereas Einstein in both of his theories *seems* to place his priority not only with measurement but with a type of measurement which was often thought to be incompatible with simple measurement in terms of tapes and rods, *as if he were changing the rules of the game*. This author is not competent to judge which priorities physicists should follow or even to go very far into physical details, but when Adler's book on space and time was finally published in 1920 which seems to have been hightide both of worldwide admiration of Einstein and for severe criticism and opposition to his views especially in Germany and Austria, criticisms which were so profuse as in *Hundert Autoren gegen Einstein* that it is doubtful even today if all of them have been carefully and definitively studied.

Most people have been taught that it is more rational to reserve judgment on bizarre matters or those we do not fully understand, than to accept them merely on the basis of authority, even if it may seem wise to act *as if* many not-understood things were correct. But this means that as matters stand it is not rational for most people to accept many theories in modern physics, including quantum and relativity theory, except on a conditional or *as if* basis! Adler was quite rational in looking for a theory which seemed easier to understand, even if he couldn't eliminate all of the "bugs" from his own point of view.

In addition to proofreading and writing on the foundations of physics Fritz was busy during 1917 and 1918 in his prison cell constructing a written defense of Mach's philosophy. He put his previous articles on Mach together, made some revisions, and published them as a book with a new preface but repeated an article title to be: *Ernst Machs Ueberwindung des mechanischen Materialismus* (1918). And in retrospect this probably became Adler's most influential book (See Section III).

Concerning political action, the ex-prisoner became the hero of the hour, especially among left-wing socialists at just the right time to prevent Austria from going communist as Hungary to the east and Bavaria to the west were doing in early 1919. Indeed, had Adler not argued worker's groups out of going communist and perhaps forming a three-nation communist block in Central Europe, one hesitates to think of what might have happened such as civil war. In short, democratically-minded people even on the right owe a debt to Friedrich Adler, which has only gradually emerged as important.

Furthermore, Austrian socialists like their comrades elsewhere in Europe were becoming split into hostile camps, a Bolshevik wing which favored revolution and a moderate wing which preferred parliamentary government and evolution. But Austrian socialism with its odd Machist and Marxist combination as advocated by Friedrich Adler and some others helped reduce the communist faction in the country to impotence.

Adler became a member of the Austrian Constitutional Convention, was elected to the Austrian parliament, emerged as leader of a worker's council, and became Party secretary again.[45] These were heady days, but his great service to the First Austrian Republic at this time has often been overlooked, particularly because the situation changed so quickly. Even as early as March 1919 Karl Renner became Chancellor of Austria, and while a socialist he was no friend of Fritz Adler. On June 11, 1920 an all-party coalition took power, and by May 1922 the socialists were out of power altogether. Nor did the new government of Ignaz Seipel look with favor on the under-punished assassin. The social democrats did, however, continue to rule "Red Vienna" for many years, though in spite of the label it was far from being communist. In fact, led by Adler and others *Austro-Marxismus* gradually became more Machist than Marxist at least in matters of epistemology and ontology, that is, by identifying matter with what could be directly noticed they were closer to Mach's position than to what Lenin believed Marx's position to be.[46]

After helping to both establish and save the First Republic, Adler and his colleagues in Vienna attempted to reconstitute the Second International of socialist parties which had existed for many years before the War while also including the Bolsheviks, but Lenin tried to forestall this by setting up the so-called Third International which was exclusively communist in character. The result was that socialist parties all over Europe split their allegiance with communists looking toward Moscow and moderate socialists toward Bern and later Zürich. Adler's vain effort to keep a reconstituted 2nd International broad enough to include all socialist factions not only failed but earned that body the reputation of being "The Second-and-a-half International".

In 1923 Adler left Vienna and became founder and co-secretary of *The Socialist Workers International* which was largely a continuation of the 2nd International, whose headquarters shifted from London to Zürich and in 1935 to Brussels. After a year, Adler became sole secretary and spent most of his time in Zürich with his family while continuing to work hard to reconcile various European socialist parties. Perhaps the best explanation for the almost complete dearth of non-socialist publications written during the last 35 years of his life would be his extreme work habits as Party secretary for the 2nd International. Julius Braunthal, a longtime friend and biographer, has described how Adler behaved:

> He knew no holidays or off-days, no real vacations, since he took his work with him. His luggage never lacked a typewriter and a pile of documents since as he would say he wanted to "work in peace". He would even telephone his office daily to get a report on correspondence,

> and then answer it as best he could.
>
> He was an honest, pedantic worker; he devoted his attention to every administrative and technical detail. He wrote all of the letters, memos, and manuscripts himself as well as all of the resolutions and minutes of executive sittings. He was a victim of his own conscientious sense of duty. He would address and put stamps on important letters and take them to the mailbox himself. He was the trusted advisor of a growing mass of frightened socialist refugees from both Fascism and also Bolshevism. He devoted a considerable amount of time to helping illegal emigrants and providing them with passports, fake addresses, contacts . . . and support for the families of political prisoners and refugees.[47]

During the 1930's he spoke out sharply against both Hitler and Stalin. He even called the Moscow public attacks on old communists "witch trials". He resigned as Secretary when World War II broke out, presumably because of the International's failure to condemn all totalitarian movements, whether from the left or the right. With the German invasion of Belgium in May 1940, Adler and his family moved to New York where he helped organize a committee to help refugees from Europe. He returned to Zürich in 1946 while his wife stayed behind for a period in America. During his last years he published much correspondence by his father to other socialists and planned a biography of him as well but which he was unable to finish before his death in 1960.

Perhaps his thoughts would occasionally wander back toward his early interest in physics and Mach's philosophy, but the absence of publications on them after the mid-1920's makes such reminiscing seem speculative. But of course verbal reticence, far from changing earlier assumptions or hiding shame, may well reinforce those assumptions.

2. FRIEDRICH ADLER AS PHYSICIST
(*Vossische Zeitung*, Berlin, May 23, 1917)[48]

An Interview with A. Einstein

During his detention period in penitentiary awaiting trial, Dr. Friedrich Adler has been intensely occupied with questions of epistemology relating to the discipline of physics. Professor A. Einstein, member of the Berlin Academy of Sciences and co-creator of the theory of relativity, has offered us his opinion on the significance of Adler's investigation into classical and ideal mechanics. He has also had the goodness to help a member of our editorial staff gain insight into Adler's work and to briefly enlighten us about his train of thought.[49]

Dr. Friedrich Adler, an eager student of Ernst Mach, studied physics in Zürich at the same time as Einstein, where both men studied almost the same physical problems, while becoming closely acquainted with each other. Adler worked for several years at the University of Zürich as their only *Privatdozent* in theoretical physics. His teaching activity led him above all into epistemological questions and into the analysis of basic physical concepts in the sense of Mach. His studies showed an objectivity which also dominated his actions. Professor Einstein has informed me of the following case which showed how selfless he was.

After Dr. Adler had been *Privatdozent* for theoretical physics for several years, a special chair for an associate professor was created at which time Dr. Adler came under serious consideration. The outlook for the man who had been the only teacher in that subject in the University was especially favorable, since the officials who were to make the decision were already well-disposed toward him.

Nevertheless, the young teacher for whom such a position might even have been a matter of survival for him and his growing family reasoned that if scientific capacity were the highest consideration, then another physicist should come first. Thus Dr. Adler turned the position down with selfless objectivity and the other man received it. Professor Einstein also reported other examples of his objectivity and selflessness, deeds which would have been impossible for a doctrinaire man of academic ambition.

Then the discussion turned to Adler's latest work which had originated in prison. Generally speaking, it has followed in the same path as most of his earlier scientific publications, particularly in his eager support of Ernst Mach. The following lines are characteristic:

> "Ernst Mach has reminded us that no essentially new principle has been promulgated in mechanics. Everything done after Newton in mechanics has been merely formal, deductive, or mechanistic . . . . Ernst Mach's entire contribution to this field has been devoted to critical analysis, enlightenment, and insight into the elementary laws of classical mechanics. But already as early as 1868 he knew and spoke out about another possible mechanics, which we will designate in short as 'ideal' in which we 'recognize the dependence of the locality or condition of bodies on each other'. Mach saw in the beautiful ideal picture of mechanics which Hertz had developed twenty years later very essential progress toward this new mechanics. Albert Einstein's general theory of relativity is even more significant. Hertz and Mach both realized that the 'ideal program' was more beautiful and unified but that for practical purposes the customary mechanics was still needed. Nevertheless, both methods of representation - the classical and the ideal - can be acknowledged as equally legitimate.
>
> Adler has attempted to show in addition that classical mechanics represents the real path which bodies take in their various forms, but always possesses two components which are characterized by the previous history of the body and by its momentary relation to the rest of the world. Both of these fundamental components can be expressed in different ways using different designations. Suitable definitions can be given either in terms of speed and acceleration [kinematics] or when considering the mass of bodies in terms of inertia and force [dynamics]. The difference between classical and ideal mechanics is that in ideal mechanics neither of these "basic" components is present, but instead the real path of bodies is [empirically] described. In ideal mechanics there is neither velocity nor acceleration and neither [intrinsic] inertia nor the force of classical mechanics. In the latter as we have seen there is a kind of synthesis in which the real path of bodies is [merely] a description of the dependence of bodies on each other . . . According to Hertz and Einstein the real path of bodies is the "straightest path" [in terms of measurement which uses light and seems to presuppose the

> constant velocity of lig'.t in a vacuum], while in classical mechanics the real way is a superposition of two straightest paths [a historical and an immanent one].

One can already see from these few lines that the work which Adler with enthusiasm and unshakable eagerness has created during his days in prison custody, wants to establish the epistemological foundations of classical mechanics and their relations to the modern theory of relativity. According to Professor Einstein, the co-creator of relativity theory, Adler represents the conception - in contrast to special and general relativity - that nature itself represents or is closely connected with a preferred system of coordinates which is related to natural appearances, especially to the apparent movement of ponderable matter. According to his conception the center of gravity determines the system of bodies which he would select as the starting point of a particular coordinate system and whose movements would be described.

If the question, for example, is about describing movements on the Earth, then the origin of the coordinates would be in the center of gravity of the Earth or of the Earth-Moon system. Similarly, the origin of the coordinates of the Sun-system would lay in the center of gravity of the whole Sun-complex, and so forth. Friedrich Adler calls this kind of coordinate system generally "Copernican coordinates". The path which was followed by Copernicus had led to the center of gravity of the Sun-system and by that means Copernicus arrived at the enlargement of dynamics into something universal.

Dynamics is primarily applicable in all Copernican coordinate systems, but in each of them in the realm of mass whose gravitational center lies at the origin of the coordinate system. The realm of validity of each Copernican system fully includes that of included systems, but not [necessarily] of parts of systems. A realm of validity which includes another realm of validity represents a Copernican system of higher order. Thus dynamics in each coordinate system only holds for its realm of validity . . . .

Kurt Joël

3. A MACHIST FORWORD

by Friedrich Adler[50]

Concerning the main questions of principle in physics, Ernst Mach is a lifetime ahead of his colleagues. He could also feel satisfied that the boldest thoughts which he had conceived in his youth could by the evening of his life become widely acknowledged after a hard fight even beyond his own expectations. But this development which is connected with him is still not finished. One could even say that his basic insights only stand at the end of the first phase. This treatise which is before you would help make them better understood. It is not designed to spare you from having to read Mach's own works, but to stimulate the reader to try harder to help make the central problems easier to master.

Perhaps, the shortest path to help clarify the inner connections in the following representation is to show you how I myself came upon these problems and how Mach showed me the way to solve them. When I was sixteen I read the work by Friedrich

Engels *Herrn Eugen Dührings Umwälzung der Wissenschaft* [On Mr. Eugen Dühring's Scientific Revolution]. This work, which is of top importance for every socialist even today, was decisive for my intellectual development. This was because it forced me for the first time to face the problem of how to relate mechanistic materialism with the historical materialism of Marx and Engels.

I was at that time a persuaded follower of both theories, but for the next ten years I saw no way either to reject one or to make them consistent with each other. Thus, I was conscious of a contradiction which caused me to plunge ever deeper into science for a solution. This led me first of all to chemistry. But when after a short time I saw that chemistry while indeed a field of immense variety and practical importance rested on a disappearingly small foundation of relevant principles and could offer nothing in the way of epistemological enlightenment to help solve my problem, I transferred to physics. I worked all the way through the normal course-plan, carried out independent laboratory research, lectured in physics, but concerning my difficulty about understanding mechanistic materialism I was no better off than before.

I had naturally come across Mach's books and had enjoyed most of the interesting details in his historical representation and insightful asides, but I had made no more headway into what he was driving at than most other people. This reaction to Mach was not something special to him, but had its source in his critical and historical approach. By concentrating on making details clear in order to overcome my psychological resistance to his ideas made understanding his main point of view more difficult. Mach resisted every representation of his views as a system, because according to him every system is only valid in science for a moment and in time will fall away. But emphasizing the lack of duration of systems by being itself part of a methodological system has lasting value, and thus makes it easier for the beginner to penetrate into Mach's standpoint, since we are always in danger of seeing the trees but not the forest.

Nevertheless, I had the experience of being able to understand Mach's philosophy only after many twists and turns and in particular only after studying the work of Richard Avenarius whose ideas were as systematic in appearance as Mach initially seemed unsystematic. At the end of my studies at the Technical University of Zürich, I read Ernst Mach's *Mechanik* again and became obsessed by the sections on Newton.

In fact, I started to write down material in order to refute Ernst Mach's definition of mass, which seemed to be the core of these chapters, but it all ended two years later [1904] with my complete adoption of Mach's point of view, and the knowledge that his proposed foundation of science is more certain and more complete than mechanistic materialism. (If one wants to know more about my earlier perspective, then look at my paper "On the Metaphysics in Ostwald's Energeticism".)

The solution of my problem was now clear. I now recognized the source of the contradiction between mechanistic and historical materialism and the way to overcome it. Later, I would learn that in his youth Mach himself had faced a similar contradiction between two ways of thinking, namely, mechanical materialism and Kant's approach.

At the same time as I solved my problem I noticed that the position of physics within the scope of the sciences as a whole was not being properly taught. In contrast to other disciplines not enough attention was being paid to how physics students were initiated into physics. Indeed, it was superficially supposed that all physical fields were not only separated from psychology in a fundamental way but that this was "quite natural".

The book you are reading now has as its goal a systematic introduction to physics from a Machian standpoint. This is both possible and necessary. We would present the problems but will not solve them all. Indeed, your author would be happy if he could add a single foundation stone since many will be needed.

From 1907 to 1910 in the University of Zürich I lectured in a course which was titled "Epistemological Foundation for the Foundations of Physics" and part of that can be found in Chapter Three of this book which has the name "Introduction to Mach's Theory of Elements", The other chapters are taken from articles written from a Machian perspective which were written by me over the last fourteen years. I have only revised those articles to the extent needed to avoid repetition. Since I am currently engaged in much more pressing work, I cannot at this moment provide fully new material, desirable as that admittedly might be. Concerning the fifth chapter and the second part of the sixth, they contain polemics [largely in defense of Mach against criticisms by Planck and Lenin] which it would take too much trouble to remove hence they remain unaltered.[51]

I have been guided in my choice of what material to include in this book by ease of comprehension. No special expertise will be needed from natural science or mathematics. Every reader who is more familiar with historical materialism than with the natural sciences will grasp what is intended if he quickly reads the first chapter and then the sixth one.

In this book I can only touch on how Mach's ideas are related to those of Kant, but since I have been working on a more detailed analysis for some time, I hope to have an occasion to finish it. As an introduction [*Vorarbeit*], I have been studying the issue of space and time in physics for the last two years while in prison. In this work, which I hope will appear in print soon, Mach's position with respect to relativity theory will be treated.

On the relation of Mach to Kant a series of treatises exist but they are all from a Kantian standpoint and they include much misunderstanding of Mach's ideas. Most of these stem from the school of Alois Riehl. On the basis of [Hermann] Cohen's interpretation of Kant, Max Adler has attempted to refute Ernst Mach in his paper "Marx und Mach" which was printed in *Marxistische Probleme* in 1913. But in spite of this rather major difference between us, we have arrived at essentially the same conclusion about Marx's "materialism" . . .

Originally, I had wanted to include a short sketch on the significance of Mach's ideas for the theory of relativity in the book you are now reading which reflected what I had taught in my lectures at the University of Zürich. But recent study has indicated to me that I can no longer simply accept Mach's standpoint about a distinctive reference system; a critical discussion is needed about it. Since these investigations have been growing in length to twice the size of this book, and since it will also presuppose some advance acquaintance with mathematics which will narrow the circle of readers, I have decided to keep it completely independent from this introductory work on Mach's philosophy.

Nothing contradicts the spirit of Mach's approach more than a dogmatic recitation of his ideas. Only a critical to and fro can do justice or be fruitful. At the end of the second chapter of his *Mechanik* he decisively emphasizes this point: "It would only be a subjective, short-sighted conception of science were I to expect that my views would be accepted by my contemporaries in the absence of vigorous contention. The history of science teaches us that subjective, scientific world pictures from mere individuals will always be corrected in detail by others and at least partly submerged in the process, and that the world picture which is finally adopted by mankind will after a long period only include selected highlights from the most significant human beings. We as individuals can do nothing better than to make the leading features of our picture clear."

If the work before you helps us become conscious that Ernst Mach's world picture is already in need of minor corrections and additions and that this is the best that natural science can offer, then it has already achieved its purpose.

Stein Penitentiary on the Danube
Early October 1918

Friedrich Adler

4. SELECTED MACH-ADLER CORRESPONDENCE

#38[52]

Friedrich Adler to Ernst Mach Zürich, July 23, 1909

Highly Respected Herr Professor:

My wife[53] wants to thank you for being so friendly as to send her the Russian edition of Mach's *Mechanik*. She has read the different translations through and found that they keep getting better. While the translation of *Analyse* left much to be desired, that of *Mechanik* is quite good.

In Russia, "the fight over Mach" has continued. The enclosed copy of a letter from Kautsky[54] which I had published in *Der Kampf* should interest you. Unfortunately, it won't help much. Plechanov[55] has already declared that Kautsky understands nothing of the situation, and in the Bolshevik wing the situation has become violent. Lenin, who is a colleague of Bogdanov,[56] has published a book of 440 pages under the title of *Materialismus und Empirio-Kriticismus- Kritische Bemerkungen über eine Reaktionäre Philosophie* in which you, Avenarius, and all of your many followers are soundly insulted [*beschimpft*]. All of the arguments they can find which reveal how little they understand of the matter are included. Lenin had not considered philosophy earlier, and in order to learn "why people seem so crazy about it" has concentrated on it [here in Zürich] for a year in order to understand it. He has certainly worked hard and struggled his way through the literature, but naturally has not had the time to think matters through. He considers your theory of elements to be just a sophistical trick! I only know the book in segments which my wife has translated for me, but that has been enough to clearly see that there is no serious argument to be found in it.

Best wishes from my wife and hearty greetings from yours truly. Your servant

Sincerely Yours

Dr. Fr. Adler

#41

Ernst Mach to Friedrich Adler Vienna, September 20, 1909

Highly Respected Herr Colleague:

Thank you very much for taking the trouble to give Planck a detailed answer. I don't know when your treatise will appear, nor can I judge how the editors of that journal will respond. You remember how Ostwald responded to what you sent him. Should it turn out that the weekly journal stands on Planck's side, then they may not accept your essay. Such a response would neither wound nor frighten me. There is it seems a right dubious philosopher in Berlin, Ferdinand Jakob Schmidt,[57] who has at once seconded Planck with an article in *Preussischer Jahrbücher* called "Der Kampf der Mechanik gegen den Psychologismus" [The Fight of Mechanics against Psychologism], which is so dumb that I certainly will not answer it.

Petzoldt[58] has also sent a reply to Planck which will appear in Verworn's journal, but which also has not yet appeared. But in this case, the Berlin clique is hardly capable of interfering.

If you could occasionally tell me a bit about what Valentinov[59] is driving at in his two books [in Russian], I would be very grateful..

I only talked with Robert Seidel[59.1] for a half-hour, but he made a good impression on me. We only spoke about school-reform. I am pleased about your *Einleitung in die Physik*. My *Einleitung in die Optik* will be similar. I have been waiting some time for Cornelius's book on epistemology to come out.[60] I don't really know why he still denies a relationship to Avenarius. Petzoldt polemisizes against what I call "economy of thought" in favor of what he calls "stability". But I don't see that the difference means much. I refuse to fight over words. With heartiest greetings, your servant

Dr. Ernst Mach

Do you need the copy [of your reply to Planck]?

#46

Friedrich Adler to Ernst Mach Zürich, January 15, 1910

Highly Honored Professor:

Concerning my book *Einleitung in die Physik* on which I am currently working I still have to fight with a series of difficulties. In order to make it as systematic in the sharpest possible way I need systematic connecting links, and I don't have them. It would help me a lot if I could learn your position on an issue which has concerned me a lot for several years, but a presupposition is that you will not find discussing the matter inconvenient.

I am talking about something I wrote in my Ostwald *Schrift*,[61] and above all a comment on pages 14-15. Here and in other places I represented the standpoint that elements do not require or necessarily have a connected self, ego, or subject and that they are superfluous for our worldview, or are [in independence from elements] even metaphysical. My question is: Do you think it is necessary to *assume* that there are elements which exist as objects in the absence of subjects? Obviously, such elements are not directly given [in consciousness], but what if one assumed them hypothetically as you seem to do in your *Analyse der Empfindungen* in the the places I have cited?

Clifford[62] answered the question in his *Natur der Dinge an sich* [63] opposite to how I answered it at the time. "Can a perception exist by itself without being conscious at all?" He answered "Yes",[64] and indeed as if the answer were conditioned by the theory of evolution. But this opinion seems to create the greatest difficulties for the theory of elements in the discipline of biology. Obviously, science can provide an economical representation of the given (which identifies the self with a group of interelated elements), but long ago before the existence of human beings, where no self or ego could be associated with elements, wouldn't the elements have had to have been unconscious objects, and identification of the self with groups of elements a later product of evolution? That is Clifford's solution, while Avenarius as far as I know would not have agreed, still, the difficulty seems to reappear in a different place.

In my Ostwald-Treatise § 7 & 8 I tried to resolve the difficulty by means of "the constant possibility of connecting links", which neither I nor anyone else would be likely to reject. I therefore did not assume the presence of individual [sensory] elements [before there were human beings] but of hypothetical connecting links. But today this solution seems untenable to me, and is actually a kind of solipsism. It also has other curious consequences. Willy[65] once assumed that we have to think at an early time in the process of evolution that the nervous system had not developed very far [such that there was no consciousness]. Berkeley alleges that God is watching [what we are not]. Gerardus Heymans[66] adds the view that there is: "an ideal, universal, and immanent observer who has sense organs like ours but of infinitely greater sensitivity."

These are all rather ghastly conclusions, but if we don't try something the situation becomes worse. Clearly bounded experience will blur. The barrier against metaphysics will dissolve. If one allows elements which are not experience it will be harder to resist other non-experience. A question of economy thus arises from a question of principle and with that at least provisionally part of the certainty of my standpoint disappears.

In any case, if we succeed in constructing the entire given world out of elements such that we do not need a world "in-itself" which is behind or parallel, then it should be most economical to construct the world before there were human beings also out of elements. But it is precisely this insight into the time before man which makes the solution of this task so difficult and pressing concerning the world of experience. This concern raises a whole series of problems which I can no longer avoid. Above all, how far should the world picture be enlarged by elements which are outside the realm of experience. When should one give precedence to the standpoint of naive realism[67] that the greenness of leaves, for example, is a lasting quality contrasted with what is actually being experienced and when is this merely to be assumed in a hypothetical manner and limited to spatial factors?

At least for today I would very much appreciate it if I could learn more exactly how you think about these matters, that is, especially about my views at that time concerning these issues. Also, the transition to a spatial way of thinking has not been entirely worked out in my article against Planck. The principal distinction is between perceptions of heat on the one side and spatial characteristics on the other side. The first consist of individual elements, but the last of whole groups of elements of which, however, only individual ones can be observed. But the significance of the latter consists precisely in their relative stability.

I am now reading Gerardus Heymans. Your remark in the *Mechanik* that you "follow him with great pleasure" is something I can agree with with my whole heart. He is truly surprisingly clear. I had previously avoided his *Metaphysik* because of the title, but I now see with agreeable surprise that he defines "metaphysics" in a completely different way[68] and that this book is at least as useful as *Elemente und Gesetze*.

Incidentally, I would like to relate to you something about a work by a *Ferdinand Reinhold* on Mach's epistemology which you didn't know. I now see that it is identical with a Leipzig doctoral dissertation by a *Reinhold Schultz* - It is a bibliographical curiosity - It is word for word identical up to the name of the author himself. The work raises certain problems in theory of elements, but contributes little to their solution.

With best greetings, your thankful servant

Dr. Fr. Adler

#48

Ernst Mach to Friedrich Adler — Vienna, January 23, 1910

Highly Honored Herr Doctor:

I cannot delay my answer any longer. First of all, please ignore its incompleteness. If one has the purpose of philosophizing about a scientific world picture it matters: 1) which phase in the development of a world view one starts with, and 2) which goal one pursues. If one starts with the organization and interrelation of diverse parts, then one has already assumed a rather developed worldview about nature.

My starting point, however, when I wrote my *Analyse der Empfindungen* [1886] was different and more primitive. I tried to model my outlook on the standpoint of a child, who is just beginning to distinguish his body from his environment. It is in this way that we learn the mutual dependence of elements on each other. This in turn leads to concepts such as subject and object and from a consideration of the self to mental organization of these factors, but not to anything which is very fundamental. If one already recognizes the elements but not yet the self or ego, then one still lacks the ground to distinguish those elements linked to the self from those which are not linked to it.

First comes the analogy based on recognizing the bodies of other people that we are a body too; nor is this something metaphysical but a scientific conclusion [even if it is not as certain as direct observation]. According to your point of view, Elements which do not belong to me don't exist. Also, how could one explain how elements might be connected both to a central nexus or theory and to a side or opposed theory as well. Your theory totally changes the position and significance of elements.

Because I only seek to rationally construct a standpoint in physics, I have no problem with the double dependence of elements, for such has remained at any time experimentally provable, and has nothing to do with philosophy or metaphysics. Given such a standpoint I can also easily assume on the basis of analogy that there are probably other selves or egos with their own elements; I can even extend this conception to animals, plants, and inorganic bodies. This hypothesis can serve as a provisional rounding out of the world picture and provides hope for the future development of biology . . .

I also understand W.K. Clifford in this sense, and even if the elements in themselves are treated as things-in-themselves, which W.P. Preyer[69] [pure theory of Perception (nach Grassmann)][70] also seems to share, none of us is sympathetic with premature speculation about what happened in the world before there were human beings. The value of a hypothesis is determined by sound biological investigation. Nevertheless, speculation can produce for us a hypothesis. Provisionally it seems to me that we ignore an entire side of experience if we try to do without hypotheses.

About your hypothetical relational nexus, while it might be regarded as a kind of molecule in physical terms, in biology it could well be regarded as a living cell. Is that what you meant to suggest?

But I have not cultivated these kind of issues further, because I feared that I was moving too close to metaphysics where there is no solid ground of experience at all. Concerning elements, Petzoldt has a tendency to lean toward the standpoint of naive realism, which at least provisionally I cannot do.

I think it is most important for our goal to relate elements with concepts of space and time. This will not satisfy you, but more can be discovered in *Erkenntnis und Irrtum*, Chapter 1, particularly noting Page 12 - "Space and Time Physically Considered" P. 434 - "The Meaning of Natural Laws" p. 449 and following - "Sur le rapport de la physique avec la psychologie" (in Binet's *L'Année Psychologique* **12** (1906) 303).

Meanwhile, Hearty Greetings

Dr. Ernst Mach

#49

Friedrich Adler to Ernst Mach — Zürich, February 11, 1910

Highly Respected Herr Professor:

As always I have forgotten to thank you for your friendly letter and the information which is so valuable for me.

When you say "that we ignore a side of our experience when we decline to consider hypotheses (such as the subject not belonging to elements)", that is confirmation for me that when I was concerned with the problem at the time in my Ostwald treatise that I had a different standpoint than you. You say: "I have not cultivated these kind of issues further, because I feared that I was moving too close to metaphysics where there is no solid ground of experience". But it seems to me that one must clearly announce where this chasm [*Abgrund*] starts, and that is in biology. For physics, a parallel "thing-in-itself" to experience is superfluous, but outside of physics we are dependent on something "in-itself" because experience is lacking.[71]

The only thing we can do is to construct what is not experienced as economically as possible, and that is done through elements. I believe that this question can be handled without either philosophy or metaphysics and also without speculation about the world before human beings [*die Urwelt*], but only if we take steps to link the different branches of science with each other. You introduced elements into physics in order to make that field compatible with psychology. The question is how the relation of physics to biology currently stands, after the introduction of elements. The mechanistic way of thinking alleges that there is no insurmountable obstacle to finding knowledge of the world before there were human beings to experience it. If we abandon their approach for elements, then we must also consider the possibility of applying elements to biological problems. it seems to me that the theory of elements can also be applied to biology.

You have differently interpreted the hypothetical relational nexus [*Centralglieder*] which I assumed at that time at least as I understood it. I conceived of a relational nexus within experiences, a hypothesis, which I today as already said, can no longer hold as correct. Indeed, from my current point of view I absolutely cannot understand what you could mean by "the relational nexus in physical terms is a molecule for other people." For the possibility seems self-evident to me that what belongs at the same time to a subject and an object would belong at the same time to two [incompatible] objects, which seems unacceptable to me.

I am primarily interested in the problem of the logical-empirical genesis of physical concepts. And it appears to me that we are still faced with a gap. Your representation of mechanics, thermodynamics, and so forth starts from characteristics such as speed and temperature and then arrives at higher concepts such as mass and capacity. The basis of the whole system consists of elements. But it seems to me that the transition from elements to these characteristics is lacking, that is, that the analysis of this transition is in no way complete. I believe that I have succeeded in making clear this transition in the case of heat perception to temperature. But I believe I have failed to make the transition clear from elements to velocity, important as that is. This is indeed much more difficult because further development is needed in order to overcome psychological resistance.

I feel certain that you meant the same thing when you wrote: "I think that it is most important for our goal to relate elements with concepts of space and time." I am sending you at the same time an essay which I have published in *Neue Zeit*, which will probably interest you concerning the discussions which are going on in the socialist camp. In a few weeks I will probably come to Vienna for a few days and hope to meet you in your best health.

With Respectful Greetings, Your Servant

[Friedrich Adler]

#51

Ernst Mach to Friedrich Adler — Vienna, March 16, 1910

Highly Honored Dear Herr Doctor:

I thank you and your dear wife heartily for your friendly visit. Reading your manuscript and hearing your plans has been very worthwhile for me. You attack your problems seriously and it seems to me from the right scientific angle. If you were not able to benefit much from what I said, that is because since 1898 I have been clumsy and unskilled in verbal expression. But in any case, I hope to send you soon a draft of my *Einleitung in die Optik.*[72] In this work, I no longer think any more about philosophers or physicists, who from the very beginning have followed other paths, but only about reasonable people. Indeed, I am mindful that I should no longer use the expression metaphysical or anti-metaphysical. That you and I have different march plans could well be beneficial.

With heartiest greetings to you and your wife from my wife and me, your old friend who values you highly

Dr. Ernst Mach

Please give my best compliments to Herr Professor Einstein.[73]

#53

Friedrich Adler to Ernst Mach Zürich, April 7, 1910

Highly Honored Herr Professor:

Please excuse me for not having answered your friendly letter and card and for sending me Dingler's publication[74] which interests me a lot and for which I have not yet thanked you.

At the beginning of this month I took over the editing job of our local socialist party organ, *Volksrecht*, which appears daily. This provisionally means that I will have no time to occupy my mind with epistemology. Also, unfortunately my work [*Einleitung in die Physik*] has not made as much progress as hoped and has to be reworked before I can send it to the press.[75]

Our discussion in Vienna was extremely worthwhile for me, and my wife also heartily enjoyed making your personal acquaintance. If you could send me your draft of *Einleitung in die Optik*[76] that would please me greatly.

I have finally received Riehl's book,[77] but after skimming through it there is not as much there as I had expected. Einstein's work has already appeared and you could well already have it. With the heartiest wishes and greetings for your wife and yourself from us both

Your True Servant

F. Adler

NOTES

1 Rudolf Haller & Friedrich Stadler (eds.), *Ernst Mach -Werk und Wirkung*, Hölder-Pichler-Tempsky: Wien, 1988, p. 259.

2 Friedrich Adler, *Ernst Machs Überwindung des mechanischen Materialismus*, Wien, 1918, p. 137. It is not clear whether Adler distinguished between conscious experience and unconscious experience.

3 For Carnap's feeling that "he was the first person to take Mach's doctrine [phenomenalism] seriously" see his autobiographical remarks in *The Philosophy of Rudolf Carnap*, edited by Paul Arthur Schilpp, Open Court: La Salle, Illinois, 1991 (1963), p. 16. For his view two pages later that all epistemological and ontological positions for him were mere "modes of speech" see page eighteen.

4 *Die Metaphysik in der Ostwaldschen Energetik*, Leipzig, 1905. For an article with virtually the same name see: "Bemerkungen über die Metaphysik in der Ostwaldschen Energetik", *Vierteljahrschrift für wissenschaftliche Philosophie*, **29**. N.F. 4 (1905) 287-333.

5 Fritz Tischler (pseud.), "Materialistische Geschichtsauffassung und Mathematik", *Die Neue Zeit*, **24**, issue 27, 2 (1905/06) 25-30.

6 *Die Physik als phänomenologische Wissenschaft*, [ohne Ort], 1907.

7 "Friedrich Engels und die Naturwissenschaft", *Die Neue Zeit*, **25**, issue 19, 1 (1906/07) 620-638.

8 "Die Entdeckung der Weltelemente (zu Ernst Machs 70. Geburtstag)". *Der Kampf* **1** February, issue 5, (1907/08) 231-240 (auch ins Russ. übersetzt). See also "The Discovery of the World

Elements", *International Socialist Review*, **8**, April, issue 10 (1908) 577-588.

9 "Wozu brauchen wir Theorien?" *Der Kampf*, **2**, issue 6 (1908'09) 256-263.

10 "Die Einheit des naturwissenschaftlichen Weltbildes", *Naturwissenschaftliche Wochenshrift*, **8**, issue 52 (1909) 817-822. (included in Adler (1918) 112-132.)

11 Pierre Duhem, *Ziel und Struktur der physikalischen Theorien*, Übersetzung von Friedrich Adler, Mit einem Vorwort von Ernst Mach, Leipzig, 1908. The French original was titled: *La théorie physique, son objet et sa structure*, Paris, 1906.

12 "Der 'Machismus' und die materialistische Geschichtsauffassung, *Die Neue Zeit*, **28**, 1 (1909 /10), Feuilleton, issues 25/26, 671-682.

13 See Richard Avenarius, *Prolegomena zu einer Kritik der reinen Erfahrung* (1876), *Kritik der reinen Erfahrung* (1888 & 1890), and *Der menschliche Weltbegriff* (1891).

14 Edmund Husserl, *Logical Investigations*, Volume One, translated by J.N. Findlay [with some changes] , Humanities Press: New York, 1970, selected short parts from Sections 55 & 56, pp. 204-211.

15 See Alexander Bogdanov, "Ernst Mach und die Revolution", *Die Neue Zeit*, **26** (1908) 695-700. This was originally published as a foreword to a Russuan translation of Mach's *Analyse der Empfindungen* in which first appeared in 1907.

16 Rudolf G. Ardelt, *Friedrich Adler*, Bundesverlag: Wien, 1984, pp. 159-161 & 166.

17 Ibid., pp. 164-165.

18 Ibid. p. 166.

19 Friedrich Adler, *Ortzeit, Systemzeit, Zonenzeit, und das ausgezeichnete Bezugssystem der Elektrodynamik. Eine Untersuchung über die Lorentzsche und die Einsteinsche Kinematik*, Wien, 1920.

20 Boltzmann was widely thought of as the champion of atomism in physics at the time, even though in fact his own philosophical position was a complicated one which evolved considerably. See John Blackmore, *Ludwig Boltzmann - His Later Life and Philosophy 1900-1906*, two volumes, (Boston Studies in the Philosophy of Science #168 & #174), Kluwer: Dordrecht, 1995.

21 Ardelt 1984, op. cit.., pp. 163-164. The original of the letter can be found in the *Adler Archiv* in the library of the *Verein für Geschichte der Arbeiterbewegung* in Vienna.

22 V.I. Lenin, *Materialism and Empirio-Criticism*, International Publishers: New York, 1927, pp. 34-35 & 45-46.

23 See Miodrag Cekic, "The Influence of Mach's Philosophy in Yugoslavia" in *Ernst Mach - A Deeper Look*, edited by J. Blackmore, Kluwer: Dordrecht, 1992, pp. 379-402.

24 F. Adler to E. Mach, Zürich, July 23, 1909. (See letter #38 in Section IV.)

25 "Die Taktik der österreichischen Sozialdemokratie", *Die Neue Zeit*, **28**, issue 9, 1 (1909/10) 306-315.

26 "Der Wert des Parlamentarismus", *Der Kampf*, **4**, issue 9 (1910/11) 413-415.

27 "Minderwertig im Internationalism", *Der Kampf*, **4** , issue 11 (1910/11) 495-499.

28 Friedrich Adler, "Ernst Machs Überwindung des mechanischen Materialismus", *Die Neue Zeit*, **35** (1916) 111. This article should not be confused with the book of the same title which came out in 1918, though the article was included virtually unchanged in the book.

29 Ernst Mach, "The Leading Thoughts of My Scientific Epistemology and Its Acceptance by Contemporaries" in *Ernst Mach - A Deeper Look*, edited by J. Blackmore, Kluwer: Dordrecht, 1992, pp. 138-139.

30 Joachim Thiele, "Ernst Mach-Bibliographie", *Centaurus*, **8** (1963) 219-221.

31 Ronald Florence, *Fritz - The Story of a Political Assassin*, The Dial Press: New York, 1971, p. 1.

32 Ibid., p. 2.

33 Ibid., p. 3.

34 While events like these may be common knowledge to military historians and need not be cited for them, most of us interested readers often do need assistance. A good place for us to start would be William L. Langer, "The First World War" in *An Encyclopedia of World History*, edited by William L. Langer, Houghton Mifflin: Boston, 1962 (1940), pp. 913-954. Information on the Serbian offensive and collapse of Bulgaria can be found on page 248.

35 Rudolf Haller & Friedrich Stadler (Hrsg.), *Ernst Mach - Werk und Wirkung*, Hölder-Pichler-Tempsky, Wien, 1988, p. 259.

36 Ernst Mach, *Die Principien der Wärmelehre. Historisch-kritisch entwickelt*, dritte Auflage, Barth: Leipzig, 1919; *Die Principien der physikalischen Optik.. Historische und erkenntnispsychologisch entwickelt.* Barth: Leipzig, 1921.

37 Friedrich Adler to Ludwig Mach. Vienna, September 6, 1917. An English translation can be found in Blackmore (1992), op. cit., p. 159. The original is located in the Austrian National Library *Handschrift Abteilung*.

38 Friedrich Adler to Albert Einstein, Bad Nauheim, June 25, 1927. The original is located in the *Verein für Geschichte der Arbeiterbewegung - Bibliothek*, Vienna.

39 Mach's foreword to his *Optik* was dated June 1913. For remarks by Ernst Mach from other sources during the same period which are similar to or compatible with his position in the foreword please see John Blackmore (ed.), *Ernst Mach - A Deeper Look*, Kluwer: Dordrecht, 1992, pp. 23, 91, & 148. For Friedrich Adler's and Ludwig Mach's opinion about Ernst Mach's position on relativity theory see pp. 159-162. For articles which suggest reasons why Ernst Mach seems to have been so reserved or sceptical about Einstein's theory of relativity see 263-276 (Holton) and 277-295 (Itagaki). For reasons given in Mach's correspondence see letters from Joseph Petzoldt to Ernst Mach dated September 22, 1910 and June 1, 1911. They have been published in John Blackmore & Klaus Hentschel (eds.), *Ernst Mach als Aussenseiter*, Braumüller: Wien, 1985, pp. 84-85 and 91-92.

40 See Enstein's correspondence with Friedrich Adler and Joseph Petzoldt. These are largely included in *The Collected Papers of Albert Einstein, Volume 8, The Berlin Years: Correspondence, 1914-1918*, translated by Ann M. Hentschel, Klaus Hentschel consultant, Princeton University press: Princeton, 1998; and in Joachim Thiele, "Briefe Albert Einsteins an Joseph Petzoldt ", *NTM*, Leipzig, **8** (1971) 70-74.

41 Albert Einstein to Friedrich Adler, Berlin, September 30, 1918. An English version appears in*The Collected Papers*, Ibid, pp. 660-661. For Einstein's criticisms of Adler's physics see his letters of August 4, 1918 and September 29, 1918 in *Colected Papers*.

42 Hans Israel, Erich Ruckhaber, & Rudolf Weinmann (hrsg.), *Hundert Autoren gegen Einstein*, R. Voigtländer Verlag: Leipzig, 1931. [It has been alleged that this book was actually published in 1921 during the height of the controversy over Einstein, but it says 1931 on the cover and books were mentioned which were published later than 1921.] For a later picture of the controversy over Einstein see: Klaus Hentschel, *Interpretationen und Fehlinterpretationen der speziellen und der allgemeinen Relativitätstheorie durch Zeitgenossen* Albert Einsteins, Basel, 1990, p. 131ff.

43 Laszlo Tisza, "Philipp Frank and Physics" in *Philipp Frank 1884-1966*, p. 8. This was a thirty-one page memorial pamphlet with an annonymous editor. We have included several short sections from this pamphlet in the chapter on Philipp Frank.

44 More literally, Mach often expressed himself as if Newton had not always been faithful to his empirical intentions. In fact, the first edition of Newton's *Principia* was clearly in the double-dualist tradition of Galileo, Descartes, Boyle, and Locke in that mind is distinguished and separated from matter and we have only indirect or represen-tational acquaintance with matter. Isaac Newton's "empirical" remarks in later editions of his great work were apparently intended to help pacify objections from Leibniz, Berkeley, and other critics without in fact changing his basic assumptions or his scientific views. For Mach's interpretation of Newton see his *Science of Mechanics*, Open Court: La Salle Illinois, 1960, pp. 226-317. For Newton's "empiricism" see his queries and side-

remarks in the last two editions of the *Principia* which were published during his lifetime (1713 & 1726).

45 Haller and Stadler (1988), op. cit. p. 259.

46 Another strong supporter of Mach in the Austrian Social Democratic Party was David Josef Bach (1874-1947) who had studied under Mach directly and who later became a journalist, friend of Arnold Schönberg, and Party librarian. (See Kurt Blaukopf, *Pioniere empiristischer Musikforschung*, Hölder-Pichler-Tempsky: Wien, 1995, pp. 159-167 & 197-199.)

47 Julius Braunthal, *Viktor und Friedrich Adler*, Verlag der Wienerbuchhandlung: Wien, 1965, p. 299.

48 Kurt Joël, "Friedrich Adler als Physiker", *Vossische Zeitung*, Berlin, May 23, 1917, No. 259, Morning Edition, p. 2.

49 In the absence of Adler's manuscript of the time and the fact that he later revised his views to meet many of Albert Einstein's objections before publishing his book: *Ortzeit, Systemzeit, Zonenzeit und das ausgezeichnete Bezugsystem der Elektrodynamik. Eine Untersuchung über die Lorentzsche und die Einsteinsche Kinematik*, Wien, 1920, it can be very difficult to fully reconstruct Adler's position. The journalist who interviewed Einstein (Kurt Joél) only gives an overview which largely represented his understanding of Einstein's understanding of Adler's position at that time. A year later, Einstein would severely criticize Adler's position. See footnote #41 for reference to the exact letters in which the criticisms were made.

Philipp Frank in his biography of Einstein relates a story that Adler's manuscript was sent by prison authorities to psychologists in order to determine if he was sane. His father and mother apparently hoped that a negative judgment might free their son. It didn't work, though his death sentence was commuted. Frank received a copy of the manuscript but the author of these lines has not located it yet. See Philipp Frank, *Einstein: His Lfe and Times*, Alfred A. Knopf: New York, 1947, pp. 174-175.

50 Friedrich Adler, "Vorwort", *Ernst Machs Überwindung des mechanischen Materialismus*, Verlag der Wiener Volksbuchhandlung Ignaz Brand & Co.: Wien, 1918, pp. 5-11. We have not included Adler's footnotes.

51 The article criticizing Planck which was included in Adler's book was: "Die Einheit des physikalischen Weltboldes", *Naturwissenschaftliche Wochenschrift*, **8** (1909) issue page 52.

52 The numbers before each letter are those originally given in the Adler Archiv Mappe 130 which is now part of the library of the *Verein für Geschichte der Arbeiterbewegung* in Vienna. Mach's letters in this article were first published in German in John Blackmore & Klaus Hentschel (Hrsg.), *Ernst Mach als Aussenseiter*, Braumüller: Wien, 1985. Adler included the text of a letter from Mach from April 1913 on page 28 of *Ernst Machs Überwinding des mechanischen Materialismus* which is not included in the Adler Archiv Mappe 130 or the publications based on it. The most complete publication of Mach-Adler letters can be found in Haller & Stadler (1988), op. cit, pp. 258-305. (See footnote #35 above.)

53 Katharina ("Kathia") Jakovlevna Adler. Her birthname was Germanischakaya.

54 Karl Kautsky (1854-1938) was a prominent Austrian socialist and was at that time editor of the journal *Die Neue Zeit*, where Adler was then publishing many of his articles. While not in full agreement with Adler on philosophical matters Kautsky nevertheless defended his colleague against Lenin's criticisms.

55 Georgi Plekhanov (1857-1918) while in Switzerland and as co-editor with Lenin of the Russian journal *Iskra* put forward a dualistic epistemology which identified the external physical world with what lay outside all sensation and consciousness. Even after he more or less joined with the Mensheviks, Lenin still used his ideas to construct his own version of Marxist theory of knowledge, a theory incompatible with Machist and Positivist presentism which Lenin identified with Berkeleyan and Humean "idealism".

56 Alexander Bogdanov (1873-1928) was a leading Bolshevik who became the object of Lenin's

wrath for supporting major aspects of Mach's philosophy while too cleverly denying it.

57 Ferdinand Jacob Schmidt (1860-1939) would be appointed Professor of Philosophy and Education at the University of Berlin somewhat belatedly (1913). It is possible that his defense of Planck against Mach did not hurt him with the university and government authorities in the Prussian capital.

58 Joseph Petzoldt (1862-1929) after the death of Richard Avenarius (1896) came into increasingly close intellectual relations with Ernst Mach. He doubted the light constancy axiom in Einstein's special theory while otherwise favoring it and later believed that the general theory removed that objection. Einstein, however, in his letter to him of August 23, 1919, denied that Petzoldt correctly understood general relativity , even alleging that "Your misunderstanding is a very fundamental one". Petzoldt later alleged that Mach would have supported Einstein's theory if his son Ludwig had let Petzoldt visit him in August 1914, a visit which was allegedly denied because of difficulties connected with the outbreak of the Great War. For Petzoldt's defense of Mach against Planck see: "Die vitalistische Reaktion auf die Unzulänglichkeit der mechanischen Naturansicht", *Zeitschrift für allgemeine Physiologie*, **10** (1910) 69-119.

59 V. Valentinov, *The Philosophical Constructions of Marxism*, Moscow, 1908. Lenin quotes Valentinov as writing: "We would not consider the 'kinship' of Mach to the idealist views of Berkeley as a philosophical crime, even if this actually were the case", Lenin (1927), op. cit., p. 31. But in spite of such blatant "Machism" which Lenin criticized severely, Lenin would appoint Valentinov to be minister of education after the Bolshevik revolution.

59.1 Robert Seidel (1850-1933) ran a Zürich bookshop, was a leader of the socialist party in the city, and from 1905 was a *Privatdozent* for Education at the Technical University.

60 The book was indeed slow in coming out. See his: *Transzendentale Systematik. Untersuchungen zur Begründung der Erkenntnistheorie*, München, 1916. Hans Cornelius (1863-1947) was a strong ally of Mach but turned increasingly toward art and aesthetics in his later years.

61 It is not clear whether he is referring to the book or the article i.e. to *Die Metaphysik in der Ostwaldschen Energetik*, Leipzig, 1905 or to "Bemerkungen über die Metaphysik in der Ostwaldschen Energetik", *Vierteljahrschrift für wissenschaftliche Philosophie und Soziologie*, **29**. N.F. 4 (1905) 287-333.

62 William Kingdom Clifford (1845-1879) taught mathematics at Trinity College, Cambridge from 1868 to 1871 and from then until his death at University College London. He is perhaps best known in mathematics for his use of biquaternions "to study motion in non-Euclidean spaces" and in philosophy for his phenomenalistic rejection of causes. Having apparently read Mach's *Erhaltung der Energie* (1872) he may be considered one of the very earliest of Mach's philosophical sympathizers. Clifford also strongly influenced Mach's leading English ally, Karl Pearson, who became well-known for his application of statistics to biology and Eugenics. (See John T. Blackmore, *Ernst Mach*, University of California Press: Berkeley & Los Angeles, 1972, pp. 124-125.)

63 This work which was published in German in 1909 seems to have been derived from an article by Clifford which had appeared in *Mind* under the title of "On the Nature of Things in Themselves" (1878).

64 It is interesting that Bertrand Russell with his theory of "sensibilia" also accepted Clifford's conclusion, namely, that sensations do not have to be conscious to be real. But this expanded *ontological* type of phenomenalism stands in considerable contrast with Mach's *epistemological* phenomenalism and the resulting narrowed ontology, where at least in his opinion only what could be *absolutely* certain such as conscious sensations deserved to be designated as real in science. Unlike many scientists and most people, Mach was not willing to use probability or even high probability as a sufficient test for what is real in science. He thought of science as a way to examine what is uncertasin in such a way as to yield certainty. Theories and hypotheses could help, but being uncertain themselves they could never belong to the end- or final-stage of science which

consisted of mathematical functions which directly describe conscious elements *sans theory*.

65 Rudolf Willy (1855-1920) was a Swiss philosopher with views similar to those of Mach and Avenarius even to the point of writing an article titled: "Der Empiriokritizsmus als einzig wissenschaftlicher Standpunkt" in *Vierteljahrschrift für wissenschaftliche Philosophie*, **18** (1896) 75-. He was one of Lenin's favorite whipping boys.

66 Gerardus Heymans (1857-1930) was a professor of philosophy in Groningen in the Netherlands. He is known for his book: *Einführung in die Metaphysik auf Grundlage der Erfahrung*, 1905.

67 The pejorative term "naive realism" as Adler employs it here has traditionally been used to denigrate direct realism in epistemology, but recently, some people have started to identify it with indirect or representational realism. Switches of this kind can always be expected with prejudicial terms, which is one reason why they should be avoided.

68 Adler like Mach and most presentists tended to define *Metaphysik* in a prejudicial way as the study of what allegedly existed beyond sensory or conscious bounds and hence was illegitimate for them as if incompatible with science. But the term is used in many other ways. Perhaps the most neutral is to make it identical with ontology as the study of reality such that far from being illegitimate or incompatible with science it helps provide basic assumptions for science.

69 William Preyer (1841-1897) was a German-English physiologist, psychologist, and philosopher who taught physiology in Jena starting in 1869 and in Berlin from 1888. He published *Elemente der reinen Empfindungslehre* in 1877.

70 Hermann Grassmann (1809-1877) was a mathematician, physiologist, and Sanscrit scholar. He had also written on Helmholtz's color theory.

71 Instead of making a mere hypothesis about what is beyond sensations, Adler seemed to think he was obliged to postulate entities there. In short, Adler seemed not to have a well-devoloped sense of relative certainty and conditional acceptance. Furthermore, he seemed to treat space and time as universal parameters rather than as mere ways to relate sensations as Mach did. Adler alleged that such "absolutism" was his old view and that he had a new one, but Mach was not persuaded. Nevertheless, it is clear from Adler's homage to Mach in 1918 that he still considered himself a dedicated Machist. But Adler's philosophical position after 1925 has remained a mystery, though further research may eventually solve it.

72 Mach did send to Adler an epistemological introduction to his book on optics, but Mach used it instead as part of two articles and not in the book. The title of the book went through phases. While in manuscript form it was normally called *Lichtlehre* (Theory of Optics), and eventually it was titled *Die Principien der physikalischen Optik. Historisch und erkenntnispsychologisch entwickelt*, Leipzig 1921.

Though the last of Mach's books to be published, *Die Optik* was one of the earliest to be partly written and was largely based on lectures he had given on geometrical optics in Prague during the 1860's and 1870's. Sometime around 1912 or 1913 he left it to his son Ludwig to finish and printing began in 1916, but because of the war it was only published five years later. Ernst Mach feared that the book had been bypassed by more recent theories and discoveries, but his son who had co-invented an optical interferometer took his task seriously and did his best to bring the book up to date by adding suitable chapters as much as possible from other work by his father. Both father and son also mentioned in correspondence from time to time that a second volume would appear which included Ludwig's attempt to measure the constancy of the velocity of light in a vacuum to see if that "absolute" aspect of Einstein's theory was true. Ludwig, however, lacked the financial means and perhaps the physical know-how to properly carry out such a delicate experiment. The second volume never appeared.

73 Since Adler and Einstein lived in the same building in Zürich and were friends it is natural that Mach would express his greetings. He had became interested in Einstein and his ideas in 1909. They briefly corresponded in August of that year. After then, he wrote to the physicists Wilhelm Wirtinger and August Föppl to give him their opinion of Einstein's work. They responded

respectively on October 5, 1909 and January 11, 1910, but in a very non-committal way. Since this took place during the period of the Mach-Planck polemics, one suspects that both Mach and Planck would have liked to draw the young genius into their own philosophical camp. But after a disturbing visit from Einstein (September 1910) and Planck's even more disturbing second article attacking Mach (November 1910) Mach seems to have given up on both physics and Einstein. Though Einstein sympathized with Mach during the controversy, by 1914 Planck would successfully entice Einstein to Berlin and he would after 1916 increasingly share the older man's opposition to Machism, positivism, and later to logical positivism.

74 Ernst Mach had befriended young Hugo Dingler (1881-1954) who was then a Privatdozent in Munich. Dingler was a mathematically-minded philosopher of science and Mach anticipated great things from him.

75 The title seems ambiguous since it could mean an introduction to a book or the book itself, but neither one seems to have been either finished or published. Adler returned to Vienna in 1911 to become Secretary of the Austrian Social Democratic Party

76 Mach sent the introduction to *Die Optik* to Adler, but soon wanted it back. Later, Adler proof-read those parts of the book which had been printed in 1916.

77 Alois Riehl (1844-1924) was an Austrian thinker who taught psychology and physiology. This book is titled simply: *Zur Einführung in die Philosophie der Gegenwart* (Introduction to the Philosophy of the Present). The third edition appeared in 1908.

REFERENCES

Adler, Friedrich. *Die Metaphysik in der Ostwalschen Energetik.* Leipzig, 1905.

Adler, Friedrich. "Bemerkungen über die Metaphysik in der Ostwaldschen Energetik". *Vierteljahrschrift für wissenschaftliche Philosophie* 29. N.F. 4 (1905) 287-333.

Tischler, Fritz (pseud.). "Materialistische Geschichtsauffassung und Mathematik". *Die Neue Zeit* 24 (1905/06) 25-30.

Adler, Friedrich. "Friedrich Engels und die Naturwissenschaft". *Die Neue Zeit* 25 (1906/07) 620-638.

Adler, Friedrich. *Die Physik als phänomenalogische Wissenschaft.* [ohne Ort], 1907.

Adler, Friedrich. "Die Entdeckung der Weltelemente (zu Ernst Machs 70 Geburtstag)". *Der Kampf* 1 (1907/08) 231-240.

Adler, Friedrich. "Discovery of the World Elements". *International Socialist Review* 8 (1908) 577-588.

Adler, Friedrich. "Wozu brauchen Wir Theorien?". *Der Kampf* 2 (1908/09) 256-263.

Adler, Friedrich. "Die Einheit des naturwissenschaftlichen Weltbildes". *Naturwissenschaftliche Wochenschrift.* 8 (1909) 817-822.

Adler, Friedrich. "Die Taktik der österreichischen Sozialdemokratie. *Die Neue Zeit* 28 (1909/10) 306-315.

Adler, Friedrich. "Der 'Machismus' und die materialistische Geschichtsauffassung". *Die Neue Zeit* 28 (1909/10) 671-682.

Adler, Friedrich. "Der Wert des Parlamentarismus". *Der Kampf* 4 (1910/11) 413-415.

Adler, Friedrich. "Minderwertig im Internationalism". *Der Kampf* 4 (1910/11) 495-499.

Adler, Friedrich. "Ernst Machs Überwindung des mechanischen Materialismus". *Die Neue Zeit* 35 (1916) 111- .

Adler, Friedrich. *Ernst Machs Überwindung des mechanischen Materialismus.* Wien, Verlag der Wiener Volksbuchhandlung Ignaz Brand & Co., 1918

Adler, Friedrich. *Ortzeit, Systemzeit, und das ausgezeichnete Bezugssystem der Elektro-*

dynamik. Eine Untersuchung über die Lorentzsche und die Einsteinsche Kinematik. Wien, 1920.
Ardelt, Rudolf. *Friedrich Adler*. Wien: Bundesverlag, 1984.
Avenarius, Richard. *Prolegomena zu einer Kritik der reinen Erfahrung*, Leipzig: 1876.
Avenarius Richard. *Kritik der reinen Erfahrung*. zwei Bände. Leipzig: 1888 & 1890.
Avenarius, Richard. *Der menschliche Weltbegriff*. Leipzig: 1891.
Blackmore, John. *Ernst Mach*. Berkeley: University of California Press, 1972.
Blackmore, John and Klaus Hentschel (Hrsg.). *Ernst Mach als Aussenseiter*. Wien: Braumüller, 1985.
Blackmore, John (ed.). *Ernst Mach - A Deeper Look*. Dordrecht: Kluwer, 1992.
Blackmore, John. *Ludwig Boltzmann - His Later Life and Philosophy 1900-1906*, two volumes. Dordrecht: Kluwer, 1995.
Blaukopf, Kurt. *Pioniere empiristischer Musikforschung*. Wien: Hölder-Pikler-Tempsky, 1995.
Bogdanov, Alexander. "Ernst Mach und die Revolution". *Die Neue Zeit* 26 (1908) 695-700.
Braunthal, Julius. *Viktor und Friedrich Adler*. Wien: Verlag der Wienerbuchhandlung, 1965.
Carnap, Rudolf. "Autobiographical Remarks". *The Philosophy of Rudolf Carnap*. edited by P.A. Schilpp. La Salle: Open Court, 1991 [1963], pp. 3-84.
Cekic, Miodrag. "The Influence of Mach's Philosophy in Yugoslavia". *Ernst Mach - A Deeper Look*. edited by J. Blackmore. Dordrecht: Kluwer, 1992, pp. 379-402.
Clifford, William Kingdom. "On the Nature of Things in Themselves". *Mind*, 1878.
Cornelius, Hans. *Transzendentale Systematik. Untersuchungen zur Begründung der Erkenntnistheorie*. München: 1916.
Duhem, Pierre. *Ziel und Struktur der physikalischen Theorien*. Übersetzung von F. Adler. Vorwort von Ernst Mach. Leipzig, 1908.
Einstein, Albert. *The Collected Papers of Albert Einstein*. Volume 8. translated by Ann M. Hentschel. Princeton: Princeton University Press, 1998.
Florence, Ronald. *Fritz - The Story of a Political Assassin*. New York: The Dial Press, 1971.
Frank, Philipp. *Einstein: His Life and Times*. New York, Alfred A. Knopf, 1947.
Haller, Rudolf, and Friedrich Stadler (eds.). *Ernst Mach - Werk und Wirkung*. Wien, Hölder-Pichler-Tempsky, 1988.
Hentschel, Klaus. See John Blackmore (1985).
Hentschel, Klaus. *Interpretationen und Fehlinterpretationen der speziellen und der allgemeinen Relativitätstheorie durch Zeitgenossen. Albert Einsteins*. Basel: Birkhäuser, 1990.
Heymans, Gerardus. *Einführung in die Metaphysik auf Grundlage der Erfahrung*. Leipzig: 1905.
Husserl, Edmund. *Logical Investigations*. Volume One. translated by J.N. Findlay. New York: Humanities Press, 1970 [German 1901].
Israel, Hans, Erich Ruckhaber, and Rudolf Weinmann (Hrsg.). *Hundert Autoren gegen Einstein*. Leipzig: R. Voigtländer Verlag, 1931.
Joël, Kurt, "Friedrich Adler als Physiker". *Vossische Zeitung*. Berlin, May 23, 1917. No. 259. Morning edition, page 2.
Langer, William L.. *An Encyclopedia of World History*. Boston: Houghton Mifflin, 1962 [1940].
Lenin, V.I.. *Materialism and Empirio-Criticism*. New York: International Publishers, 1927.
Mach, Ernst. *Erhaltung der Arbeit*. Prag: Calvé, 1872.
Mach, Ernst. *Die Principien der Wärmelehre*. dritte Auflage. Leipzig: Barth, 1919 [1896].
Mach, Ernst. *Die Prinzipien der physikalischen Optik*. Leipzig: Barth, 1921.
Mach, Ernst. *Science of Mechanics*. La Salle: Open Court, 1960 [1893].
Mach, Ernst. "The Leading Thoughts of My Scientific Epistemology and Its Acceptance by Contemporaries". *Ernst Mach - A Deeper Look*. edited by J. Blackmore. Dordrecht: Kluwer, 1992, pp. 133-139.
Petzoldt, Joseph. "Die vitalistische Reaktion auf die Unzulänglichkeit der mechanischen Naturansicht". *Zeitschrift für allgemeine Physiologie* 10 (1910) 69-119.
Planck, Max. "Die Einheit des physikalischen Weltbildes". *Physikalische Zeitschrift* 10 (1909)

62-75.
Planck, Max. "Zur Machschen Theorie der physikalischen Erkenntnis: Eine Erwiderung". *Physikalische Zeitschrift*. 11 (1910) 1186-1190.
Preyer, William. *Elemente der reinen Empfindungslehre*. Leipzig: 1877.
Riehl, Alois. *Zur Entführung in die Philosophie der Gegenwart*. Leipzig: 1903.
Ruchhaber, Erich. See Hans Israel.
Stadler, Friedrich. See Rudolf Haller (1988).
Thiele, Joachim. "Ernst Mach-Bibliographie". *Centaurus* 8 (1963) 219-221.
Thiele, Joachim. "Briefe Albert Einsteins an Joseph Petzoldt". *NTM - Schriftenreihe für Geschichte der naturwissenschaften, Technik, und Medizin*. Leipzig. 8 (1971) 70-74.
Tisza, Laszlo. "Philipp Frank and Physiks". *Philipp Frank 1884-1966*. Cambridge Mass.: 1966.
Valentinov, V.. *The Philosophical Constructions of Marxism* [in Russian]. Moscow: 1908.
Weinmann, Rudolf. See Hans Israel.
Willy, Rudolf. "Der Empiriokritizsmus als einzig wissenschaftlicher Standpunkt". *Vierteljahrschrift für wissenschaftliche Philosophie* 18 (1896) 75- .

Chapter Three

PHILIPP FRANK

largely by himself

Section Guide
1. T.S. Kuhn's Interview
2. Recollections of Frank: (Fürth, Tisza, Bergmann, Holton and Bernstein)
3. Frank's Philosophy
4. Correspondence

1. T.S. KUHN'S INTERVIEW[1]

TSK: Tell me, when did you actually begin at Vienna as a student?

F: I think it was 1904 or something like that or a little less even. I graduated from Vienna, I think, in 1907. And since I studied about four years, it must have been 1903. I graduated in 1907. I took the regular program until the PHD. Of course at that time Vienna was dominated by Boltzmann [who died in September 1906]. I was of course mostly a student of Boltzmann. You see Ernst Mach was already retired at that time. He was no longer teaching when I was a student. He retired rather early. He was not so very old [63] but he retired early. Thus to help clarify the situation: I studied with Boltzmann. And the interesting thing about Boltzmann and quantum theory was that he had already formed a conceptual anticipation,[2] a fundamental conception, about the probability of a certain state phase space. He had long had the idea that phase space must be fundamentally quantized. It must be that the cells in phase space must have a certain finite size because otherwise the probability would be zero or infinite. But anyway, the whole idea of probability really presupposes that there are finite cells in phase space. As Boltzmann liked to express himself, because as you know he was a follower of atomistics, he said: 'Maybe there are atoms of time as well.' Phase space of course contains time. It is energy times time'. And as the great scientist always said: 'Probably there will be atoms of time too.'

TSK: I had not known that . . . .

F: I would say that Vienna was a big school of physics. And therefore, since I myself come from the city of Vienna, I never thought of studying anywhere else. It was generally thought that everybody who was interested in science studied at Vienna, and therefore, I never thought of going elsewhere, except for just a little at another place. I studied mostly with Boltzmann. But Exner, you see, became connected with the basis of quantum theory by one thing, by one lecture. I think it was his inaugural lecture as *Rektor* -- It was about the role of statistics in physics.[3] And then he said one thing which became rather famous;

J. Blackmore, R. Itagaki, and S. Tanaka (eds.). Ernst Mach's Vienna 1895-1930. 61-83.

he said that it may be that the basis of physics will be statistical. And it may not be that every statistical law can be derived from dynamic laws. Because generally, of course, the [reigning] idea had been that statistics only gave an average. The basic laws were the Newtonian laws. But it was Franz Exner who already said that maybe this was not true. It could be that the basic laws are statistical. I think that this concept had a great influence on Schrödinger.[4]

TSK: I haven't read Exner's address but I did read Schrödinger's essay on this. But was this something the rest of you also knew? Or has that statement only become famous since Schrödinger?

F: Oh no, no. It was already known at that time. It was widely discussed even when I was a student . . . . All the physicists in Vienna were interested in the philosophy of science. Hence, if there was anything connected with philosophy of science [such as the status of statistical laws] it would be widely discussed. Admittedly, Franz Exner was not well-known for his interest or publications in philosophy of science, but in this matter he did clearly lead the way. Exner was an experimental physicist who usually left philosophical matters to Mach and Boltzmann who was the theoretical physicist. [But after Boltzmann's death, Exner took it upon himself not to let Boltzmann's concern with statistical laws die.] At this time, I think that Exner's main field was connected with electricity, atmospheric electricity.[5] It was the beginning of ion theory - - of ions and electrons at this time. Oh yes, it was much discussed whether the foundation of this was statistical or not, because everything was connected with philosophy[6]

TSK: There are some other things which if I may, I'd like to ask you somewhat more about this. Did you get to know Boltzmann fairly well?

F: Yes, fairly well.[7]

TSK: Did he talk at all about the Planck quantum?

F: Yes, you see, [but] that was [mostly in terms of his own] anticipation [of quantum theory] . . . .

TSK: But he definitely did speak about it?

F: Oh yes, he talked about it. He definitely did . . . . [8]

TSK: Did anyone you remember in Vienna have a feeling [that it might be a big breakthrough]?

F: I can't think of anyone. Oh yes, the only man of course would be Ehrenfest. He had a suspicion.

TSK: Did he already have it then?

F. Oh yes, certainly. Ehrenfest always looked for queer things, even then. But he wasn't so much in Vienna. He lived in Russia and he came only occasionally. I remember that Ehrenfest was regarded as generally crazy.

TSK: Was Hasenöhrl in Vienna when you were a student?

F: Yes, but by the time I was a student he was what one called a *Privatdozent*. He was essentially a student of Lorentz, and he was studying electro-magnetic theory and like things, and later of course at least to a certain degree relativity theory. But also more in the way of Lorentz. He died very young. Hasenöhrl was killed in World War I. You see, he became famous for [partly anticipating] Albert Einstein's formula E equals mc^2 which he had interpreted in the sense of Lorentz.[9] He was in fact quite an interesting man, but not a particular genius. He could have said something about Planck, but I don't remember that he did. In general, he was interested in Planck's type of work. Hasenöhrl it seems once knew a brother of Paul Ehrenfest in Geneva, and once when Ehrenfest came to Vienna he said to Hasenöhrl: 'Tell my brother that I am not crazy. It will be a great help if you will tell him.' But then Hasenöhrl replied: 'I'm sorry, but it wouldn't help. The only result would be that he would believe that I was crazy.' . . . . .

TSK: Before I let you get there, still back in this period, had the influence of Mach vanished in so far as scepticism about the atom was concerned?

F: No, it did not vanish. There was always this interesting point: What was the relation between Mach and Boltzmann. This is a point which of course plays a great role, and has become rather political in Russian literature where we find it sometimes said that Mach was responsible for Boltzmann's suicide, as if Ludwig Boltzmann were quite desperate [about opposition to his theories], but as a matter of fact this was not really the case, because Boltzmann was himself philosophically speaking rather a follower of Mach.[10] Boltzmann once said to me: 'You see, it doesn't make any difference to me if I say all atoms are merely pictures. I don't mind this. I don't require that they be absolute. I don't say this. 'An economical description', Mach said. Maybe the atoms are an economic description. This doesn't hurt me so much. From the viewpoint of the [anti-metaphysical] physicist this doesn't make a difference.' Ludwig Boltzmann was rather philosophical about it. He did not require that you believe in the existence of atoms. And there wasn't I would say any opposition from Mach's viewpoint. This opposition only existed, so to say in the philosophical [i.e. metaphysical] realm. Yes, it did exist there. Also strange as it might seem, Vienna physicists were all followers of both Mach *and* Boltzmann.[11] It wan't the case that people would hold any antipathy against Boltzmann's theory of atoms because of Mach. And I don't even think Mach had any antipathy.[12] The situation wasn't like that. Nor did it play any role [in my case]. I was of course very interested in the whole problem, but it never occured to me that just because of Mach's theories that one shouldn't pursue the theories of Boltzmann.

TSK: Yet, Ernst Mach was a person who thoroughly believed in a phenomenological interpretation of thermodynamics and was strongly opposed to the statistical.

F: No, I wouldn't go quite that far. I wouldn't say that. He was not against the statistical interpretation [per se but only when it rested on assumptions about the reality of atoms]. The views of Mach can be found in the presentation of [your own] American philosopher [John Bernard] Stallo. He believed that there were not any sufficient proofs, I mean experimental proofs, in order to justify the assumption of atoms. From the perspective of philosophy Mach rejected them for reasons which are similar to those of Stallo. I think that his principal argument had always been that if the atom were like a mechanical object, then you could not explain for instance the great complication of the spectral lines.

The answer was not known in his time, thus another approach was needed. This apparent view of Ernst Mach somehow seems to have influenced Schrödinger, since his conception of atoms is not at all that of the [mechanical] atomic theory.

TSK: No.

F. It was not a theory of the atom [as a substance or indivisible particle], because as you see, the theory kept [evolving]. But of course individually and privately there could be cases where people were somehow diverted from the study of the atomistic theory because they believed in phenomenological theory, but I don't think it played a great role among the truly productive physicists. All were more or less followers of Ernst Mach in the philosophical sense.[13] Nor was it so different during Einstein's younger years. He was also a follower of Mach.[14] You can see this too with Heisenberg, and probably even Bohr. It is strange that it didn't play a [more central] role. There were some people of course who interpreted it in this sense [Mach's epistemology as outlawing discontinuity in science],[15] and they were for this reason opponents of Mach in the philosophical sense [where continuity and discontinuity were not philosophical issues].

TSK: Was Ernst Mach much discussed among the physicists in Vienna?

F: Oh yes. He was very much discussed among the physicists. Mach played a very great role.[16] Boltzmann always spoke about Mach.

TSK: And which of Mach's works were particularly widely read by the physicists?

F: Mostly of course his main books: *Mechanik* and *Analyse der Empfindungen*, plus a book which doesn't even exist in English: *Erkenntnis und Irrtum*.. The latter book was also widely read in Vienna. [See *Knowledge and Error* (1976).] The chemist and President of Harvard University, J. B. Conant, had a completely wrong idea about Mach. I often discussed the matter with him. Later he recognized this. Conant liked to claim that Mach had said that physics is merely descriptive and therefore that it was not a method for progress in physics. But Conant was wrong. In his *Erkenntnis und Irrtum* Mach clearly and repeatedly points out it is precisely simple description which produces progress in science. But it is true that in some of his earlier books he had not emphasized this point so much.

Of course I frequently discussed these problems with Albert Einstein [Frank had succeeded Einstein in Prague]. Einstein was very strong. There is also a certain interesting perspective which concerns quantum theory. This idea which was developed years later about the relation between physics and philosophy was first put forward by Einstein's own assistant Lanczos. He held that the [general] theory of relativity is based on a metaphysical interpretation of science, while the quantum theory is based on a positivistic interpretation of science.[17] When I first read this I was astonished of course, since originally the [special] relativity theory was based on positivism. But the interesting thing is that this notion of Lanczos had a certain truth in it which one sees in the whole history of Einstein.

It was Albert Einstein who later more or less abandoned the positivistic conception of science. And this is one of the reasons I think why we have this antipathy against the Bohr theory and the Copenhagen school. He regarded the Copenhagen school as something which was radically connected with positivism. Of course originally and in his first obituary of Mach [1916] that Einstein wrote, he still bluntly [asserts] that the relativity theory is grounded on the ideas of Mach.[18] But in later relativity theory [the general theory which was published at almost exactly the same time], Einstein emphasized another point of view. It is said that the general heory is based more on field theory and not so much on the physical interpretation of measurement as it was earlier. But at their birth both [the Copenhagen] quantum theory and relativity theory were connected with positivism. This is not perhaps as well known as it should be.

I would like to speak about Heisenberg. In his first paper in which he introduces his matrix theory he says that the basis of his theory, both in terms of philosophy and science, is that we no longer look for what kinds of motions produce spectral lines; nor do we try to reduce this to mechanics. Rather we investigate radiation by itself the frequencies and try to give a mathematical presentation of those frequencies. And then comes Heisenberg's theory. But Heisenberg in the first paper had already mentioned explicitly that [his approach] is an application of the general ideas of Mach on physics. In short, that quantum theory is also connected with Mach. On the other hand, Sommerfeld was, as many German physicists have been, a great opponent of the so-called Machist theories, and he believed in the real existence of things - - I remember a speech at a meeting where Sommerfeld said: 'The whole theory of Mach, the metaphysical theory of Mach, that physics is only a description, is completely sterile. It is quite hostile to progress in physics.' Then I replied to him: 'There are only two exceptions: relativity theory and quantum theory.'

TSK: When did Sommerfeld say this?

F: He said this in a meeting in Prague. In Germany, all German speaking countries, there was already a society the: *Deutsche Naturforscher Versammlung*. This was like the A.A.A.S. here. And there was also a special meeting of physicists later.[19] One of these meetings was held in Prague, and I was the chairman. Sommerfeld also spoke here. In addition there was a special meeting, which was one of the first meetings of the Unity of Science Congresses about the epistemological basis of physics. I also gave a speech, which was reproduced in my book.[20] I discuss there what Sommerfeld said and my answer.

Also in this book, I related Mach to American philosophy, that is, to pragmatism. Sommerfeld responded: 'I really didn't know anything of American philosophy but always suspected that it must be disgusting, [but thanks to you] I am now sure it is.' Sommerfeld was a man with a very . . . character. Of course he was a very good physicist in many respects. But he always had this thing, this terribly nationalistic belief, as if German philosophy was indigenous to the German nation, and as if not believing in Kant would make you a traitor. He was locked into a nationalistic or religious notion of philosophy. So to a certain extent was Planck, but not so much, though one can still detect vague Kantianism.

Yes, I knew Planck a little. Concerning science, the most I ever had to do with him related to this [early] book which he wrote about the first law of thermodynamics, which had won some kind of prize.[21] His main point was to refute Poincaré and similar authors who had held that the first law was tautological. He formulated it in such a way that it was not, and [supporting Poincaré] I rejected this. His main argument was, and this is often repeated, that the first law has to be formulated in such a way that if you have a cyclic process, then by a cyclic process no weight can be lifted. And I said this is not sufficient, because what is a cyclic process? It is one which will restore the same state. And what is the same state? Something remains in the same state if no weight has been lifted. And this is also a kind of tautology.

Planck was a great enemy of Mach. He even wrote out two articles against him [1909-11] which were often published and republished.[22] Nor do I think it was right what Planck said in this respect. He was a strange character, this Planck. Although he was a man who discovered some good ideas, he didn't think very logically. He often let popular ways of thinking influence him. But he was a good writer, that is, a good popular writer. He was always writing in the *Berliner Tagblatt*. Ehrenfest was also impressed by his way of writing. But all of his writing of course has many mistakes. All those gaps concerning radiation theory and a lot of other things are not very precise, though he had good ideas. But he very often confused other people by his arguments, which were not correct. But he had to use statistics in both thermodynamics and radiation theory, but there was a certain rift between him and Boltzmann. And in Planck's last book when I believe he must have been over 90, his *Scientific [Auto]biography* says: 'There was always something which I regretted in my life, namely, that I was never recognized by Boltzmann.'[23] But I think that Boltzmann was partly at fault . . .

TSK: When did you first get acquainted with Einstein? When you replaced him at Prague?

F: No. My first contact came from the philosophical side. In my very first paper -- which is in the volume -- *Kausalgesetz und Erfahrung*.[24] I tried to follow the line of Poincaré but in a very radical way. I said that the principle of causality is a tautology. Its just a convention. I was perhaps twenty or so and had not even graduated, and here was a letter from Einstein in which he said it was very good, and at least from a logical point of view it was very true. 'I like this paper very much', he said to me. He then turned to the type of objections which he always made and which (I make also) . . . We can give a few simple causal laws in some cases, and physics contains nothing else. But while it is true that we cannot say more, if we want to take a general or philosophical course, then of course Poincaré is very right, and you are very right, this is only a tautology . . . .

TSK: In retrospect, how would you say doing science in Vienna [circa 1900-1910] was different from, say, if you had done it at Munich or Göttingen?

F: I would say that one of the [most important] things was that in Vienna [the professors and students] were more interested in the philosophy of science, and definitely more than elsewhere. It was very strong. It was probably the influence of Mach at that time, as well as other people who followed a similar line. Yes, it was strong, very strong, I would say . . . .

2. RECOLLECTIONS

A

I [R. Fürth][1] have known Philipp Frank since 1912 when I started my study of physics and mathematics at the German University of Prague and he had just been appointed to the chair of Theoretical Physics as successor of Einstein. At 28 years of age Frank was by far the youngest professor in the university, a fact unknown to his students who were much impressed by the mastery of his teaching and his full beard.

The first lecture course which he gave to us was on Clerk Maxwell's theory of electromagnetism and some of us, including Charles Löwner as well as the late Arthur Winternitz, had selected this course although we were still only beginners. At first Frank did not realize that the course would be attended by first years students, who had no knowledge of partial derivatives, differential equations, and vectors, but on being told so he took great trouble in helping us overcome these difficulties.

His style of lecturing was, on the face of it, rather easy going and slow, and he normally devoted the first quarter hour of each lecture to a repetition of the contents of the preceding lecture. But in this manner he helped make us understand the subject thoroughly, in spite of our inexperience, and to arouse our enthusiasm for it.

During the time of my undergraduate studies Frank never attempted to cover the field of theoretical physics systematically and apparently selected his lecture courses according to his inclinations. I remember in particular the courses he gave on probability theory, statistical mechanics, and kinetic theory, subjects that I found so fascinating that they have occupied my main interests ever since.

Frank also liked to teach specialized and topical subjects by the way of seminars, where the more senior students were given the task of reporting in turn on chapters of books or papers. He did, of course, most of the talking himself, interrupting the students frequently in order to emphasize salient points and to make comments. I especially remember a seminar on relativity theory, Frank's main field of interest at that time. We found these seminars most enjoyable because they were conducted in an informal and friendly atmosphere and brought us in closer contact with Frank . . . .

The outbreak of the first world war in 1914 brought about great changes in the Departments of Experimental and Theoretical Physics in Prague which were housed in the same building. The former became completely depleted of scientific staff and, although still an undergraduate, I took on a great deal of teaching and laboratory work there. I also transferred my optical research to the experimental department and hence spent most of my time in it. Nevertheless, Frank offered me, some time later, the post of "Assistant" in his Department which I held for twelve years until I was appointed to a professorship . . . .

Privately Frank's wide interests and acute observations made him a lively and witty companion. After his lecture in the morning he invariably went to the room of George Pick, the well-known mathematician, who was head of the Mathematics Department (and by the way, together with Einstein and the experimental physicist [Anton] Lampa had made the proposal for Frank's appointment). He would sit there for a while, in later years also to be joined by Pick's colleague Ludwig Berwald, the successor of Kowalewski. Judging from the occasions when I was present myself, these talks were far from being confined to academic matters. Who would have thought at that time that Pick and Berwald, like so many of our friends, would perish in German concentration camps!

With his students Frank was on easy and friendly terms and on numerous occasions joined them in departmental outings and parties. In fact, he was completely free of any racial or national prejudice; when after the end of the first world war the relations between the German and Czech universities in the new Czechoslovak state were in general rather tense he was always on the best of terms with his Czech colleagues and we used to meet them frequently on scientific and social occasions. Often we also met some of the Russian émigré scientists, who had come to Prague after the war and were made to feel at home by Hania and Philipp Frank. I remember in particular the physicists Eichenwald and Rashevsky . . . .

I was Frank's colleague in the Faculty of Science for eleven years. In contrast to many of my older colleagues he did not like to be involved in academic affairs, and when I happened to sit next to him at board meetings, he used to make amusing and sarcastic asides to me about some of the proceedings. On the other hand, he never approved of the conduct of some other colleagues, who preferred to keep silent for the sake of peace, but spoke up courageously whenever anything seemed wrong to him, which sometimes did not make him particularly popular. He would strongly resist any attempts of the Nazi sympathizers amongst staff and students of the University to apply racial doctrines to the admission of students or the appointment of staff, and he vigorously supported the appointment of the late E. Finlay-Freundlich to the Chair of Astronomy. Frank was also responsible for the creation of a Chair of Philosophy of Science within the Faculty of Science, against considerable opposition, and for the Appointment of Rudolf Carnap to this Chair.

During the last years before the invasion of Czechoslovakia. Frank, who seemed to understand much better than most of us what was going on, had become very pessimistic about our future prospects. When in the autumn of 1938 he went with his wife to the U.S.A. to give lectures at various universities I had the feeling that we would not see him again for a long time to come. Indeed, twelve years had to elapse before we had a happy re-union in London.

B

Around 1930, I [Laszlo Tisza][2] came across two papers which were by Philipp Frank on philosophical implications of modern physical theories. For the novice that I was, impressed and confused by the theories in question, the lucid reasoning of Philipp Frank was a sheer delight. From this moment on he was established in my mind as a philosopher of science, or perhaps the physicist's philosopher of science. This impression was strengthened some dozen years later when I had the good fortune to meet him in person here in Cambridge.

Until quite recently I was not acquainted with Frank's early work as a physicist, and I confess it was not without trepidation that I agreed to say a few words about this aspect of his activities. His professional competence was, of course clear enough from his authoritative article on Hamiltonian optics and mechanics in "Frank-Mises", a classic of mathematical physics of which he was co-editor. Yet this paper is so austerely written that it hides more than it reveals about its author's personality.

It was therefore an interesting experience for me to scan through some of Frank's technical papers of the period from 1904 to the early 1930's. These papers range over a wide field in pure and applied mathematics and physics. Some of them were written in collaboration with renowned mathematicians like [Georg] Pick and [Charles] Löwner. They deal with variational calculus, Fourier series, function spaces, convex bodies, and with applications to analytic dynamics and hydrodynamics. About half a dozen papers of the period 1908-1911 are devoted to problems of the special theory of relativity. They are reasoned with clarity, elaborated in precise mathematical terms, and are always conceived to clarify a fundamental problem or difficulty. In one interesting paper he and Hermann Roth were able to derive the Lorentz transformation, without assuming the constancy of light velocity, by a skillful exploitation of the fact that the transformations form a group. Recent years witnessed a revival of interest in considerations of this sort.

Frank made several contributions to Hamiltonian geometrical optics and developed an early interest in Schrödinger's wave mechanics. Again he revealed himself as the philosopher interested in the interrelation of theories. He also established a connection between wave mechanics and Boltzmann's statistical mechanics, and in a side remark he even conjectured that in a hypothetical future interpretation of wave mechanics the concept of a point particle inherited from Newton and Boltzmann might have to be abandoned.

Philipp Frank was certainly different from the physicist whose attention was focused on the questions of current interest. In an unguarded moment he gave himself away as the benevolent outsider; for example, in connection with a relativistic reformulation of

Maxwell's equations, he writes in 1911: "It would be possible to render the calculation considerably shorter and more transparent by using consistently Grassmann's 'Ausdehnungslehre', and thus combine Sommerfeld's intuitive approach with the elegance of Minkowski. Nevertheless, I shall spare my physicist reader from exposure to unfamiliar formalisms; he has had to contend with more than his share of mathematical abstraction in recent years already . . ."

It is quite apparent from these papers, which I could review only in the sketchiest of terms, that Frank was not only the popularizing philosopher, but also a creative scientist involved with some of the fundamental living problems of the physics of his times. In his relation to Einstein he seems to have been not only the biographer, but a colleague with whom Einstein might have, and probably has, exchanged ideas on matters of mathematical and conceptual structure. There emerges thus, an image of the young frank, an image that is different from that most of us may have formed of him in later years, but that is nonetheless entirely believable.

C

It was in the fall of 1933 that I [Peter G. Bergmann][3] came to the city of Prague as a refugee from my native Berlin, where I had been unable to register as a student. I was but eighteen years old. I had suffered no physical injuries, no imprisonment, and I was not destitute. But my parents had lost all possibility of existence in my native country, and I found myself in a strange country of whose language I knew not even one word. I had picked Prague as the site of my further studies for one simple reason: Among all the foreign universities offering instruction in German - the only language I had a speaking command of - Prague was the least expensive by a considerable margin; and that was all I knew about the city of Prague. And I was certainly not the only person appearing at Prague under those far from auspicious circumstances. There were refugees from Germany, there were exiles from various Balkan countries, from the Soviet Union, and even from the then independent Baltic republics, all attempting to complete their formal training for the doctorate at the various Prague institutions of higher learning.

In this overheated and jittery atmosphere there was one fatherly figure who represented all that was best at the University, Philipp Frank. As Albert Einstein once said of him: "His shirt might show grease spots, but his mind was always in apple-pie order." Philipp Frank never in the three years I was at his Institute began a class on time; he might be late anywhere from twenty minutes to over half an hour, and that for a forty-five minute period. But he was full of sparkle, full of stimulus. After class he would drift into the Institute library where Wallauschek and I took care of the collection of books, and he would talk about politics, about physics, about anything that he might have picked up at the "Kaffeehaus". He would encourage all of us students, and he gave us the feeling of a wide-open intellectual window, open to things that happened in and out of physics, and open to things that happened outside of the country as well. Philipp Frank saw to it that there was close contact with philosophy of science, presided over by Rudolf Carnap, with experimental physics at Fürth's Institute downstairs, and with pure mathematics, which was taught by professors Berwald, Löwner, and Winterits.

Philipp Frank never lost his ebullient good spirits, and I found out, many years after I had left Prague, how he had helped along my own professional career by recommending me to Einstein, without letting me know. All of us who have been his students will value his guidance as a teacher, and his friendship as an elder colleague.

D

I, for one, was a student when his collection of essays, *Between Physics and Philosophy*, was published back in 1941. The effect upon me [Gerald Holton],[4] as on so many of Professor Frank's readers and students, was deep and lasting. Here was a link in the charismatic chain leading back to such predecessors as Ernst Mach, Ludwig Boltzmann, and Henri Poincaré. Some years later I had the chance to take a course in the philosophy of science which Professor Frank was giving at Harvard, and shortly after that to help out in the same course as an assistant. In this way was started a treasured association which brought me a little bit closer to the person whose scientific and philosophical writings I had already been studying, whose power and even whose style comes through on almost every page that he has written. Thus, the introduction, entitled "Historical Background - Discussion in a Vienna Coffee House", which is to the second edition of Professor Frank's collection of essays, begins with these lines:

At the time when the first chapter of this book was written (1907) I had just graduated from the University of Vienna as a doctor of philosophy and physics. But the domain of my most intensive interest was the philosophy of science. I used to associate with a group of students who would assemble every Thursday night in one of the old Viennese coffee houses. We stayed until midnight and even later, discussing problems of science and philosophy. Our interest was spread widely over many fields, but we returned again and again to our central problem: How could we avoid the traditional ambiguity and obscurity of philosophy? How could we bring about the closest possible rapprochement between philosophy and science? By 'Science' we did not mean 'natural science' only, but we always included social studies and the humanities.[5]

This paragraph illustrates what has been the lasting formula: deepest involvement in central intellectual issues; construction of intellectual unities rather than mere iconoclastic revolt; and - equally important - pursuits of these discussions in a humane and deceptively informal setting rather than in a stuffy ex-cathedra manner.

This apparently paradoxical combination of serious power and effortless style made Professor Frank the effective and beloved teacher of unnumbered readers and students who came from all over the world. Each new essay, lecture, or conversation left his audience with new recognitions. That this should have been felt equally by Professor Frank's beginning students and long-term colleagues is the measure not only of a quality of mind but equally of spirit; for there seemed to be in and around him a sort of calm equilibrium that allowed problems and answers to be seen with new clarity. And in this, as much as in his writings stretching over nearly sixty years, Professor Frank taught us perhaps the most profound meaning of the word "philosopher".

E

As a freshman at Harvard in 1947, not interested in making science a career, I [Jeremy Bernstein][6] took Natural Sciences 3. About December we encountered very briefly, Einstein's relativity theory, which fascinated me enormously but of which I understood next to nothing. Fortunately, just at that time, I was advised that Professor Frank was about to give a somewhat advanced General Education course in the ideas of modern physics which included relativity, and I enrolled at once.

We would meet twice a week in the large lecture room in Jefferson, and Professor Frank spoke to us - "lecture" would be too formal a term - for about an hour after which he would, as he put it, "make a certain interval", followed by a second hour of discussion. Nothing pleased him more than sharp disagreement with his own points of view in these discussions. One of his students was a very intelligent Jesuit father, and for long afterwards Professor Frank, who had himself read widely in Thomistic philosophy, told me how much he had enjoyed debating about the interpretations of modern physical ideas with him.

We were taught the pivotal ideas of modern physics, the principle of relativity, the photon concept, the uncertainty principle, through the vivid, almost pictorial, images that are at the center of the really profound theories. The lectures had a deceptive simplicity, often I would walk home, after them, and try to reconstruct the argument, only to realize that what had seemed so transparent in class really concealed a subtle logical framework that had to be assembled in just the right order to make it work right. For Professor Frank these ideas were part of his instinct. he had grown up with 20th century physics, as a student of Boltzmann, and was a close friend and associate of most of the men who have created our discipline. He could explain so simply only because he understood so clearly.

He loved telling jokes and anecdotes in class. These like the rest of his lectures, were rendered in an inimitable German accent that seemed to bear traces of the various countries whose languages Professor Frank had learned during his academic voyages. When I first came to know him his recall was nearly total. He remembered details of conversations he had had with Einstein dating back half a century, some of which he would then tell us. One of his favorites concerned the perils of popularizing science. Sometime in the early twenties Professor Frank went to see Albert Einstein in Berlin. Einstein had recently published his small popular book on relativity. He was very pleased with the results and remarked that any twelve-year-old upon reading the book, could readily grasp the theory. In fact, he said his daughter, then about 12, had read the manuscript and understood it. Einstein left the room and Professor Frank took the opportunity to ask Miss Einstein whether she really had understood her father's book. "Oh, yes," she answered, "I understood everything in it except what is a 'coordinate system?"

I think that all of us who attended these classes were constantly awed, although this was never Professor Frank's intention, by his almost incredible erudition. He seemed to have read and digested the great philosophical, literary and scientific works in an enormous variety of languages. He once told me that he had studied Arabic, as a young student, in order to be able to read the great texts in that language and, fifty years later, he remembered it sharply enough to be able to write it out, which I once saw him do in a discussion with an Iranian student, some of the passages in Arabic that had intrigued him. This vast general culture was also worn instinctively, without pretense, and with the same mastery that characterized his scientific cultivation. If someone pressed him in discussion, he would dig one level deeper in his store of knowledge to respond. As a student I had the feeling that what he taught to us represented only the pure distillation of a vast reserve beneath.

When I got to know him personally through reading courses and many conversations in his office and at home with Mrs. Frank, there was another quality about him that impressed me as much as his erudition. With many philosophers, scientists, or religious leaders, one often has the impression that there is a contradiction between what they profess and how they live; often philosophers do not appear philosophical, scientists not scientific, and religious leaders not religious. Professor Frank, as a philosopher, had the equilibrium and serenity of the truly philosophical. His view of the world was realistic, humorous, and affectionate. he had lived in many countries, had traveled widely, and had witnessed, especially as a European living during the rise of Hitler, the ebb and flow of the political and social events of our time. The result of his observations was to make him somewhat sceptical of the drama of the present and optimistic for the future.

Those of us who had the chance to study with him, if were fortunate, may have absorbed some of his outlook along with the philosophy and science that he gave us. It has been a privilege for me to be able to dedicate these few remarks to the memory of Professor Frank.

3. SELECTIONS FROM PHILIPP FRANK

"The Law of Causality and Experience"[1]
(1907)

The thesis that we shall attempt to prove states that the law of causality, the foundation of every theoretical science can be neither confirmed nor disproved by experience, not, however, because it is a truth known a priori, but because it is a purely conventional definition.

The law states that if, in the course of time, the state *A* of the universe is once followed by the state *B*, then whenever *A* occurs *B* will follow it.

If no state *A* could ever be repeated, then the law would be meaningless theoretically, since it refers only to recurring states.

Editor's Comment s: It is interesting that Ludwig Boltzmann briefly refers to Frank and his ideas about causes in his Nachlass: "Ursache und Wirkung nach Philipp Frank", which is probably from 1905, that is, two years before Frank published the article and a year before Boltzmann's suicide. Frank's change in the title of course reflects his identification of causes here with laws or with what is subordinate to laws, whereas Boltzmann at least in this context identifies causes with forces. Mach as a phenomenalist identified causes like Hume with constant conjunctions among appearances, a view which has recently been called "functional explanation". For Boltzmann's reference to Frank see Ilse Fasol-Boltzmann, Karl Heinz Fasol, and Walter Höflechner, "Ludwig Boltzmann - Notizen und Gedanken aus seinen letzten Lebensjahren". (Hrsg.). *Österreichische Gesellschaft für wissenschaftsgeschichte - Mitteilungen* 18 (1998) 112.

"The Importance of Mach's Philosophy for Our Times"[2]
(1917)

There are indeed but few thinkers who can induce such sharp differences of opinion, that is, who are so inspiring to some and so utterly repugnant to others.

In recent years, especially among creatively active physicists and mathematicians, there has become noticeable a reaction against the conceptions of Ernst Mach. When one of the most outstanding theoretical physicists of our time, Max Planck, and one of the foremost living geometricians, F. Study, characterize these conceptions as being partly misleading, partly incapable of being carried out, and partly actually harmful for science, the fact provides food for thought and cannot be lightly brushed aside.

According to Ernst Mach, on the other hand, physics is nothing else but a collection of statements about the connections among perceptions and theories are nothing but economical means of expression for summarizing these connections. "The aim of natural science," says Mach, "is to obtain connections among [sensory] phenomena. Theories, however, are like withered leaves, which drop off after having enabled the organism of science to breathe for a time." This phenomenalistic conception, as it is called, was already familiar to Goethe.

Mach himself saw the real value of theories in the fact that they permitted setting up a connection as free from contradictions as possible, between physics on the one hand and physiology and psychology on the other.

Wherein then, lies the value of the doctrines of Ernst Mach for physics? My view is that their main value is not that they help the physicist to go forward in his physical work, but rather that they provide the means for defending the edifice of physics against attacks from outside.

A person who will examine dispassionately the concepts which today are the basis of the system of hypotheses of physics will hardly be able to assert seriously that the atom, the electron, and the action-quantum form really satisfactory ultimate building blocks. Every thinker who is somewhat inclined toward logical thoroughness can find many obscurities in these concepts. Into these nebulosities boring doubt can penetrate and try to shake the whole system of physics as the foundation of our scientific world picture. Here Mach steps in and says: "All these concepts are only auxiliary concepts in physics. What is significant is the connection among [sensory] phenomena. Atoms, electrons, and quanta are only links which represent a connected system of science; they make it possible for us to derive logically the immeasurable system of connected [sensory] phenomena from a few abstract principles. But these abstract principles are then only the means to an economical representation. They are not the epistemological basis. The reality of physics can never be shaken by any criticism of auxiliary concepts."

I will not deny that Mach allowed himself to be misled by this argument into attacking the use of atomistics in physics more sharply than can be justified. After all, the usefulness of atomic theories in this limited realm is certainly indisputable. His followers, as is generally the case, often saw in this weakness of the master his greatest sttrength, and wished to banish the atom entirely from physics. I believe that one can completely free the nucleus of Ernst Mach's teachings from this historically and individually conditioned aversion to atomistics. The atoms are auxiliary conceptions just like the many others which can be employed advantageously in a limited domain. They are not suitable for an epistemological foundation. Once we have adopted this point of view, we are all the freer in employing the concept of atoms wherever it is admissible.

In general, phenomenalism will neither particularly further nor hinder the physicist in his field of work. Thus, Maxwell, who doubtless thought positivistically, wrote the work that laid the physical foundation for the molecular theory of gases. The phenomenalistic conception becomes a danger only in those cases where the requirement of economy is not realized with equal intensity. The most noteworthy historical example is perhaps Goethe's doctrine of colors.

It will perhaps be instructive, as a comparative case, to recall a theoretical physicist who, as an immediate student of Mach, really tried to construct a system of physics and chemistry in which no hypothetical corpuscles, whether atoms or electrons, occur, and which embraces all the phenomena known at present. It cannot be denied that Gustav Jaumann in numerous works has undertaken this task with great constructive force. I do not believe, however, that the result has turned out to be really in the spirit of Ernst Mach's teachings. To be sure, it does correspond to the surface requirement that all atomistics be omitted, but his result hardly corresponds to the requirement of economy.

"Ernst Mach and the Unity of Science"[3]
(1938)

I shall speak about Mach's activity only in so far as he may be considered as one of the spiritual ancestors of the Unity of Science Movement and particularly as the real master of the Vienna Circle.

According to Mach this desire to make use of a unified mode of expression in all fields of science is a result of the economical design of science. This design implies the comprehension of as many facts as possible by means of the simplest possible system of propositions.

Hence Ernst Mach's chief tendency may be described by the slogans "unification (i.e. economical presentation) of science" and "elimination of metaphysics".

If we try to find out what he really thought when he maintained that our world is constructed from perceptions or complexes of perceptions, we shall soon see that this phrase "complexes of perceptions" is in Ernst Mach's eye not in any sense a statement concerning a property of the real world, but only a useful means for the unification of science and also for the elimination of metaphysics.

Mach never maintained that our world consisted of complexes of perceptions, but that every scientific proposition was a statement about perceptions.

We can see that, according to Mach, the unification of science is possible, but only by formulating all scientific propositions as propositions about complexes of perceptions in the widest sense of this word.

Hence to Mach the expression "elimination of metaphysics" means the elimination of all sentences that are not reducible to sentences containing only perception terms as predicates.

The elimination of metaphysics from science was not for Mach, as we understand, a demand which merely arose from some anti-metaphysical mood, but the only means of making possible the unification of science. According to Mach, metaphysics had to be eliminated "because it is contradictory to the economical function of science."

Editor's comments: Mach wrote: "The assertion then is correct that the world consists only of our sensations. In which case we have knowledge *only* of sensations". In short, there is *no* reference to statements or propositions as such. This quotation is from Mach's *Analysis of Sensations* (1959) p. 12. Frank's defense of *propositions* about perceptions, while perhaps compatible with Frank's own earlier views and Carnap's theory of language does not seem to have a firm or consistent foundation in Mach's writings.

Comments on Mach in *Einstein - His Life and Times* (1947)

The French philosopher/sociologist Auguste Comte advanced the sociological theory that the "metaphysical" stage in the development of a science will be or is already succeeded by a "positivistic" one. This means that the demand for the use of a specific analogy such as the organismic and mechanistic views is abandoned and after that a theory is judged only as to whether it presents "positve" experience to people in a simple, logically unobjectionable form. This approach was most widely and profoundly developed by the Austrian physicist Ernst Mach who became one of Einstein's immediate forerunners.[4]

.Mach raised the demand which is now frequently described as the positivistic criterion of science; namely, that only those propositions should be employed by us from which statements regarding observable phenomena can be deduced.[5]

Even though Einstein greatly admired Ernst Mach's ideas on the logical structure of physics, there were many things he could not accept. According to Einstein's judgment Mach did not give enough credit to the creative mind of the scientist who imagines general laws in science beyond a mere economic description of facts. Mach's opinion that the general laws of science are only a means by which individual facts can be remembered more easily, did not appear satisfactory to Einstein. To him the phrase "remembered more easily" could in this connection apparently mean only "remembered with less effort." Mach's *economy* seemed to be *economy* in a psychological sense.

Hence, after conversing awhile with Mach, Einstein raised the following question: "Let us suppose that by assuming the existence of atoms in a gas we became able to predict an observable property of this gas that could not be predicted on the basis of a non-atomistic theory. Would you then accept such a hypothesis even if calculations of its consequences required very complicated computations, comprehensible only with great difficulty? I mean, of course, that from this hypothesis one could infer the interrelation of several observable properties that without it would remain unrelated. Is it then 'economical' to assume the existence of atoms?"

Mach answered: "If with the help of the atomic hypothesis one could actually establish a connection between several observable properties which without it would remain isolated, then I should say that this hypothesis was an 'economical' one; because with its aid relations between various observations could be derived from a single assumption. Nor should I have any objection even if the requisite computations made were complicated and difficult."

Einstein was exceedingly satisfied with this statement and then replied: "By 'simple' and 'economical' you mean, then, not a 'psychological economy' but rather a 'logical economy'. The properties which are observable to us should be derived from as few assumptions as possible, even though these assumptions appear arbitrary and the computation of the results might be difficult."

With *economy* interpreted in this *logical* sense, there was no longer any conflict between Mach's standpoint and Einstein's as to the criteria to be filled by a physical theory. Although Mach made the concession in conversation, yet Einstein saw in his writing only a demand for "psychological economy". Thus for the moment Einstein was satisfied, but he retained a certain aversion to the "Machian philosophy".[6]

Editor's comments: 1) It is no longer acceptable for historians to re-create conversation à la Thucydides or Frank, even when the general content is known, though of course one can paraphrase or give the point or gist. 2) Einstein's support of atoms and Planck's second attack on Mach, both in late 1910 (published in early 1911) seem to have helped persuade Ernst Mach that he should leave what he called "the Church of Physics", which he seems to have done in early 1913 by means of resigning from the Austrian Academy of Sciences[7] even as a corresponding member and by re-emphasizing his interests in anthropology and psychology.

"Einstein, Mach, and Logical Positivism"[8]
(1949)

In the original "positivistic conception of science" as advocated by Ernst Mach, the concepts of which the principles consisted were very near to direct observation and, therefore, very near to possible scientific experiments. The road from the experiments to the principles was short and easy to understand.[9]

Einstein's conception of modern science departs from Mach's "positivistic requirement" in the following point: According to Mach and his immediate followers, the fundamental laws of physics should be formulated so that they would contain only concepts which could be defined by direct observations or at least by a short chain of thoughts which were connected with direct observations. Albert Einstein, however, recognized that this requirement is an oversimplification.[10]

Under the name "Logical Positivism" a new school of thought has appeared, which can be regarded as an attempt to develop Ernst Mach's philosophy of science according to the new developments in theoretical physics.[11]

Einstein speaks here almost completely in the line of the logical empiricists; which should not surprise us, in as much as logical empiricism is, at least to a considerable extent, a formulation of the very way in which Einstein envisaged the logical structure of his later theories, e.g. the theory of gravitation. Occasionally even Einstein himself uses the term "metaphysics" in exactly the same [abusive] sense in which it has been employed by the logical empiricists. He speaks of "metaphysics or empty talk", meaning by it any set of principles from which no conclusion - i.e. no statement about sense-experience can be derived.[12]

Roughly speaking, we may distinguish, according to Planck, two conflicting conceptions in the philosophy of science: the metaphysical and the positivistic conception. Each of these regards Einstein as its chief advocate and most distinguished witness.[13]

Editor's comments: Alas, the rhetorical side of Philipp Frank has clearly come to the fore. 1) Max Planck obviously does not mean "metaphysics" in the pejorative positivist sense as "empty talk" but in a traditional sense close to ontology or theory of reality. 2) Most non-positivists use a theory of reference where it is possible to "meaningfully" refer to what is beyond sensory and conscious bounds, such as past events. 3) Einstein in his 1944 article in the Schilpp volume on Bertrand Russell wrote: "'a fateful fear of metaphysics' arose which has come to be a malady of contemporary empiricistic philosophizing"[14] and "I am particularly pleased to note that, in the last chapter of the book [by Russell], it finally crops out that one can, after all, not get along without 'metaphysics'".[15] In short, what Frank says about "metaphysics" and its relation to Mach, Planck, Einstein, and science does not seem to be free of unfair suggestion in the passages quoted above.

Comments about Mach in *Philosophy of Science* (1957)

Ernst Mach's suggestion amounts to the following hypothesis that the fixed stars exert upon the material bodies on earth an influence that cannot be accounted for by Isaac Newton's law of gravitation. This suggestion seemed at that period very daring and even absurd. The great German physicist, Max Planck, the originator of the most revolutionary hypothesis of the twentieth century, rejected Mach's reformulation of the Newtonian mechanics emphatically. He had definitely approved Einstein's theory of relativity, but he regarded Mach's theory according to which the rotation of the Foucault pendulum is due to an action emanating from the ficed stars, as a fantastic assertion which has its source in Mach's theory of knowledge. Planck regarded Mach's theory of the Foucault Pendulum as a definite proof of the absurdity of his philosophy. However, Albert Einstein began a new analysis of Newtonian mechanics which eventually vindicated Mach's reformulation.[16]

Albert Einstein's principle of relativity would never have been accepted if the metaphysical belief in absolute space and time had not been shaken by the empirical philosophy of men like Ernst Mach.[17]

The criteria of Reichenbach and Carnap, which are based, like Stuart Mill's inductive logic, upon agreement with observations, have to be complemented by the criterion of economy and simplicity which was advanced in the history of science by men like William Ockham, Isaac Newton, and Ernst Mach. In our twentieth century, the importance of criteria other than mere agreement with observation was stressed by both von Mises and Bronowski.[18]

4. CORRESPONDENCE

(1)

Gustav Jäger[1] to Ernst Mach: Vienna, June 5, 1910

It is not my fault for only first answering you today concerning [Paul] Gerber's treatise [*Gravitation und Elektrizität*]. In the belief that Professor [Friedrich] Hasenöhrl who has a much better judgment on such matters than I, would look through the work, I gave it to him since he stated a willingness to read it. After fourteen days, however, he gave it back - after having briefly looked through it and declaring it worthy of attention - but stated that at present he had no time to study it. I also believe that the work deserves notice . . . . [Still] the good agreement between the calculated constant of gravitation and what has been measured is purely accidental . . . .

[But] as a precaution I would like to allow *Privatdozent* [Philipp] Frank to look through the work, a man who I believe is the best qualified of any one of the Vienna physicists to render a judgment on the matter.

With Greatest Respect Your Servant

Dr. Gustav Jäger

(2)

Philipp Frank to Ernst Mach: Vienna, June 13, 1910

Concerning Gerber's work[2] which you were friendly enough to send me: It is somewhat confused, such that in spite of considerable effort I have not been able to understand it . . . . The basic thought is that while the electrical and magnetic strength of the field is dependent on fluctuations of the vector potential, the value of this vector potential is itself the gravitational constant, which agrees with observation. Nevertheless, the deductions which lead to this result are so unsatisfactory that no exact conclusion can be rendered as to the validity of the result.

With Highest Respect, Your Servant

Dr. Philipp Frank

(3)

Philipp Frank to Ernst Mach: Vienna, undated [1910?]

I would very much appreciate it if you could do me a great favor. I am currently negotiating with Johann Ambrosius Barth Publishing House in Leipzig about translating a book by P. Duhem: *L'evolution de la mechanique* into German. I would be very grateful if you would be friendly enough to send a card to J.A. Barth with some words of recommendation or to me so that I can send them.

I would like to mention further, I am now working on a representation of the theory of relativity which is understandable to non-mathematicians as you requested in your letter Herr Hofrat and as Herr Professor Lampa[3] has also asked for. I will especially try to represent Minkowski's thoughts[4] on space and time in an understandable way.

Thanking you in advance and with the very highest respect

Augustengasse 4

Your Servant

Dr. Philipp Frank

NOTES

Section One

1 We have been allowed to use up to 1/3rd of: "Transcript of a Tape-recorded Interview", by Thomas S. Kuhn, 16 July 1962, Archives for the History of Quantum Physics, copy at Niels Bohr Library, Amerian Institute of Physics, College Park, Maryland USA. Frank was born in 1884 and died in 1966.

2 See Ludwig Boltzmann, "Über die Beziehung zwischen dem zweiten Hauptsatze der mechanischen Wärmetheorie und der Wahrscheinlichkeitsrechnung respektive den Sätzen über das Wärmegewicht", *Sitzungsberichte der kaiserlichen Akademie der Wissenschaften*. Wien, 1877, pp. 379-435. Also: Ludwig Boltzmann *Wissenschaftliche Abhandlungen*, ed. F. Hasenöhrl, Vol. II,

Chelsea Pub. Comp. (reprint): New York, 1968, pp. 164-223.

3 Franz Serafin Exner became *Rektor* of the University of Vienna in the Fall of 1907. His address was titled: "Gesetze der Naturwissenschaften und Humanistik". It was included in his late book: ***Vorlesungen über die physikalischen Grundlagen der Naturwissenschaften,*** Deuticke Verlag: Wien, 1. Auflage 1919; 2. Auflage 1922.. For details on the lecture and mention of the controversy it created see B. Karlik & E. Schmid, ***Franz S. Exner und sein Kreis,*** Verlag der Österreichischen Akademie der Wissenschaften: Wien, 1982, pp. 80-83.

4 This is also the opinion of Karlik and Schmid, ibid., p. 82. Schrödinger had become one of Exner's laboratory assistants about this time.

5 Karlik & Schmid, Ibid., "Wissenschaftlichen Wirken von Franz S. Exner", pp. 66-80.

6 Philipp Frank gave lectures or directed discussion sessions five times in the University of Vienna Philosophical Society between 1907 and 1915. He was Secretary of the Society [*Schriftführer*] during the 1911-12 schoolyear. Many physicists were members of the Philosophical Society including Philipp Frank, Arthur E. Haas, Alois Höfler, Gustav Jäger, Ernst Mach, Stefan Meyer, Karl Przibram, Egon von Schweidler, and Marian von Smoluchowski ("Mitglieder-Verzeichnis - Stand im October 1913", in ***Ruckblick auf die ersten fünfundzwanzig Vereinsjahre der Philosopischen Gesellschaft an der Universität zu Wien,*** Selbstverlage der Philosophischen Gesellschaft: Wien, 1913, pp. 43-47.)

7 We have omitted Frank's view that Boltzmann knew nothing about philosophy, since this has been refuted by recent research which points out that Boltzmann had taken nine courses in the subject while a student. The notes for his lectures in philosophy 1903-1906 also contradict Frank's claim. Boltzmann in his Nachlass (1905?) does refer to "Ursache und Wirkung nach Philipp Frank" which may be a reference to what later became an article by Frank: "Kausalgesetz und Erfahrung", *Ostwald's Annalen der Naturphilosophie,* **6** (1907) 445-450. See Ilse M. Fasol-Boltzman, Karl Heinz Fasol, und Walter Höflechner. (Hrsg.) "Ludwig Boltzmann - Notizen und Gedanken aus seinen letzten Lebensjahren". *Österreichischen Gesellschaft für Wissenscaftsgeschichte - Mitteilungen.* **18** (1998) 112.

8 Lise Meitner, who studied under Boltzmann from 1902 until 1906, did not remember Boltzmann ever talking about quantum theory, either his own anticipation or Planck's formulation of it. Patricia Rife, *Lise Meitner - Ein Leben für die Wissenschaft,* Claassen Verlag: Düsseldorf, 1990, p. 41.

9 Philipp Frank writes in his biography of Einstein as follows: "It had long been known that light falling on a surface exerts a pressure on it as if particles were being hurled against the surface. In 1904 the Austrian physicist Hasenöhrl had concluded from this knowledge: If light radiation is enclosed in a vessel, it will exert a pressure on the walls. Even if the vessel itself does not have any material mass, yet because of the pressure of the enclosed radiation it would under the impact of a force behave like a body with material mass. And this 'apparent' mass m will decrease according to $E = mc^2$. This principle is obviously a special case of Einstein's law. If radiation is already contained in a body, its mass will decrease when the radiarion is given off. Einstein's law, however, is much more general. He says that the mass of a body, no matter what in nature, decreases if the body loses energy in any manner whatsoever." (Philipp Frank, ***Einstein - His Life and Times,*** Alfred A. Knopf: New York, 1947, pp. 165.

10 For the phases which Ludwig Boltzmann seemed to pass through in his philosophical understanding while Frank was his student see: John Blackmore, ***Ludwig Boltzmann - His Later Life and Philosophy, 1900-1906,*** Book Two, Kluwer: Dordrecht, 1995.

11 Some students such as Philipp Frank and Schrödinger did indeed pick up ideas from both Mach and Boltzmann without seeing any apparent contradiction, but others such as Meitner, and Smoluchowski noticed it, and some like H. Mache and Gabriele Rabel were deeply bothered by it. For some quotations from the parties themselves please see: John Blackmore, ***Ernst Mach - His Life, Work, and Influence,*** University of California Press: Berkeley & Los Angeles, 1972, pp. 208-209. See also Blackmore (1995), Book One, pp. 75-76 & 128-132.

12 Mach seems to have had several closely related opinions concerning atomism. First, he often

thought the theory might be useful in physics as long as one did not believe in the reality of atoms. Second, he believed it was difficult to apply the atomic theory to both psychology and physics. Third, his own theory of elements is often called an "atomistic theory" in psychology in contrast with both gestalt theory and a continuum theory of experience. Fourth, when critical of the reality of atoms, he normally meant the Greek sense of "indivisible substance" and thought Boltzmann was being evasive by advocating divisible atoms or "corpuscles" such as would become normal after J.J. Thomson and the distinction between electrons and nuclei. Fifth, he normally called physical atoms "things of thought" and was very happy when Ostwald seemed to refute the reality of atoms in 1905. And sixth, after Ostwald returned to atomism in 1908, Mach continued to defend Ostwald's "energeticist" alternative to atomism (See Mach's autobiographical manuscript of 1913 in John Blackmore, "Three Autobiographical Manuscripts by Ernst Mach", *Annals of Science*, **35** (1978) 401-418.)

13 This is of course an exaggeration, but it is true that most physicists were against what they called "metaphysics", and that some of them understood that this meant that they were supposed to be opposed either to the search to understand reality or efforts to try to understand trans-sensory reality.

14 This needs qualification. Einstein did agree that scientific ideas had an empirical origin and had to be tested by empirical means. He also read several of Mach's books and said nice things about him, especially in his obituary of Mach (1916), but Einstein also supported Boltzmann, his work on Brownian motion helped prove the reality of atoms, and he used multi-dimensional geometry in physics in his general theory in a sense which Mach had opposed. By 1922 he was denouncing Mach's epistemology and by the 1930's he was criticizing logical positivism. But some Machist empiricism can still be detected after that.(1936). It is difficult to attribute a consistent philosophical position to Einstein for any period of his life.

15 Philipp Frank is probably referring to Mach's longtime assistant, Gustav Jaumann, who competed with Einstein for a position in Prague in 1910. See Frank (1947), op. cit., pp. 78-79.

16 For a recent book on Mach's influence in America, see Gerald Holton, *Science and Anti-Science*, Harvard University Press: Cambridge, 1993, Chapter One. For an older and somewhat more inclusive account see John Blackmore (1972), op. cit., chapters nine and thirteen or on an even larger scale the UCLA doctoral dissertation (1970) on which the 1972 book is based.

17 For a later version of his position see: Cornelius Lanczos, *The Einstein Decade (1905-1915)*, Academic Press: New York & London, 1974, pp. 15, 25, & 57.

18 For an English translation of Einstein's obituary of Mach see John Blackmore (ed.), *Ernst Mach - A Deeper Look*, Kluwer: Dordrecht, 1992, pp. 154-159.

19 1929.

20 It is probable that Frank is referring to *Modern Science and its Philosophy*, Harvard University Press: Cambridge, 1949. Another influential book of his was simply titled *Philosophy of Science*, Prentice Hall Incorporated: Englewood Cliffs, New Jersey, 1957.

21 Max Planck, *Der Prinzip der Erhaltung der Energie*, Leipzig, 1887, fifth edition 1924. Planck had won second prize for his book in a competition in Göttingen. No one won first prize.

22 Planck's first article was "Die Einheit des physikalisches Weltbildes", *Physikalische Zeitschrift*, **10** 1909) 62-75. It was also published separately as a pamphlet. His second article was "Zur Machschen Theorie der physikalischen Erkenntnis: Eine Erwiderung", *Physikalische Zeitschrift*, **11** (1910) 1186-1190 & *Vierteljahrschrift für wissenschaftliche Philosophie*, **34** (1910) 497-507.

23 Planck died in 1947 at the age of 89. Philipp Frank is probably referring to an English translation titled: *Scientific Autobiography and Other Papers*, which appeared in 1949.

24 Philipp Frank, "Kausalgesetz und Erfahrung", *Annalen der Naturphilosophie*, Volume 6, 1907. An English translation is in Philipp Frank, *Modern Science and Its Philosophy*, Harvard University Press: Cambridge, 1950. An expanded book version of the article can be found in Philipp Frank, *Das Kausalgesetz und Seine Grenzen*, Julius Springer: Vienna, 1932. For a translation see: *The Law of Causality and Its Limits*, edited by R.S. Cohen, translated by Marie Neurath, (Vienna Circle

Collection), Kluwer: Dordrecht and Boston, 1992.

Section Two

1 R. Fürth, "Reminiscences of Philipp Frank at Prague", *Boston Studies in the Philosophy of Science*, Volume **Two**: *In Honor of Philipp Frank*, edited by Robert S. Cohen and Marx W. Wartofsky, Proceedings of the Boston Colloquium for the Philosophy of Science, 1962-1964, Humanities Press: New York, 1965, pp. xiii-xvi. This book includes a number of other interesting reminiscences of Philipp Frank by Peter G. Bergmann, Rudolf Carnap, Gerald J. Holton, Edwin C. Kemble, Hilda von Mises, Raymond J. Seeger, and Kurt Sitte.

2 Laszlo Tisza, "Philipp Frank and Physics", *Philipp Frank 1884-1966*, The editor of this short obituary brochure from Harvard University is not listed, Cambridge, Mass., 1966, pp. 7-9.

3 Peter G. Bergmann, "Philipp Frank in Prague", ibid., pp. 4-6.

4 Gerald Holton, "In Remembrance of Philipp Frank", ibid., pp. 1-3.

5 This quotation has helped give birth to the notion of a "first Vienna Circle" before World War I. For more details see: Thomas E. Uebel (ed.), *Redoscovering the Forgotten Vienna Circle*, (Boston Studies in the Philosophy of Science, Volume **133**), Kluwer: Dordrecht, 1991.

6 Jeremy Bernstein, "Philipp Frank as a Teacher in America", *Philipp Frank 1884-1966*, op. cit., pp. 23-26. Other interesting papers in this brochure on Philipp Frank include: Laszlo Tisza, Ernest Nagel, Joseph T. Clark, S.J., Edwin C. Kemble, and Robert S. Cohen. One should also add that these short papers are different from the ones listed in footnote 1 starting this section.

Section Three

1 This paper originally appeared as: Philipp Frank, "Kausalgesetz und Erfahrung", *Ostwald's Annalen der Naturphilosophie*, **6** (1907) 445-450. Our selection is taken from *Between Physics and Philosophy*, Harvard University Press: Cambridge, Massachusetts, 1941, pp. 16-27.

2 Philipp Frank, "Die Bedeutung der physikalischen Erkenntnistheorie Machs für das Geistesleben der Gegenwart", *Naturwissenschaften*, **5** (1917) 65-72. Our selection is from *Between Physics and Philosophy*, ibid., pp. 28-53.

3 Philipp Frank, "Ernst Mach - The Centenary of His Birth", *Erkenntnis* **7** (1938). Our selection is from *Between Physics and Philosophy*, ibid., pp. 211-226.

4 Philipp Frank, *Einstein - His Life and Times*, translated by George Rosen and edited by Shuichi Kusaka, Alfred A. Knopf: New York, 1947, p. 39.

5 Ibid.

6 Ibid., PP. 104-105.

7 Ernst Mach to Generalsekretär, Österreichische Akademie der Wissenschaften, Vienna, April 28, 1913.

8 Paul Arthur Schilpp (ed.), *Albert Einstein: Philosopher-Scientist*, Volume **One**, Open Court: La Salle, 1969 (1949), pp. 271-286.

9 Ibid., p. 273.

10 Ibid., p. 274.

11 Ibid., p. 275.

12 Ibid., p. 279.

13 Ibid., p. 271.

14 Albert Eunstein "Remarks on Bertrand Russell's Theory of Knowledge", The Philosophy of Bertrand Russell, edited by P.A. Schilpp, Open Court: La Salle, 1989 (1944), p. 289.

15 Ibid., p. 291.

16 Philipp Frank, *Philosophy of Science*, Prentie Hall: Englewood Cliffs, 1962 (1957), p. 153.

17 Ibid., p. 299.

18 Ibid., p. 350.

Section Four

1 Gustav Jäger (1888-1938) had studied physics under Loschmidt and Stefan, was a great admirer of Ludwig Boltzmann and like him worked on kinetic gas theory, and became a full professor of Physics at the Technical University of Vienna in 1918. This letter and the other two to Mach can be found in the Ernst-Mach-Institut, Eckerstr. 4, Freiburg im Breisgau in Germany.

2 Paul Gerber is known for his attempt to account for some of the same phenomena that Einstein did but without using relativity theory. See Klaus Hentschel, *Interpretationen*, Birkhäuser: Basel, 1990, p. 323 and Helge Kragh, *Quantum Generations*, Princeton University Press: Princeton, 1999, pp. 96-97.

3 Anton Lampa (1868-1938) was a physicist first in Vienna and then in Prague. Later he became a director of Viennese adult education. He was a strong supporter of Ernst Mach's philosophy.

4 It seems possible that Mach's reaction to Minkowski's ideas was not positive. Jeremy Bernstein has written: 'It happened, rather, that the first real course I took in physics, [at Harvard] in the spring of 1948, was taught by Philipp Frank who had, as Einstein once remarked, 'imbibed Mach with his mother's milk' . . . . I recall Professor Frank's telling me that Mach [circa 1910] was not very enthusiastic about it' [i.e. Hermann Minkowski's 'brand new' four-dimensional formulation of the theory of relativity]. Please see Jeremy Bernstein's article: 'Ernst Mach and the Quarks' in his: *Cranks, Quarks, and the Cosmos*, Basic Books: New York, 1993, p. 28.

REFERENCES

Anon. [Alois Höfler?]. *Rückblick auf die ersten fünfundzwanzig Vereinsjahre der Philosophischen Gesellschaft an der Universität zu Wien.* Selbstverlage der Philosophischen Gesellschaft, 1913.

Bergmann, Peter G. "Philipp Frank in Prague". *Philipp Frank 1884-1966.* Cambridge: 1966. pp. 4-6. [part of a brochure with no editor or publisher listed]

Bernstein, Jeremy. "Philipp Frank as Teacher in America". Philipp Frank 1884-1966. Cambridge: pp. 23-26. [part of a brochure with no listed editor or publisher]

Bernstein, Jeremy. "Ernst Mach and the Quarks". *Cranks, Quarks, and the Cosmos.* New York: Basic Books, 1993.

Blackmore, John, *Ernst Mach - His Life, Work, and Influence.* Berkeley & Los Angeles: Berkeley, 1972.

Blackmore, John. "Three Autobiographical Manuscripts by Ernst Mach", *Annals of Science* 35 (1978) 401-418.

Blackmore, John (ed.). *Ernst Mach - A Deeper Look.* Dordrecht: Kluwer, 1992.

Blackmore, John. *Ludwig Boltzmann - His Later Life and Philosophy 1900-1906. Book Two The Philosopher.* Dordrecht: Kluwer, 1995.

Boltzmann, Ludwig. "Über die Beziehung zwischen dem zweiten Hauptsatze der mechanischen Wärmetheorie und der Wahrscheinlichkeitsrechnung respektive den Sätzen über das Wärmegewicht". *Sitzungsberichte der kaiserlichen Akademie der Wissenschaften.* Wien, 1877, pp. 379-435. Also in Ludwig Boltzmann. *Wissenschaftliche Abhandlungen.* Band II. Hrsg. Friedrich Hasenöhrl. New York: Chelsea Pub. Co., 1968, 164-223.

Einstein, Albert, "Ernst Mach". *Physikalische Zeitschrift* 17 (1916) 101-104.

Einstein, Albert. *Albert Einstein - Philosopher.* Volume One. edited by P.A. Schilpp. La Salle: Open Court, 1969 [1949]. pp. 271-286.

Einstein, Albert. "Remarks on Bertrand Russell's Theory of Knowledge". *The Philosophy of Bertrand Russell,* edited by P.A. Schilpp. La Salle: Open Court, 1989 [1944], pp. 277-291.

Exner, Franz Serafin. *Vorlesungen über die physikalischen Grundlagen der Naturwissenschaften.* Wien: Deuticke Verlag, 1919.

Fasol-Boltzman, Ilse M., Karl Heinz Fasol, und Walter Höflechner. (Hrsg.) "Ludwig Boltzmann -

Notizen und Gedanken aus seinen letzten Lebensjahren". *Österreichischen Gesellschaft für Wissenscaftsgeschichte - Mitteilungen.* **18** (1998) 87-124.
Frank, Philipp. "Kausalgesetz und Erfahrung". *Annalen der Naturphilosophie,* 6 (1907) 445-450.
Frank, Philipp. "Die Bedeutung der physikalischen Erkenntnisstheorie Machs für die Geistesleben der Gegenwart". *Naturwissenschaften,* 5 (1917) 65-72.
Frank, Philipp. "Ernst Mach - The Centenary of his Birth". *Erkenntnis* 7 (1938).
Frank, Philipp. *Between Physics and Philosophy.* Cambridge: Harvard University Press.
Frank, Philipp. *Einstein - His Life and Times.* New York: Alfred A. Knopf, 1947.
Frank, Philipp. *Modern Science and Its Philosophy.* Cambridge: Harvard University Press, 1949.
Frank, Philipp. *Philosophy of Science.* Englewood Cliffs: Prentice Hall, 1957.
Fürth, R. "Reminiscences of Philipp Frank in Prague", *Boston Studies in the Philosophy of Science.* Volume Two. In Honor of Philipp Frank, edited by Robert S. Cohen and Mark W. Wartofsky, Proceedings of the Boston Colloquium for the Philosophy of Science, 1962-1964. New York: Humanities Press, pp. xiii-xvi.
Hentschel, Klaus. *Interpretationenen.* Basel: Birkhäuser, 1990.
Holton, Gerald. "In Remembrance of Philipp Frank". *Philipp Frank 1884-1966.* Cambridge: 1966. pp. 1-3. [part of a brochure with no editor or publisher listed]
Holton, Geralf. *Science and Anti-Science.* Cambridge: Harvard University Press, 1993.
Karlik, Berta, und Erich Schmid. *Franz S. Exner und sein Kreis.* Wien: Verlag der Österreichischen Akademie der Wissenschaften , 1982.
Kragh, Helge. *Quantum Generations.* Princeton: Princeton University Press, 1999.
Kuhn, Thomas S. "Transcript of a Tape-recorded Interview with Philipp Frank". [a manuscript located in the Niels Bohr Library, College Park, Maryland] 1962.
Lanczos, Cornelius. The Einstein Decade (1905-1915). New York & London: Academic Press, 1974.
Planck, Max. "Die Einheit des physikalischen Weltbildes". *Physikalische Zeitschrift* 10 (1909) 62-75.
Planck, Max. *Der Prinzip der Erhaltung der Energie.* Leipzig: fifth edition 1924 [1887].
Planck, Max. "Zur Machschen Theorie der physikalische Erkenntnis - Eine Erwiderung.". *Physikalische Zeitung* 11 (1910) 1186-1190. This article also appeared in: *Vierteljahrschrift für wissenschaftliche Philosophie* 34 (1910) 497-507.
Planck, Max. *Scientific Autobiofephy and Other Papers,* translated by Frank Gaynor, 1949.
Rife, Patricia. *Lise Meitner - Ein Leben für die Wissenschaft.* Düsseldorf: Claassen Verlag, 1990.
Tisza, Lazlo. "Philipp Frank and Physics". *Philipp Frank 1884-1966.* Cambridge; 1966. pp. 7-9. [part of a brochure with no editor or publisher listed]
Uebel, Thomas E (ed.). *Rediscovering the Forgotten Vienna Circle.* Boston Studies in the Philosophy of Science. Volume 113. Dordrecht: Kluwer, 1991.

Chapter Four

ERWIN SCHRÖDINGER

by Henk W. de Regt

Section Guide

1. BIOGRAPHICAL SKETCH

The Viennese scientist Erwin Schrödinger became world famous at a relatively late age, that is, for a theoretical physicist active during the 1920's. In 1926, when he was already a doddering thrty-eight, he was still capable of developing a new quantum theory of atomic structure. His new theory was presented as an alternative to the theory of matrix mechanics, which had been formulated by Werner Heisenberg along with Max Born and Pascual Jordan some months previously. Matrix mechanics was a rather abstract and predominantly mathematical theory, which had deliberately refrained from putting forward claims about the unobservable world of atoms. The great advantage of wave mechanics, Erwin Schrödinger repeatedly emphasized, was that it appeared to promise a visualizable, realistically interpretable description of atomic structure. Thus, in his first paper on wave mechanics, Schrödinger spoke of his theory as representing: "some vibrational process in the atom, which would more nearly approach reality than the electronic orbits, the real existence of which is very much questioned today".[1]

As these quotations illustrate Erwin Schrödinger freely spoke of "reality" and "real existence" when discussing the unobservable structure of atoms in his scientific writings. It is therefore not surprising that it has been assumed that Schrödinger was a realist who held that a scientific theory is (or at least aims hard to be) a literal description of an independently existing reality.[2] Nonetheless, I will argue in this chapter that this assumption is false, and that contrary to being a naive correspondence realist of sorts, Erwin Schrödinger adhered to an anti-realist philosophy.

Schrödinger became strongly interested in philosophy. During the First World War, for example, he made a serious study of many philosophical works. Consequently as he was to write in the preface to a philosophical treatise *Meine Weltansicht*, around 1918 he was "deeply imbued with the writings of Benedict Spinoza, Arthur Schopenhauer, Ernst Mach, Richard Semon, and Richard Avenarius.[3] Indeed, Schrödinger's philosophical writings reveal the influence of some of these philosophers. Undoubtedly, Ernst Mach is the most important example. In particular, Mach's opinions concerning epistemology and ontology appear to have made a strong and lasting impression on the young Schrödinger for he keeps referring to them through his entire career. As will be shown in this chapter,

J. Blackmore, R. Itagaki, and S. Tanaka (eds.). Ernst Mach's Vienna. 1895-1930. 85-104.

Ernst Mach's idealistic monism, his phenomenalism according to which "sensations" are the only reality, was adopted by Erwin Schödinger and never relinquished. At the epistemological level, Ernst Mach's position leads to an anti-realist view of scientific theories, which refuses to ascribe to unobservable entities (e.g. atoms and electrons) the status of really existing objects. Erwin Schrödinger followed Mach also in this epistemological anti-realism.

There is, however, one major subject on which Schrödinger seriously disagreed with Mach. In contrast to Mach's radically empiricist view, which asserts that the aim of science is to be *description* of observable phenomena in as economical a way as possible, Schrödinger held that science aims at *understanding* of these phenomena. According to Schrödinger, a scientific theory can provide understanding only if it is visualizable in space and time. In other words, he gladly accepted what the Germans call *Anschaulichkeit* as a necessary criterion for the construction of theories.[4] This preference for visualizable theories can be easily traced back to the influence of Ludwig Boltzmann. It is this strain in Schrödinger's philosophy which is the source of confusion and misunderstanding as regards the issue of realism. His demand for visualizable theories might easily encourage one to jump to the conclusion that Schrödinger also thought that these visualizations are direct representations of reality. However, this is not a necessary consequence, and I will show that he refused to make this step to realism. The abundance of references to "atomic reality" in Erwin Schrödinger's work can be explained as being of merely pragmatic or methodological origin.

Thie opening chapter presents a systematic analysis of the influence of Mach's ideas on Schrödinger's thought. By way of introduction, Section 2 briefly describes Erwin Schrödinger's early life and his scientific career. Subsequently, in Section 3, Ernst Mach's influence on Erwin Schrödinger is analysed in detail. The anti-Machian tendencies in Schrödinger's thinking, which were mainly inspired by Boltzmann, are discussed in Section 4. In the next section it is shown how Erwin Schrödinger attempted to achieve a synthesis of the Machian and Boltzmannian elements of his philosophy. Finally, Section 6 will discuss the relation between Schrödinger's philosophy, his discovery of wave mechanics, and his views concerning the interpretation of quantum mechanics.

Erwin Schrödinger (1887-1961) was born in Vienna and would enter the University of Vienna to study physics in 1906. The memories of Boltzmann who had just taken his own life, and Ernst Mach, who had retired in 1901, still lingered in the air, and their heated debates over the existence of atoms and the philosophical foundations of physical science had been far from forgotten. Boltzmann's immediate successor for 1906-07 was his longtime assistant Stefan Meyer. The permanent successor, Fritz Hasenöhrl, came the next year and it was very largely from him that Erwin Schrödinger learned theoretical physics. Hasenöhrl's inaugural lecture was a brilliant exposition of Boltzmann's ideas and achievements. The contents impressed Schrödinger indelibly and directed his own thoughts towards a path from which he would never depart. Schrödinger later testified that "no other human being had a greater influence on him than Fritz Hasenöhrl, except perhaps my father Rudolph".[5] But this influence, however, consisted above all in the ideas of Boltzmann. Thus in his 1929 inaugural speech, he remarked: "[Boltzmann's] line of thought may be called my first love in science. No one has ever thus enraptured me or will ever do so again."[6]

The influence of Franz Exner, who taught experimental physics, can be found in Erwin Schrödinger's ideas on causes and indeterminism.[7] Furthermore, being involved in experimental physics also had a noticable effect on young Schrödinger's philosophy of science. Although he was chiefly interested in theoretical physics, Schrödinger worked in both fields until 1920, and his doctoral dissertation of 1910 was an entirely experimental study. After he obtained his degree he was offered a one-year university assistantship under Exner in experimental physics. During this time, Schrödinger discovered that he was not well suited to be an experimentalist, but he later expressed his gratitude for having been able to experience directly the experimental side of physical research.[8] In his biography of Schrödinger, Moore states that "his work as a laboratory assistant had helped him to determine the philosophical framework that he was willing to accept for a physical theory", namely the view that "physics is not based upon mathematical fantasies but is based on the solid ground of experimental observations".[9]

Just before the war, in 1914, Schrödinger completed his *Habilitation* in theoretical physics and obtained an unpaid position as *Privatdozent* at the University of Vienna. During the war he served in the Austrian army, but he still managed to carry out some theoretical research, especially on statistical dynamics. He also carefully studied Albert Einstein's general theory of relativity and, moreover, immersed himself more in philosophy. After briefly teaching in Jena, Breslau, and Stuttgart Erwin Schrödinger In September 1921 settled in Zürich, having been offered a professorship in theoretical physics. But there he worked primarily on statistical physics. Quantum theory did not have his special attention, although he made minor contributions to it. The main reason for this lack of interest was that he regarded the old quantum theory of Bohr and Sommerfeld as unsatisfactory, primarily because of "its inherent contradictions".[10]

In 1925, however, his work on gas dynamics led him to examine Albert Einstein's paper on the quantum theory of the monatomic gases and Bose statistics. Through this paper he would become acquainted with Louis de Broglie's hypothesis of matter waves. In late December in 1925, with an outburst of creativity, Schrödinger developed the essential foundations of wave mechanics, a theory of atomic structure within which only wave concepts were used. During the first half of 1926 he further elaborated his theory and presented a proof of its mathematical equivalence with matrix mechanics, a rival theory put forward by Heisenberg, Born, and Jordan. Matrix mechanics had repelled Erwin Schrödinger because of its absence of *Anschaulichkeit*. However, the physical, visualizable interpretations that he proposed for his wave mechanics all ran into difficulties, and he did not succeed in defending them against the attacks from the Copenhagen school of quantum mechanics. The manner in which this school then incorporated his theory was completely unacceptable to Schrödinger. He considered Max Born's statistical interpretation of the wave function to be a complete misinterpretation of his theory. Furthermore, in Schrödinger's opinion Niels Bohr's notion of complementarity was nothing more than a "marvously soothing [empty] expression".[11] While this magical word led physicists to believe that philosophically everything was settled, actually it did not solve any of the conceptual problems linked to quantum mechanics. First and surely foremost among these was the so-called measurement problem, highlighted by the famous thought-experiment of Schrödinger's cat.[12]

In 1927, Schrödinger succeeded Max Planck in the chair of theoretical physics at the University of Berlin. But the advent of the Nazi regime in 1933 induced him to emigrate to Oxford. The same year would bring him the Nobel Prize in physics for his discovery of wave mechanics. But Erwin Schrödinger longed for Austria and returned shortly to teach in Graz. After the 1938 *Anschluss*, however, he fled again, this time to Dublin, where he stayed in exile until 1956. His last years were spent in his birthplace, Vienna.

Erwin Schrödinger's scientific work in his last years was primarily concerned with discovering a "unified field theory", which would combine Einstein's general theory of relativity with the theory of electromagnetism. His own involvement with quantum theory would be restricted to questions about interpretation. Erwin Schrödinger became very disappointed when the Copenhagen conception of quantum mechanics seemed to become the dominant view. On numerous occasions he voiced criticism of its theses.[13] In the post-war period Schrödinger devoted much of his time to philosophy and related subjects. Best known among these books is *What Is Life?* (1944) in which he explores the physical basis of biological processes and which has had a great impact on the development of molecular biology. In addition, the volume *Science and Humanism* (1951) focused mainly on the philosophical implications of modern physics. In his 1954 book *Nature and the Greeks* Erwin Schrödinger traces the fundamental principles of Western science back to their ancient origins. Finally, he expounded his views concerning the mind-body problem in *Mind and Matter* (1958).

2. SCHRÖDINGER AND MACH

On March 22, 1940, in a letter to Arthur Eddington, Schrödinger recalled the period of his intellectual formation and wrote about the most important influences on his thinking. This letter is worth quoting at some length:

> I was born and educated in Vienna with Ernst Mach's teaching and personality still pervading [much of] the atmosphere. I was devoted to his numerous writings, which I went through almost completely. Just as strong or even stronger than Ernst Mach's [influence] during that period was the after-effect of the great Boltzmann, whose splendid pupil and admirer Hasenöhrl had just taken over his chair in physics (which had been so cruelly evacuated the year before in 1906). Both Boltzmann and Mach were, as you know, just as much interested in philosophy, more especially in epistemology, as they were in physics; in fact all of their later publications were pervaded by perspectives which were based on epistemology. Their views were not the same. But being filled with great admiration of the candid and incorruptible struggle for truth in both of the men, we who were in Vienna did not consider them irreconcilable. Ludwig Boltzmann's approach had consisted of forming absolutely clear, almost naively clear and detailed "pictures" - mainly in order to be *extremely* certain of avoiding contradictory assumptions. Ernst Mach's ideal was a cautious synthesis of observational facts that can, if desired, be traced back to the plain, crude, sensual perception (pointer reading). He was anxious not to contaminate absolutely reliable timber with anything else of more doubtful origin.
>
> However, we decided that for ourselves these were just different methods of attacking problems, and that we were just as permitted to follow one as the other provided that we did not lose full sight of the main principles that were being strongly emphasized by the followers of the other approach.[14]

The seemingly opposite positions of Mach and Boltzmann lie at the root of Erwin Schrödinger's own philosophy of science. As will become clear, Schrödinger held on to these views all of his life and attempted to reconcile them into a coherent philosophy. In the present section the influence of Ernst Mach's ideas will be analysed. In the next sections I will proceed to discuss Boltzmann's influence and Erwin Schrödinger's reconciliation of Mach and Boltzmann.

From the letter to Eddington we learn that Erwin Schrödinger made a serious study of Mach's philosophy; he claims to have read almost all of Mach's writings. Probably he did this during his student days in Vienna, but as mentioned earlier, certainly during the war Mach's philosophy directly inspired Schrödinger's scientific work, especially when he was occupied with research on colour vision. In his contributions to colour theory he also attempted to construct a "colour space" by means of a direct comparison of elementary sensations.[15] More important, however, is the influence that Mach had on Schrödinger's philosophy. Ernst Mach's ideas affected first and foremost Schrödinger's ontological conceptions, and to a somewhat lesser extent his epistemological views. This clearly shows in the philosophical treatise *Suche nach dem Weg*, written by Erwin Schrödinger in 1925, in which he analysed the desperate state of metaphysics in Western thought.[16]

In the fourth chapter of this essay, Erwin Schrödinger has presented a discussion of the mind-body problem that contains an affirmation of Mach's ontological position.[17] He begins by rejecting out of hand the assumption that a soul can exist independently of a body as naive and childish, and states that both the "I" and the "world" consist of the same "empirical elements". Then he addresses the question what would happen when the "I" observes something, for example a tree. Erwin Schrödinger criticizes the well-known solution of trying to distinguish between the observation of the tree and the tree itself (the tree *an sich*), because it does not clarify in any respect what the observation of the tree consists in. Subsequently, he sketches the radical and simple answer to this problem, proposed by Mach and Avenarius: As the tree *an sich* is inaccessible for our experience, we might just as well declare once and for all that it is uninteresting and that we will abandon it. Hence, only one given aspect of the tree remains, and there is no reason not to call it "tree". He writes: "This *one* tree, then, is the one datum we have; it is at one and the same time the tree of physics and the tree of Psychology".[18] He then remarks with self-approval that this implies that the "I" and the "world" do consist of the same elements, leading to the elimination of many pseudo-problems and being a vindication of alleged "naive realism".[19] However, one problem remains, namely the question of what would happen when different people observe the same tree. The only consistent answer, according to Schrödinger, and indeed the answer that Mach and Avenarius gave, is to say that the tree, being *one* complex of elements, is a common ingredient of different "I's". Schrödinger claims that this conclusion, which may at first sight seem quite bizarre, is in fact closer to our daily human experience and to "naive realism" than the idea, which is common to Western thought, that the experiences and thoughts of different individuals rightly belong to what are mutually exclusive spheres.

The ontological doctrine which Schrödinger had adopted from Mach is known as *neutral monism*. It asserts that minds and bodies do not differ in their intrinsic nature, but that the difference between them lies in the way that a common ("neutral") material is arranged. This common material may actually consist of many seeming entities of the same fundamental kind, e.g. experiences. Neutral monism is opposed to subject-object (or mind-body) dualism. Schrödinger attributed the latter view to Immanuel Kant, and his opposition to dualism might have contributed to his rejection of Bohr's philosophy of complementarity.[20] The philosophical doctrine of neutral monism was never abandoned by Schrödinger. In *Nature and the Greeks* (1954) he endorsed a view of the mind-body problem, which reads like a standard formulation of Machian neutral monism.

> The "real world around us" and "we ourselves", that is, our minds, are made up of the same building material, the two consist of the same bricks, as it were, only arranged in a rather different order - sense perception, memories, imagination, and thought . . . We can think of these - let me call them *elements* - either as that which constitutes mind,

> everyone's own mind, or, as that which constitutes the material world. But we cannot, or can only with great difficulty, think both things at the same time. To get from the mind aspect to the matter aspect or vice versa, we have as it were, to take the elements asunder and to put them together again in an entirely different order.[21]

Erwin Schrödinger's position is a specifically Machian version of neutral monism, because like Mach he takes *sensations* (empirical elements) as the neutral building blocks of both mind and matter. In his autobiography, Karl Popper writes that when discussing philosophical questions with Erwin Schrödinger around 1954-55 he was surprised to discover that Schrödinger was inclined to defend neutral monism, while admitting that it was a form of idealism.[22] Considering the Machian roots of Schrödinger's position, this will not surprise us, because also Mach may be classified as an idealist.[23] Popper seems to have held however, that "Schrödinger had absorbed his idealism from Schopenhauer". Indeed, Schrödinger had read Schopenhauer's work with fascination, and it was an additional source of his idealistic sympathies, and through it he had encountered ancient Indian philosophy of *Vedanta* and the *Upanishads*.[24] The main characteristics of these doctrines is the belief in an undivided reality of thought, a transcendent unity of consciousness. The impact of these Eastern philosophies on Schrödinger's thought is particularly evident in *Suche nach dem Weg*. In Chapter Five of the 1925 treatise, he had developed an alternative metaphysics based on the *Vedanta*, in which the manifold of appearances appears reduced to a kind of universal undivided consciousness.[25]

Many years later in his *Mind and Matter* (1958), Schrödinger again came out in favor of an idealist view of the world. The first sentences of this book read:

> The world is a construct of our sensations, perceptions, and memories. It is convenient to regard it as existing objectively on its own. But it certainly does not become manifest by its own existence.[26]

The very initial sentence is a bold statement of a radically empiricist view with sharp idealist overtones, clearly inspired by Mach's philosophy. Additional evidence of Mach's influence in this respect can be found in Erwin Schrödinger's 1950 essay "What is an elementary particle?", where he defends "a view on matter and the material universe, to which Ernst Mach, Bertrand Russell, and a number of other people were led by careful analysis of concepts".[27]

The second sentence of *Mind and Matter*, quoted above, summarizes what Erwin Schrödinger has called "the principle of objectivation". According to Schrödinger, this principle originated from ancient Greek thought and is fundamental to all Western science.[28] A common formulation of it is the "hypothesis of the real world".[29] The principle of objectivation leads to a separation of the subject and object. On the one hand there is the objective world, and on the other hand the knowing subject. While this principle has turned out to be extremely fruitful to science, it also leads to problems, which were already acknowledged by the Greeks. A famous example, cited repeatedly by Schrödinger, Democritus' paradox of the sensory qualities:

> Democritus introduces the intellect as having a major argument with the senses about what is "real". The former then declares to the latter: 'Ostensibly there is color, ostensibly sweetness, and ostensibly there is bitterness, but actually only atoms and the void', to which the senses retort: 'Poor intellect, do you hope to defeat us while borrowing from us

> your evidence? Your victory is your defeat.'[30]

By excluding the human subject from the scientific world-picture, the principle of objectivation also abandons sensations, which are the empirical building blocks of scientific knowledge. Erwin Schrödinger's answer to this paradox is to grant the principle and the ensuing subject-object distinction a pragmatic, methodological status only: "Though we may have to accept it in everyday life 'for practical reference', we ought, so I believe, to abandon it in philosophical thought."[31]

This brings us back to Schrödinger's idealistic notion of neutral monism. Discussing the question of whether the world exists if there is no mind to observe it, he concludes:

> The world is given but once. Nothing is reflected. The original and the mirror image are identical. The world which has been extended into space and time is only our own representation of it (*Vorstellung*). Experience does not give us the slightest clue of its being anything beside that - as Berkeley was well aware.[32]

In summary, Erwin Schrödinger's opinions concerning ontology were evidently based upon Mach's philosophy. First, like Ernst Mach, he endorsed neutral monism. Second, he identified all neutral elements with "empirical elements" (Mach's sensations). However, inspired by the main doctrine of the *Vedanta*, Schrödinger transformed the Machian ontology into substantive monism, that is, the doctrine that ultimately there exists only a single entity, comprising all empirical elements. By identifying this entity as an undivided consciousness, Schrödinger explicitly took an idealist stance.

Schrödinger's ontological position seems to entail implicit confirmation of Mach's empiricist epistemology that all we know is given by experience. From empiricism, Ernst Mach had inferred a radically anti-realist view of scientific theories, one which led to his notorious opposition to the kinetic theory of gases and his denial of the existence of atoms. In *Nature and the Greeks*, when discussing the Greek atomists, Schrödinger had mentioned Ernst Mach's rejection of atomism in a quite neutral manner,[33] which is remarkable because most scientists scorn Mach for precisely this reason. Did Schrödinger also adhere to anti-realism with respect to scientific theories? I submit that indeed he did. Direct evidence for this can be found in *Science and Humanism*,[34] which contains a discussion of the nature of our models of the atom:

> The pictures [provided by models] are only a mental help, a tool of thought, an intermediary means, from which to deduce out of the results of experiments that have been made, a reasonable expectation about the results of new experiments that we are planning. We then plan them for the further purpose of seeing whether they conform to the expectations, that is, to whether the expectations have been reasonable, and whether the pictures or models we use are *adequate*. Notice that we prefer to say it is adequate, not *true*. For in order that any description be capable of being true, we must be able to compare it *directly* with actual facts. That is usually not the case with our models.[35]

It thus appears to be at the epistemological level that Schrödinger was an outspoken anti-realist. He did not regard theoretical models and pictures as candidates for being true (or false). As the quotation has indicated, this was a direct consequence of his empiricist perspective, since only the directly observable were regarded as evidence.. Already in the year 1926, in a letter to Wilhelm Wien, Schrödinger had put forward the same thesis: "There is certainly no criterion for reality if one does not want to say: the *real* is only the complex of sense impressions; all the rest are only pictures."[36] It may be concluded that Erwin Schrödinger endorsed anti-realism throughout his whole scientific career.

Whereas he agreed with Ernst Mach's philosophy with respect to ontology and epistemology, Schrödinger strongly disagreed with Mach's narrow view of scientific method. Ernst Mach had claimed that the only aim of science is to describe the complexes of sensations in the most economical manner. By contrast, Schrödinger believed that the final aim of science is not merely to describe phenomena, but also to *explain*, to gain significant *understanding* of them. Arguing against Mach's methodological views, Schrödinger stated, in the very first chapter of his book *Suche nach dem Weg*, that "elimination of metaphysics means taking the soul out of both art and science, turning them into the merest skeletons incapable of any further development", and urged the reader to:

> call to his mind that sense of misgiving, that cold clutch of dreary emptiness which comes over everybody, one may expect, when they first encounter the description given by Kirchhoff and Ernst Mach of the task of physics (or of science generally): "a description of the facts, with the maximum of completeness and the most economy of thought"; a feeling of emptiness which one cannot master, despite the emphatic and even enthusiastic agreement with which one's own theoretical reason can hardly fail to accept this prescription. In actual fact (and let us examine ourselves honestly and faithfully), to have only this purpose before one's eyes would not suffice to keep the work of research going forward in my field whatsoever.[37]

Instead of Ernst Mach's descriptive task of science, Schrödinger regarded understanding as the chief aim of science. In *Nature and the Greeks* he even claimed that, together with what he called the principle of objectivation we mentioned above, *the* central feature of Western science, is "the hypothesis that *the display of nature can be understood by us*".[38] This methodological principle, which I will call the "intelligibility thesis", had its roots in ancient Greece and initiated the rise of science. Obviously, the question immediately arises: "What does comprehensibility really mean, and in what sense if any, does science give explanations?"[39] Schrödinger subsequently discusses the Machian, positivist view which amounts to a rejection of the intelligibility thesis. Admitting that positivism has its merits, he states that even from that point of view "One ought not, so I believe, to declare that science conveys no understanding at all".[40] Schrödinger's own answer to the question of the meaning and content of intelligibility will be given here in Section 5. First, however, I will discuss Ludwig Boltzmann's influence on Schrödinger's thought, which is the source of his view of intelligibility.

3. SCHRÖDINGER AND BOLTZMANN

Erwin Schrödinger's opinions concerning methodology of science had their source in the views of Mach's opponent, Ludwig Boltzmann, whom he equally admired. In Section 2 some of Schrödinger's laudatory remarks on Boltzmann were quoted and many more can be found in his writings. Again and again, Schrödinger mentions Boltzmann as the source of inspiration in his scientific work.

Dieter Flamm, who is Ludwig Boltzmann's grandson, has informed us that Erwin Schrödinger possessed a copy of Boltzmann's *Populäre Schriften*, in which he made notes on a number of articles.[41] This fact proves that Erwin Schrödinger *studied* Boltzmann's philosophical writings, and thus had a considerably more than superficial understanding of his philosophy, and consequently, of the controversy between Boltzmann and Mach. That is to say, he was very probably aware of the fact that at the epistemological level Boltzmann was not an ordinary correspondence realist, and that it was thus rather at the ontological and the methodological level that the main struggle between Boltzmann and Mach was conducted. Furthermore certain notes Schrödinger made in his copy show that he had reservations about some of Boltzmann's stronger realist claims.

Which of Boltzmann's approaches were adopted by Schrödinger? First of all, as he states in his letter to Eddington quoted in Section 2, Schrödinger took from Boltzmann the *Bild* conception of physical theory: "Boltzmann's idea consisted in forming absolutely clear, almost naively clear and detailed "pictures" - primarily in order to be *quite* certain of avoiding contradictory assumptions."[42] In his book *Populäre Schriften*,[43] Boltzmann had developed a view of the aim of physical theory as the construction of a "picture" of the outer world, that serves as a guiding star in thought and experiment.[44] Erwin Schrödinger happily agreed with this methodological principle. As we shall attempt to point out below, he regarded visualizability (*Anschaulichkeit*) as a necessary criterion for all scientific theories.

Second, Erwin Schrödinger adopted Ludwig Boltzmann's methodological realism. As I have emphasized elsewhere,[45] Boltzmann'a epistemological position during the 1890's was rather subtle: He was cautious not to assert a simple identity relation between the theoretical picture and the real world. At the methodological level, however, he strongly advised a realist attitude and he forcefully rejected the ideas of Mach and his followers. Boltzmann's main strategy in attacking Mach were to argue, first, that his phenomenological position was no less metaphysical than his own, and secondly, that Mach's ideas were unproductive and would hinder the progress of science. Erwin Schrödinger accepted this thesis: When doing research he immersed himself in the world-picture of science (which helps explain why he persistently used the word "real" in scientific writings in spite of his Machian epistemology).[46]

Boltzmann'a theories had a direct influence on Schrödinger's scientific work. A considerable part of Erwin Schrödinger's early work was in the field of statistical mechanics and gas theory, and thus was a direct continuation of Boltzmann's line of research. An interesting influence of Boltzmann's way of thought can be discerned in Erwin Schrödinger's 1914 article: "Zur Dynamik elastisch gekoppelter Punktsysteme".[47] In the paper he dealt with controversy between the atomistic and continuum theories of matter, and building upon Boltzmann's strategy mentioned above, he then argued that the "idealized" differential equations must be replaced by difference equations, and then he set out to state precise conditions under which differential equations have actually led to incorrect results, thus attempting to prove the superiority of the atomistic theory of matter.

In many of his scientific articles, Schrödinger also defended Boltzmann's *Bild*-conception of physical theory. A nice example of this came out in 1926, when Erwin Schrödinger's theory of wave mechanics had to compete with matrix mechanics. Though these two theories are very different in character (that is, the former theory is "physical", "visualizable", and "realistic" with the latter being "mathematical", "non-visualizable", and "positivistic"). It was Erwin Schrödinger himself who proved their mathematical equivalence.[48] Having demonstrated this, he went on to compare the physical content of the two theories, and observed that:

> Today, there are not a few physicists who, like Mach and Kirchhoff, regard the task of physical theory as being merely a mathematical description (*but as economical as possible*) of the many empirical connections which exist between observable quantities, ithat is, a description which reproduces the connections, as far as possible, without the intervention of unobservable elements. On this view, mathematical equivalence has almost the same meaning as physical equivalence. In the present case there just might perhaps appear to be a certain superiority in the matrix representation, because through its *unanschaulichkeit*, it does not tempt us to form space-time pictures [*Bilder*] of atomic processes, which must then perhaps remain beyond control.[49]

Against this conclusion, however, Erwin Schrödinger put forward two arguments. First, because the mathematical equivalence is fully symmetrical, the formalism of wave mechanics does not have superfluous, "metaphysical" parts that transcend the physical (empirical) content of matrix mechanics. In Erwin Schrödinger's very own words the wave functions: "do not form, as it were, an arbitrary and special 'fleshy clothing' for the bare matrix skeleton, provided to pander to the need of *Anschaulichkeit*, [which] would establish the superiority of the matrices, from the epistemological point of view".[50] Incidentally, this argument is interesting for its Machian flavour since it assumes that those parts of a theory which transcend any empirical evidence are epistemologically irrelevant. As wave mechanics equals matrix mechanics in this respect, Schrödinger continued by stating his second argument, which was meant to prove the superiority of wave mechanics over Matrix mechanics. He has argued that mathematically equivalent theories may differ in their possible extensions, and that they can therefore differ as well in their *fruitfulness*. According to Schrödinger, atomic theory is confronted with problems that seem "extraordinarily difficult to tackle [. . .] as long as we feel obliged on epistemological grounds to repress *Anschauung* in atomic dynamics, and operate only with such very abstract ideas as transition probabilities, energy levels, etc."[51] Thus he has asserted that *anschauliche* theories, presenting visualizable pictures, are preferable because they are heuristically more powerful.[52]

In sum, Erwin Schrödinger borrowed two important methodological principles from Ludwig Boltzmann: (1) methodological realism, *i.e.* the approach of immersing one in the scientific world picture; and (2) the Bild-conception of theory, thus requiring the *Anschaulichkeit* (vizualizability) of scientific theories.

4. MACH AND BOLTZMANN RECONCILED

In his letter to Eddington, quoted in Section 3, Schrödinger stated that he did not consider the views of Mach and Boltzmann to be irreconcilable. His comment that "these were just different methods of attack" might suggest that Schrödinger regarded Mach's and Boltzmann's theses both as alternative *methodological* principles. However, I think this suggestion is misleading. I argue that Ernst Mach's and Ludwig Boltzmann's seemingly incompatible views affected different levels of Schrödinger's philosophy of science, which opens up the possibility of reconciling both views in a consistent synthesis.

These different basic levels of human thought were ontology, epistemology, and methodology. As I have shown in the previous sections, he adhered to Ernst Mach's ideas at the ontological and epistemological level, but he seemed to prefer Boltzmann's perspective concerning methodology of science. But notwithstanding Schrödinger's basic remark about his methods, the letter to Eddington reveals the distinction between levels. Schrödinger mentions that already in the year 1913 he had discussed the "terrible 'jumps' in Niels Bohr's orbits" with a friend, and had expressed the opinion "that very probably the place of an electron within an atom had no meaning because we had no means of observing it, that we may and must use pictures (Boltzmann), but with our eyes open towards their limitations, which are given by what can be observed, because the ultimate aim of our pictures is only to serve as a scaffolding for our sensory perceptions (E. Mach)".[53] Evidently, such a statement puts Boltzmann in the methodological domain, while Ernst Mach rules in the epistemological and ontological realm.

A clear example of Erwin Schrödinger's reconciliation of a Machian anti-realism at an ontological and epistemological level with methodological realism in the manner of Ludwig Boltzmann can be found in the positive response to the radiation theory of Bohr, Kramers, and Slater (BKS).[54] The BKS theory contained "virtual" elements (oscillators and radiation), which were not interpreted realistically. For many physicists the status of these virtual oscillators and virtual radiation was problematic. Not so for Erwin Schrödinger. This can be gleaned from his letter to Niels Bohr of May 24, 1924:

> I cannot go along with you completely when you continue to call [this radiation] 'virtual' [. . .] For what is the 'real' radiation if it is not that which has 'caused' transitions, that is, which created the transition probabilities? Another sort of physical radiation will surely not be assumed in addition. Indeed, if one adopts a purely philosophical standpoint, one might even dare to doubt which electronic system has a greater reality - the 'real one' which describes the stationary states or the 'virtual one' that can supply virtual radiation and which scatters impinging virtual radiation.[55]

Erwin Schrödinger had criticized the BKS for its distinction between "real" and "virtual" radiation. He argued for abandoning the distinction, not because he believed that the BKS description faithfully portrays atomic reality, but, on the contrary, because he believed that no theory could do so. Thus, there is no point in distinguishing between a "real" and a "virtual" part of theoretical description, and we might as well call all parts "real". From the philosophical (read: epistemological) point of view here, according to Erwin Schrödinger, no criterion for reality appears possible, except the Machian criterion of attributing reality only to sensory impressions. But at the methodological level, however, he sees no a priori objection to calling any theoretical terms "real". Precisely because his epistemological position amounts to a type of Machian anti-realism, Erwin Schrödinger is in a position to object to calling some terms in the theory "virtual". If he had really adhered to a hard-headed correspondence realism, then he would have dismissed the BKS theory out of hand.

The fact that Schrödinger agreed largely with the ontological and epistemological views of Mach, while basing his scientific methodology on Boltzmann's ideas, leads to a somewhat paradoxical conclusion, noted by Wessels[56] and Moore.[57] Schrödinger held that good physical theories should provide space-time pictures, but apparently he did not think such theories represented anything real. Since Erwin Schrödinger rejected any form of epistemological realism, one may wonder why he valued *Anschaulichkeit* so highly. Why, for example, did he so desperately try to find a visualizable wave mechanics, as we shall see in the next section?

A clue can be found in a letter from Schrödinger to Wilhelm Wien which is dated on August 25, 1926, in which he criticized both Born's and Bohr's interpretations of quantum mechanics. About Born's statistical interpretation of the wave function, Schrödinger remarked that

> Born [...] has overlooked that [...] the particle or the guiding field [Born favors the former, HdR] which he wants to regard as real now is only so in a manner which would depend on the taste of the observer. There is surely no criterion for reality if one does not want to say that the real is only a complex of sense impressions. All the rest are only pictures.[58]

Note that Schrödinger did not argue for the real existence of the guiding field (*i.e.* the waves). On the contrary, he maintained that no criterion for existence can be given, except for the observable consequences of a theory. This is clearly an endorsement of Mach's empiricism. And right after this statement Schrödinger went on to criticize Bohr:

> Niels Bohr's standpoint, that a space-time description is impossible is something I reject *a limine*. Physics does not consist only of atomic research, science does not consist only of physics, and life does not consist only of science. The purpose of atomic research is to adjust our empirical knowledge concerning it with our other thinking. All of this other thinking, in so far as it is about the outer world, is active in space and time. If it cannot be fitted into space and time, then it fails in its whole aim, and one does not know what purpose it really serves.[59]

These statements reveal why for Erwin Schrödinger there was no tension between his insistence on *Anschaulichkeit* and his own epistemological anti-realism. He had intended the former criterion not as restricting the possible character of reality, but the possible character of *understanding*. Scientific theories should be designed to understand nature, and only a visualizable theory can provide understanding, because:

> We cannot really alter our way of thinking in space and time, and what we cannot comprehend within it we can't understand at all. There *are* such things - but I do not believe that atomic structure is one of them.[60]

Schrödinger believed in the intelligibility of nature, and he held that *Anschaulichkeit* is a necessary criterion for understanding. Besides its heuristic power, this was his primary reason for valuing *Anschaulichkeit* so highly. It follows that Schrödinger considered it possible to understand the phenomena by means of a theory of which one, at the same time, admits that its truth (i.e. its correspondence with reality) cannot be basically justified epistemologically. While this position may at first sight appear outlandish and even inconsistent, it turns out to be tenable in view of the distinction between epistemology and methodology. The key question about whether a theory provides understanding - and thus as well whether the criterion of *Anschaulichkeit* belongs in the latter domain - should not be confused with epistemological questions. A modern defence of this position is presented by Bas van Fraassen.[61] In his own terminology, the explanatory power of a theory has been labelled a *pragmatic* virtue (as opposed to epistemic virtue). Pragmatic virtues, Bas van Fraassen states,[62] "do not concern the relation between theory and the world, but rather the applicability and utility of the theory; they provide reasons to prefer the theory independent of questions of truth".

5. PHILOSOPHY AND QUANTUM MECHANICS

In this section, the relations between Schrödinger's philosophy and his scientific work, particularly with respect to quantum theory, will now be considered.[63] In 1926, Schrödinger published four papers "Quantisierung als Eigenwertproblem" (Part I-IV), in which he developed the theory of wave mechanics.[64] In this conception atoms are characterized by a wave-function Ψ. In the first section Schrödinger had derived a time-independent wave equation, governing the behaviour of the wave function. Today this equation is known as the Schrödinger-equation, and it forms the core of quantum mechanics. At the finish of his paper Schrödinger discussed a possible interpretation of his theory:

> It is, of course, strongly suggested that we should try to connect the function Ψ with some *vibrational process* in the atom, which would more nearly approach reality than the electronic orbits, the real existence of which is very much questioned today.[65]

Schrödinger realized that this interpretation had created problems, *viz.* the dispersion of wave packets and the fact that Ψ was a function only in configuration space and not in real space. Still, he was enthusiastic about the possibilities for visualizable interpretations. He envisaged a measure of possible understanding of Bohr's *ad hoc* frequency by conceiving of the emission of light during transitions as a kind of "beat" which was caused by the changing of vibrations, and added:

> It is hardly necessary to emphasize how much more congenial it would be to imagine that at a quantum transition, energy changes over from one form of vibration to another, than to think of a leaping electron, since the changing of a vibration form can take place continuously in space and time.[66]

Wave mechanics was thus portrayed by him to be an atomic theory in which there was no need for any non-visualizable "quantum jumps" nor for the renunciation of space-time description. In other words, it was an attempt to bring back *Bilder* and *Anschaulichkeit* into quantum physics.

In his second paper, Schrödinger then compared his wave interpretation with the views of the Copenhagen school (Bohr, Heisenberg, and others). He maintained that his interpretation did allow for the apparently necessary renunciation of the concepts of "position" and "path" of the electron, but it avoided the conclusion that a space-time description of atomic structure had to be given up altogether, a grim conclusion which he regarded, from the philosophical standpoint, "as equivalent to a complete surrender".[67] Thus, according to Schrödinger, the strength of wave mechanics, in contrast to that of matrix mechanics, lies in its "guiding, physical point of view", promising "*anschauliches* understanding".[68] In a subsequent article, "Über das Verhältnis der Heisenberg-Born-Jordanschen Quantenmechanik zu der meinen", Erwin Schrödinger proved matrix mechanics and wave mechanics were formally, *i.e.* mathematically, equivalent. In a famous footnote, he would remark: "I naturally knew about his [Werner Heisenberg's] theory, but was discouraged, if not repelled, by what appeared to be difficult methods of transcendental algebra, and by the absence of *Anschaulichkeit*.[69] Also, after his equivalence proof, Erwin Schrödinger discussed the physical relation, and defended the superiority of his *anschauliche* theory (his arguments were described in more detail in Section 4).

In his fourth paper Schrödinger proposed a new interpretation of the wave function, because his original interpretation of it had turned out to be untenable, for the reasons indicated above. Schrödinger now suggested that "$\Psi \cdot \Psi^*$ is a kind of *weight-function* in the system's configuration". He then admitted that this new interpretation seemed to be something of a defeat, "since we have previously often spoken in such an *anschaulich* concrete way of the 'Ψ-vibrations' as though they were something quite real". However, he maintained that "there is also something tangibly real behind the present conception, namely, the very real effective fluctuations of the electric space-density". Although one cannot maintain that the wave function itself is able to describe vibrations in the atom, Schrödinger still believed it is in some way related to real spatio-temporal processes. He clung to his belief in the possibility of a visualizable (*anschaulich*) description of atomic structure.

Notwithstanding his persistent use of the expression "real", Erwin Schrödinger's interpretation was clearly inspired by his belief in *Anschaulichkeit* as a criterion for understanding, rather than by a realist epistemology. This can be inferred from his negative response to Born's statistical interpretation of the wave function, expressed in the letter to Willy Wien quoted in Section five. Born interpreted the product $\Psi \cdot \Psi^*$ as a "weight-function" also, but as one which could determine probabilities for the behavior of the particles. Schrödinger's criticism was not directed against the claim of the reality of particles as such (for there is no criterion for reality *at all*), but against the fact that Born's interpretation seems to imply that a visualizable space-time description is fundamentally impossible.

Bohr's Copenhagen school did accept Schrödinger's theory but not his interpretation. In October 1926 Niels Bohr invited Erwin Schrödinger to visit Copenhagen. The heated, emotional discussions on interpretational questions have been described by Werner Heisenberg.[70] Erwin Schrödinger's interpretation was severely criticized by Bohr and Heisenberg. e.g. for the fact that he was unable to account for the particle-aspect of matter. It appears Schrödinger returned quite discouraged. However, despite this apparent defeat, he did not relinquish his ideal of *Anschaulichkeit*.[71]

In his recent book *Schrödinger's Philosophy of Quantum Mechanics,* Michel Bitbol (1996) distinguishes between four principal periods in the development of Schrödinger's view of quantum mechanics. The first and second period correspond to his two successive interpretations of 1926: Ψ as directly describing vibrations within the atom, and Ψ indirectly determining charge density fluctuations. In the third period (1928-1935), Schrödinger set out to analyse the Copenhagen interpretation, in order to develop a more satisfactory alternative. According to Michel Bitbol, he was able to adopt the Copenhagen view of the wave-function Ψ as a "catalog of information", but nevertheless attempted to interpret Ψ ontologically.[72] This approach reveals the influence of Mach's philosophy, because the change or transformation of an apparently epistemic interpretation into an ontological one accords with Machian idealistic monism..

Schrödinger's efforts finally led him to a new interpretation of wave mechanics, according to which particles do not exist and reality can be exhaustively described by wave-functions only. In criticisizing the particle-concept (in order to justify the ontology of Ψ-waves) one again notices the influence of Mach. He used a phenomenalist view of matter, according to which matter should be analysed as a complex of sensations, memory images.[73] A concise formulation of Schrödinger's mature view of quantum mechanics can be found in his book *Science and Humanism* (1951). Here Schrödinger states that quantum theory confronts us with a dilemma. Whereas experimental evidence does not appear to allow continuous space-time description, we need space-time pictures (for otherwise we might be compelled to use "hazy, arbitrary unclear thinking").[74] Schrödinger's solution of this dilemma is the following:

> We do provide a complete description, continuous in space and time without leaving any gaps which conforms to the classical ideal - a description of something. But we do not claim that this 'something' is the observed or observable facts; and still less do we claim that we thus describe what nature (matter, radiation, etc,) really *is*. In fact we use this picture (the so-called wave picture) in full knowledge that it is *neither*.[75]

This statement demonstrates how even in his approach to very daunting interpretational problems of quantum mechanics, Schrödinger synthesizes a Machian anti-realism with a Boltzmannian demand for *anschauliche Bilder*. On the one hand, we have a space-time picture of Ψ-waves, a visualization which satisfies our methodological criteria. On the other hand, this picture may not be identified with anything real, neither the observable facts nor nature itself.

6. CONCLUSION

In his later philosophical publications, Schrödinger repeatedly draws our attention to his two fundamental principles of science: "the principle of the understandability of nature" and "the principal of objectivation".[76] These two principles correspond exactly with the two doctrines which he inherited from Ludwig Boltzmann. Understandability is for Erwin Schrödinger equivalent to *Anschaulichkeit*, i.e. visualizability in space and time. But the principle of objectivation leads immediately to the *Hypothese der realen Aussenwelt*, which advises us to be realists in scientific practice.[77]

Both principles appear to be anti-Machian. However, it is important to note that Schrödinger speaks of "two general principles which establish the foundation of the scientific *method* (ibid., my italics)". First, the principle of objectivation is in the opinion of Schrödinger "a certain simplification which we have adopted to master the infinitely intricate problem of nature".[78] In short, it is a pragmatic, methodological assumption. The same holds for the principle of understandability in the sense that it concerns the relation between a scientific theory and its user, the scientist who wants to acquire the state of understanding. In this respect, Erwin Schrödinger's view may be compared with that of Bas van Fraassen, who has analysed explanation as a pragmatic virtue of theories.[79] The most important difference is that in van Fraassen's opinion criteria for understandability may vary greatly, whereas Erwin Schrödinger holds that visualizability is a necessary condition for understanding by human beings.

Since both principles are methodological, Erwin Schrödinger is in a good position to reconcile these seemingly anti-Machian principles with a pro-Machian standpoint on the epistemological and ontological level. In his biography of Schrödinger, Moore states: "As a young man, Schrödinger was able to maintain a kind of pragmatic balance between Machian positivism and Boltzmannian realism, but as he grew older this compromise no longer satisfied him and he sought but probably never achieved a synthesis of the two".[80] But as I have tried to demonstrate, Moore's diagnosis is incorrect. Erwin Schrödinger did achieve a synthesis between the ideas of Mach and Boltzmann, by applying them to *different domains*: Mach's to ontology and epistemology, and Ludwig Boltzmann's to methodology of science. Moore claims that Schrödinger "had never been more than a reluctant disciple of Mach".[81] This is incorrect as well. In this chapter I have presented ample evidence that Schrödinger was deeply influenced by Mach's philosophy. In particular, his ontological position, an empiricist version of neutral monism with a touch of idealism, was partly derived from Mach. Moreover, at the epistemological level, Schrödinger adopted Mach's anti-realist attitude to scientific theories.

Erwin Schrödinger's philosophical commitments, both to Ernst Mach and to Boltzmann, dated from his student days in Vienna, and in essence they never changed. During the 1920's he had elaborated his Machian views in his manuscript on philosophy *Suche nach dem Weg*, while his Boltzmannian methodology was apparent in his approach to quantum theory. In the 1950's he still defends Machian phenomenalism in his more philosophical writings, e.g. *Science and Humanism* and the William James Lectures,[82] while his work on the interpretation of quantum mechanics was again inspired by the demand for Boltzmannian *Bilder*. In conclusion, Mach's ideas had a stronger influence on Schrödinger's philosophy while Boltzmann's methodology was certainly more important for his discovery of wave mechanics. However, when it came down to questions concerning how to best interpret quantum mechanics Mach's philosophy became of equal importance.

Acknowledgements:

These investigations were supported by the Foundation for Research in the Field of Philosophy and Theology and has been subsidized by Netherlands Organization for Scientific Research (NWO).

NOTES

1 E. Schrödinger, *Collected Papers on Wave Mechanics*, Blackie and Son: London, 1928, p. 120.

2 J.T. Cushing, *Quantum Mechanics. Historical Contingency and the Copenhagen Hegemony*, U. of Chicago Press: Chicago, 1994, pp. 106 & 124; and Rom Harré, "Bohr and Schrödinger", in *Erwin Schrödinger: Philosophy and the Birth of Quantum Mechanics*, M. Bitbol and O. Darrigol (editors.), Editions Frontières: Gif-sur-Yvette Cedex, 1992, pp. 147.

3 E. Schrödinger, *My View of the World*, Cambridge University Press: Cambridge, 1964, p. viii.

4 The word *Anschaulichkeit* can be translated as visualizability or as intelligibility, and thus seems to capture the relation between visualization and understanding endorsed by Schrödinger. When quoting from translations of Schrödinger's work, I have largely kept the original German terms *anschaulich*, *Anschaulichkeit*, and so forth because they are inadequately translated (e.g. as "intuitive"). in De Regt (forthcoming) I present a more detailed discussion of *Anschaulichkeit*.

5 Erwin Schrödinger. [Three books are in one.]: (1)*What is Life?* (2) *Mind and Matter* (3) *Autobiographical Sketches*, Cambridge University Press: Cambridge, 1992, p. 168.

6 Erwin Schrödinger, *Science Theory and Man*, Dover: New York, 1957, p. xiv.

7 P.A. Hanle, "Indeterminacy before Heisenberg: The case of Franz Exner and Erwin Schrödinger", *Historical Studies in the Physical Sciences*, **10** (1979) 225-269.

8 Schrödinger (1992), op. cit., p. 169.

9 Walter Moore, *Schrödinger - Life and Thought*, Cambridge U. Press: Cambridge, 1989, p. 59.

Erwin Schrödinger's William James Lectures also display his serious concern with experiment and reveal its noticable influence on his philosophy. See Erwin Schrödinger, *The Interpretation of Quantum Mechanics: Dublin Seminars 1949-1955 and Other Unpublished Essays*, ed. by M. Bitbol, Ox Bow Press: Woodbridge, 1995, pp. 123-149.

10 Schrödinger (1957), op. cit., p. xiv.

11 Schrödinger (1995), op. cit., p. 103.

12 Erwin Schrödinger presented his cat paradox in his 1935 paper: "Die gegenwärtige Situation in der Quantenmechanik", reprinted in: *Gesammelte Abhandlungen*, Verlag der Österreichischen Akademie der Wissenschaften: Wien, 1984, pp. 484-501.

13 Schrödinger (1995), op. cit. See especially the lecture notes.

14 Schrödinger, ibid., pp. 121-122.

15 See Walter Moore (1989), op. cit., pp. 120-129, for a summary of Schrödinger's work on colour theory.

16 But *Suche nach dem Weg* was only published in 1961 as part of the larger work *Meine Weltansicht*. An English translation of this book came out three years later under the title *My View of the World*, Cambridge University Press: Cambridge, 1964.

17 Schrödinger (1984), op. cit., pp. 12-18.

18 Schrödinger (1964), op. cit., p. 15.

19 Note that Erwin Schödinger's expression "naive realism" does not refer to a naive or an ordinary correspondence realism with respect to scientific theories, but to a direct view of the world as it presents itself to us.

20 E. MacKinnon, *Scientific Explanation and Atomic Physics*, Chicago University Press: Chicago, 1982, p. 221n.

21 Schrödinger (1996), op. cit., pp. 93-94.

22 K.R. Popper, *Unended Quest*, Fontana/Collins: Glasgow, 1976, p. 135.

23 See J. Blackmore, *Ludwig Boltzmann. His Later Life and Philosophy, 1900-1906. Book Two: The Philosopher*, Kluwer: Dordrecht, 1995, pp. 48-51.

24 Schrödinger (1992), op. cit., p. 169.

25 Schrödinger reaffirms this metaphysical belief in Chapter 4 of *Mind and Matter*, which is entitled "The Oneness of Mind", and which refers again to the *Upanishads*. See also his three-part book: (1)*What is Life?* (2) *Mind and Matter* and (3) *Autobiographical Sketches*, Cambridge University Press: Cambridge, 1992, 128-139. See also Hans Poser, "The Notion of consciousness in Schrödinger's Philosophy of Nature", in J. Götschl (ed.), *Erwin Schrödinger's World View - The Dynamics of Knowledge and Reality*, Kluwer: Dordrecht, 1992, pp. 153-168.

26 Schrödinger (1992), op. cit., p. 93.

27 Schrödinger added: "According to this view, a piece of matter is the name we give to a continuous string of events that succeed each other in time; the immediately successive ones being as a rule closely similar. The single event is an inextricable complex of sensates, of associated memory images, and of exprectations associated with the former two." (Schrödinger (1957), op. cit., p. 207.)

28 Schrödinger (1996), op. cit., pp. 92-93.

29 Ibid., p. 93. See also Schrödinger (1992), op. cit., p. 118.

30 Schrödinger (1992), p. 153; (1995), p. 134; (1996), p. 89.

31 Schrödinger (1992), ibid., p. 127.

32 Ibid., p. 135.

33 Schrödinger (1996), op. cit., p. 76.

34 Ibid., p. 126.

35 Ibid.

36 Moore (1989), op. cit., p. 225.

37 Schrödinger (1964), op. cit., pp. 3-4.

38 Schrödinger (1996), op. cit., p. 90. (The original is in italics.)

39 Ibid., p. 90-91.

40 Ibid., p. 91.

41 Dieter Flamm, "Ludwig Boltzmann's Influence on Schrödinger", in C.W. Kilmister (ed.), *Schrödinger, Centenary of a Polymath*, Cambridge U. Press: Cambridge, 1987, pp. 11-12.

42 Schrödinger (1995), op. cit., p. 121.

43 L. Boltzmann, *Theoretical Physics and Philosophical Problems*, B. McGuinness (ed.), Reidel: Dordrecht, 1974, p. 33.

44 See also Blackmore (1995), op. cit., pp. 69-74. Blackmore convincingly argues that Boltzmann's *Bild*-conception was a methodological principle which he kept and utilized even while mentally experimenting with different ontological theories about nature.

45 H.W. de Regt, "Philosophy and the Kinetic Theory Of Gases", *British Journal for the Philosophy of Science*, **47** (1996) 41-43.

46 Bitbol presents a similar view of Schrödinger's position, calling it "quasi-realist". (See M. Bitbol, *Schrödinger's Philosophy of Quantum Mechanics*, Kluwer: Dordrecht, 1996, pp. 39-41.) Also, Erwin Schrödinger's methodological realism resembles the constructive empiricist approach of Bas van Fraassen. (See Bas C. van Fraassen, *The Scientific Image*, Clarendon Press: Oxford, 1980, pp. 80-83.)

47 Reprinted in Schrödinger (1984), op. cit., vol. I, pp. 124-142.

48 Erwin Schrödinger, "Über das Verhältnis der Heisenberg-Born-Jordanschen Quanten-mechanik zu der meinen" (1926), reprinted in Schrödinger (1984), vol. 3, pp. 143-165). English translation in Erwin Schrödinger (1828), pp. 45-61. Incidentally, the validity of Schrödinger's proof has recently been contested by F.A. Muller, "The Equivalence Myth of Quantum Mechanics", *Studies in History and Philosophy of Modern Physics*, **28** (1997) 35-61.

49 Schrödinger (1928), op. cit., p. 58.

50 Ibid.

51 Ibid., p. 59.

52 In the case of wave mechanics versus matrix mechanics this indeed proved to be true. For example, as Mara Beller reports, wave mechanics "was successfully applied to a great variety of problems unamenable to matrix treatment". (M. Beller, "Matrix Theory Before Schrödinger", *Isis*, **74** (1983) 489-490.)

53 Schrödinger (1995), p. 122.

54 See Schrödinger's letter to Bohr, May 24, 1924, in *Niels Bohr, Collected Works*, vol. 5, North-Holland: Amsterdam, 1976, p. 490; See also "Bohrs neue Strahlungs-hypothese und der Energiesatz", reprinted in Schrödinger, vol. 3, (1984), op. cit., pp. 26-30. For an illuminating account of BKS theory, see M. Dresden, *H.A. Kramers. Between Tradition and Revolution*, Springer Verlag: New York, 1987, pp. 159-178.

55 Bohr, vol. 5, (1976), p. 490.

56 L. Wessels, "Erwin Schrödinger and the Descriptive tradition", in R. Aris, H.T. Davis, and R.H. Stuewer (editors), *Springs of Scientific Creativity*, University of Minnesota Press: Minneapolis, 1983, p. 271.

57 Moore (1989), op. cit., p. 217.

58 Ibid., p. 225.

59 Ibid., p. 226.

60 Schrödinger (1928), op. cit., p. 27.

61 Bas van Fraassen, *The Scientific Image*, Clarendon Press: Oxford, 1980, pp. 87-96.

62 Ibid., p. 88.

63 A more detailed account of Schrödinger's discovery of wave mechanics will appear in De Regt (forthcoming).

64 Reprinted in Erwin Schrödinger, vol. 3 (1984), op. cit., pp. 82-136 & 166-250. English translations in Schrödinger (1928), op. cit.

65 Schrödinger (1928), op. cit., p. 9.

66 Ibid., p. 10-11.

67 Ibid., pp. 26-27.

68 Ibid., p. 30.

69 Ibid., p. 46n.

70 W. Heisenberg, "Quantum Theory and Its Interpretation" in S. Rozental (eds.), *Niels Bohr as Seen by His Friends and Colleagues*, North Holland: Amsterdam, 1967, pp. 94-108.

71 See his 1928 lecture: "Conceptual Models in Physics and their Philosophical Value", which is reprinted in Schrödinger (1957), pp. 148-165.

72 Bitbol (1996), op. cit., pp. 87-90.

73 Schrödinger (1957), op. cit., p. 207.

74 Schrödinger (1996), op. cit., p. 144.

75 Ibid.

76 Schrödinger (1992), op. cit., p. 117; and (1996), op. cit., pp. 90-93.

77 Schrödinger (1996), ibid., p. 93.

78 Schrödinger (1992), op. cit., p. 118.

79 Van Fraassen (1980), op. cit., Chapter 5.

80 Moore (1989), op. cit., p. 45.

81 Ibid., p. 437.

82 Schrödinger (1995), op. cit., pp. 123-149.

REFERENCES:

Beller, Mara. "Matrix Theory before Schrödinger". *Isis*, 74 (1983) 492-508.

Bitbol, Michel. "Mach et Schrödinger de l'éleent au tout". *Fundamenta Scientiae*, 8 (1987) 345-372.

Bitbol, Michel. *Schrödinger's Philosophy of Quantum Mechanics*. Dordrecht: Kluwer, 1996.

Blackmore, John. *Ernst Mach - His Life, Work, and Influence*. Berkeley & Los Angeles: University of California Press.

Blackmore, John. *Ludwig Boltzmann His Later Life and Philosophy. 1900-1906. Book Two: The Philosopher*, Dordrecht: Kluwer, 1995.

Bohr, Niels. *Collected Works*. Amsterdam: North-Holland, 1976.

Boltzmann, Ludwig. *Theoretical Physics and Philosophical Problems*. edited by Brian McGuinnes Dordrecht: Reidel, 1974.

Cushing, J.T. *Quantum Mechanics. Historical Contingency and the Copenhagen Hegemony*, Chicago: University of Chicago Press, 1994.

De Regt, Henk W. "Philosophy and the Kinetic Theory of Gases". *British Journal for the Philosophy of Science*, 47 (1996) 31-62.

De Regt, Henk W. (forthcoming) "Erwin Schrödinger, 'Anschaulichkeit', and Quantum Theory". *Studies in History and Philosophy of Modern Physics*.

Dresden, Max. *H.A. Kramers. Between Tradition and Revolution*. New York: Springer Verlag, 1967.

Flamm, Dieter. "Ludwig Boltzmann's Influence on Schrödinger". in C.W. Kilmister (ed.).

Schrödinger, Centenary of a Polymath, Cambridge: Cambridge University Press, 1987, pp. 4-15.
Götschl, J. (ed.). *Erwin Schrödinger's Worldview - The Dynamics of Knowledge and Realit.* Dordrecht: Kluwer, 1992.
Hanle, Paul A. "Indeterminacy before Heisenberg; The Case of Franz Exner and Erwin Schrödinger". *Historical Studies in the Physical Sciences*, **10** (1979) 225-269.
Harré, Rom. "Bohr and Schrödinger", in M. Bitbol and O. Darrigol (eds.). *Erwin Schrödinger: Philosophy and the Birth of Quantum Mechanics.* Gif-sur-Yvette Cedex: Editions Frontières, 1992.
Heisenberg, Werner. "Quantum Theory and Its Interpretation". in S. Rozental (ed.). *Niels Bohr as Seen by his Friends and Colleagues.* Amsterdam: North-Holland, 1967, pp. 94-108.
MacKinnon, Edward. *Scientific Explanation and Atomic Physics.* Chicago: Chicago University Press, 1982.
Moore, Walter. *Schrödinger - Life and Thought.* Cambridge: Cambridge University Press, 1989.
Muller, F.A. "The Equivalence Myth of Quantum Mechanics". *Studies in History and Philosophy of Modern Physics*, 28 (1997) 35-61.
Popper, Sir Karl R. *Unended Quest.* Glasgow: Fontana/Collins, 1976.
Schrödinger, Erwin. *Collected Papers on Wave Mechanic. London:* Blackie and Son, 1928.
Schrödinger, Erwin. *Science Theory and Man.* New York: Dover, 1957.
Schrödinger, Erwin. *My View of the World.* Cambridge: Cambridge University Prese, 1964.
Schrödinger, Erwin. *Gesammelte Abhandlungen.* Wien: Verlag der Österreichischen Akademie der Wissenschaften, 1984.
Schrödinger, Erwin. *What is Life?* with *Mind and Matter* and *Autobiogrphical Sketches.* Cambridge: Cambridge University Press, 1992.
Schrödinger, Erwin. *The Interpretation of Quantum Mechanics: Dublin Seminars 1949-1955 and Other Unpublished Essays.* edited by M. Bitbol. Woodbridge: Ox Bow Press, 1995.
Schrödinger, Erwin. *Nature and the Greeks and Science* and *Humanism.* Cambridge: Cambridge University Press, 1996.
Van Fraassen, Bas C. *The Scientific Image.* Oxford: Clarendon Press, 1980.
Wessels, Linda. "Schrödinger's Route to Wave Mechanics". *Studies in History and Philosophy of Science*, 10 (1979) 311-340.
Wessels, Linda. "Erwin Schrödinger and the Descriptive Tradition". in R. Aris, R.T. Davis, and R.H. Stuewer (eds.). *Springs of Scientific Creativity. Minneapolis:* University of Minnesota, 1996 [1883].

Chapter Five

OTTO NEURATH 1913-1915*

by Michael Stöltzner

Section Guide

1. AUXILIARY MOTIVES

A

Otto Neurath argues in this period1 that not only for practical purposes of daily life but also in science decisions must be made even before evaluating all theoretically possible options, or even if thorough investigation ends up with equally probable alternatives, such as often appeared to be the case with the wave and particle theories of light from the 17th to the 19th centuries. In these cases Neurath has recommended the rationally-minded to decide on the basis of an "auxiliary motive". This is the way he labels procedures which do not add new content to the matter which is in question. Borrowing Kantian terminology, let me classify them as "formal". The purest auxiliary motives are simply tossing a coin or throwing dice, but also voting, instinctive action, or analogies can be applied in this way. They can help us choose between theories when the theories themselves seem equally supported by empirical data.

Usually the auxiliary motive is understood within what Rudolf Haller has termed the "Neurath Principle",[2] that is, if a protocol sentence contradicts a scientific theory, we can either change the theoretical system or reject the sentence itself. Accordingly, in principle, everything in science is a matter of decision. But such decisions cannot all be based on logical inference. Instead they are to be investigated historically and sociologically, and the resulting interpretation can be well-founded for two reasons; First, there is "Neurath's boat" (1913) which when put to sea pictures the scientific enterprise as a constant reconstruction that cannot resort to any *fundamentum inconsussum*. Secondly, Neurath's nautical holism appears to be a radicalization of a step taken by Duhem. He endorses Duhem's thesis of the underdetermination of theory by experimental fact and has joined him in rejecting an *experimentum crucis*.

Furthermore, if a system of alternatives already supports practical life by providing an ontological framework, then it would seem to be all the more important to carry through Neurath's program of science as an everlasting provisional reconstruction. Auxiliary motives can serve a twofold purpose in this enterprise: It is a tool for our establishing a system and also in selecting alternatives within a system. Taking this stance, one finds in Neurath's works on the history of optics from 1914/15 the missing discussion of the auxiliary motive in science.[3] But in examining the contents of these two papers he published then, one discovers that Neurath's early work appears much more

J. Blackmore, R. Itagaki, and S. Tanaka (eds.). Ernst Mach's Vienna 1895-1930. 105-122.

committed to Ernst Mach than is usually believed. Ultimately, the Machian reading I am going to propose will also result in the *Neurath Principle*.

Also interesting concerning Neurath's attitude toward Mach is one of two undated letters to him presumably from 1915 when Neurath had finished his early work on the history of optics. He expresses his high esteem for Mach's *Mechanics* and asks Mach for further hints on the analogy between light and sound. Moreover, he admits the influence of Mach's holism on his ideas on political economy. "It is the tendency to derive the meaning, the sense of the particular from the whole, not the whole as a summation of particulars."[4] Neurath writes he will send to Ernst Mach a copy of his article on systems of hypotheses and optics which he describes as a "further exploitation" of his paper on the foundations of optics (which while published in 1915 was apparently written earlier than the former paper).[5]

From the letter , however, one gets the impression that Neurath probably had heard of (but not seen) Ernst Mach's book manuscript *Prinzipien der physikalischen Optik*, finished around 1913, of which, however, parts had been written decades before. Though printing had started in 1915, the war and perhaps other factors delayed publication until 1921, and with a famous preface that expresses criticism against relativity theory. There are, however, in Mach's earlier works several remarks on analogy as mentioned by Neurath as an auxiliary motive, e.g. in *Erkenntnis und Irrtum* (1905). In addition, the longer 1915 letter already addressed two of the topics to be investigated in this contribution, holism and analogy. The other topics are, to my mind, the core of the notorious (sometimes quite ideological) debate about the preface of Mach's *Optik*: firstly, his attitude towards mathematics and the principles of a theory; and secondly, the ontological status of a "system" touched upon above.

B. HOLISM AND THE PRINCIPLE OF ECONOMY OF THOUGHT

Neurath already mentions in his article of 1913 on auxiliary motives[6] that the rationalist should not haughtily oppose human beings acting according to their instincts. This attitude reflects Mach's high estimation of instinct as mankind's first strategy for gaining knowledge. Instinct seems to represent a preliminary stage of rationality, not its "pseudo-rationalistic" decay, to use another expression from Neurath. According to Mach, knowledge begins by isolating particular features within the whole of the empirical world. We order them with the aid of concepts acquired already and seek an *economical organization* of the facts thus established, in order to communicate them to future generations of craftsmen or scientists, since no single human being could ever acquire a reasonable state of knowledge without availing himself of the experiences of past generations and symbolically by concepts and laws. This evolution of knowledge, from instinctive to scientific experiences, culminates in "the ideal of a *unified* world conception, which is the only one compatible with a healthy mind."[7] But this ideal is still to be reached. Yet "the highest philosophy of the natural scientist is to *tolerate* an incomplete world view and to give it preference over an ostensibly complete, but really insufficient one".[8]

Otto Neurath treats this tolerance as the key feature of rationalism which "conceives its main triumph in sharply recognizing the limits of any specific insight."[9] But he radicalizes Mach's holism by changing it from a future ideal of all mankind into a permanent necessity for every human being now. Accordingly, human decisions are not only found *in retrospect* by studying the psychological motivations of scientists, as Ernst Mach did throughout his historical-critical analyses. Instead, they are *omnipresent* as a consequence of the permanent need for a practicable world view. And it is here one finds a philological bridge between Neurath's three papers: In the earliest one (1913) on auxiliary motives he mentions the "possibility to develop different theories of light such as those of wave and emission from the same starting point"[10] as an illustration of the "capability to simultaneously live distinct theoretical lives."[11] But an argument from the1914 and 1915 articles severely restricts this option: "In order to think one hypothesis through to the end, an entire life is needed."[12]

The inescapability of our making decisions even within the natural sciences has consequences for the *fine structure* of theoretical systems. For Mach, all physical theory aims at communication or storage of facts. Accordingly, the notion of a wave in undulatory theory in optics is a very economical symbol for many facts comprised by this theory. This is also true for the emission theory of light during the period studied by Neurath. Although the latter required a number of additional hypotheses to explain all optical phenomena, it was in principle equivalent to the former because it encompassed all relevant facts.* Mach has sketched in his works many alternative histories and possible conclusions in order to prove different formulations to be equivalent. He attributed these differences to the psychological motivations and the leading intuitions of the particular scientists.

Neurath too, demands that alternative hypotheses should be formulated to be as perfect as possible. Mach and Neurath, however, slightly disagree on what an *alternative is*. For this purpose of communication one chooses among various kinds of hypotheses according to practicability and economy. The most economical hypothesis in optics may well be the one about the nature of light, whether it consists of waves or of particles. However, if one regards making a decision between theories as primary, such a representation of alternatives is inappropriate, since neither theory excludes the other. Undulation and emission are by no means*logical dichotomies*. At best the dichotomies were used *historically*, often as a human "product of scientific pugnacity".[13] Actually, the hypothesis about the nature of light is only an important part of these theories listed under the respective names which are by far more complex and contain even some identical traits. Hence, in order to rationally prepare to make a decision between theories in such a manner that the decision can eventually be determined by an auxiliary motive, our theoretical complexes have to be clearly divided up into logical alternatives. Neurath uses optics as his example: Let *a* be the undulatory, *b* the emission hypothesis, and *c* a further independent feature. Altogether there are eight possible theories not all of which were necessarily held by any scientist.

Mach repeatedly wrote about the logical structure of a theory. However, as it was common then, his use of the expression "logic" corresponded to "methodology", not to formal logic. For him, formally elaborated theories, such as Lagrange's mechanics, represented merely economical schemata that had abstracted all of the possible cases in such a manner that they could be resolved by mathematics. Over two decades later (1938)[14] Neurath would conclude that Mach was lacking sufficient logical expertise to carry through his holistic program, but from 1913 to 1915 and even later Neurath was far from being that critical toward most of his ideas.

Though we cannot pursue the matter in this paper, we should mention that Neurath presented yet another lecture on Ernst Mach's ideas before the University of Vienna Philosophical Society some months later which was titled: *Machs Stellung zum Gravitations- und Trägheitsproblem* and which was given on December 15, 1916. It was apparently about what we now call "Mach's Principle", a topic which may well have been as controversial then as it is now since two additional sessions were added to the Philosophical Society's calendar on January 8 and 22, 1917 to handle this discussion along with earlier issues raised from a lecture on Mach by Viktor Kraft.[15] Neither lecture seems to have been published by the Society.[16]

It would be a mistake to consider Neurath a crude decisionist. Although with respect to the structure of a theoretical system the principle of economy is subordinated to decidability, he still held that quick decisions between alternatives that are rationally equivalent are very economical, not only for Buridan's ass or Descartes' lost wanderer in the forest but also generally in biological and historical evolution. "If, however, an auxiliary motive were lacking [in cases of equivalent theories], the resulting vacillation will exhibit less economical behavior."[17] Neurath wrote this in a paper devoted to the conceptual system of economics that deals with analogous non-dichotomy as his works on the history of optics had done two years before. In fact, Ernst Mach himself constantly emphasized the connection between science and economics and traced back the principle of economy and his notion of practicability (*Zweckmässigkeit*) embrace both levels of economy Neurath rightly distinguishes. Only if one presupposes this distinction is one likely to view the general aspect of Mach's principle of economy as an anticipation of modern theories of rationality and decision as Rudolf Haller does.[18]

C. ANALOGY AS AN AUXILIARY MOTIVE

Aside from the principle of economy, but within its more general setting, Ernst Mach discusses his *principle of comparison*, as a leitmotiv of scientific thinking. "Most important is *analogy* which represents an abstract similarity. An analogy can sometimes remain entirely concealed from direct sensory observation, and manifests itself only by comparing the internal conceptual relations of the characteristics of one object with the conceptual relations of the characteristics of another object."[19]

Among the examples that Mach discusses, one can discern three types of analogies: First, more or less *instinctive* ones such as those between the flow of heat and the movement of a substance;[20] second, *far-reaching* and fruitful ones, such as between light and sound on which Huygen's optics is based[21] (and which is Neurath's main concern); and third, *purely formal ones*, such as "the possibility of discovering analogies of the Principle of Least Action in distinct fields of physics".[22] More generally, analogy shows the "way in which a general physical phenomenology including all fields will develop".[23] In particular, Mach frequently uses mechanical analogies to discuss, for instance, refraction of light. However, analogies do not justify the reduction of optics or the theory of heat to mechanical processes.

In Mach's view, analogies build up structures, act heuristically in theory formation as well as for the purpose of communication. But they do not constitute factual identities. Thus, their validity is limited in time, such that a hitherto fertile analogy can impede scientific progress. The analogy of light and sound, for example, prevented Huygens from discovering diffraction which seems to follow from his theory of light. Accordingly, both for Mach and Neurath analogy has a woolly fringe. Thus, without further specification it cannot be considered a satisfactory logical instrument for either abduction or induction.

Otto Neurath circumvents this logical problem again by a radical pragmatism that eventually takes analogy as an auxiliary motive. Its original purpose, however, was probabilistic: "In order to orient oneself in these systems [which are compatible with experimental data] and to select among them, again one has to operate with whatever assumptions about the probability that certain systems would be realized. This probability, however, is based above all on analogies."[24]

There are basically two general strategies to justify analogies: Either one bases their use on a belief in a certain homogeneity of nature, as Duhem[25] understood his "natural order" or one views it as no more than psychological advantage, as Mach held. Albeit rather vaguely, Otto Neurath favors the second alternative. But this is not a matter of utmost importance, if one conceives the analogy as a practical (ultimately auxiliary) rule to locate and utilize an easily decidable system of hypotheses. Analogies can also make it possible to establish logical relations despite their logical wooliness, because at least, they establish some kind of formal order or help us to orient our understanding. Moreover, Neurath's auxiliary motive represents a significant addition to scientific methodology - in a weak and pragmatic sense of course. But this can also be viewed as a step away from Mach, who conceived the principle of economy as merely descriptive and who at times insisted that there is no result of science which in point of principle could not have been discovered wholly without method.

In short, Ernst Mach's application of analogies is usually "conservative", while Otto Neurath aims at reconstruction (or even something close to a Kuhnian revolution). Mach emphasizes historical continuity where possible by giving mechanical analogies or simple schemata even for highly abstract principles. This makes Mach's *Mechanics* at some points rather embarrassing reading today. For instance, the Principle of Least Action is illustrated for him by the equilibrium of a string in order to prove its equivalence with D'Alembert's Principle. The reason for this lies in Mach's attitude towards mathematics. As a matter of fact mathematically the relation between the principles is much more complex.

D. MATHEMATICAL HIERARCHIES

The role of mathematical entities appears to me to be the main clue to understanding Einstein's famous *dictum* of 1922 that what Mach has done is "to make a catalogue, not a system".[26] Although this phrase is sometimes used quite unfairly against Mach, one cannot deny that it touches a sensitive core issue. For if one reads the wrong definition of "realism" into Mach, one easily inverts the front lines in the classical dispute between Einstein and the Machians by turning Einstein into a Positivist and Mach into the realist, as Paul Feyerabend has shown.[27] Understanding Einstein's statement properly involves two issues: one about the logical structure of theories and one about ontologies therein.*

But while Otto Neurath in both cases at least partially takes steps which seem to be required for progress toward a modern standpoint, nevertheless, he seems to blur two important distinctions. First, when he emphasizes that he "does not intend to further investigate, whether these elements represent elementary experiences or elementary concepts for the particular thinkers,"[28] he moves away from Ernst Mach's conception of physical theory which distinguishes carefully between them. And second, by putting a mere formal principle on the same level, he even deviates from Mach ontologically, even though he does not draw all of the consequences which he could from them. The basis of this shift by Otto Neurath is his entirely different conception of mathematics.

For Mach himself, elementary *experiences* constitute the only basis for physics. Nevertheless, "intuition" is by no means devoid of theory, but it marks the endpoint of a lengthy process of discovery. As for mathematics, it "is the method of replacing in the most comprehensive and *economical* manner possible, *new* numerical operations by old ones done already with known results."[29] Mathematics never adds anything to physical fact. Nevertheless, Mach emphasized the importance of mathematics and was well aware of genuinely mathematical results. In later editions of the *Mechanics* he touches upon Jacobi's notion of conjugate points, which can be found in hardly any modern textbook. Emphasizing himself that science consists in discovering functional relations among our experiences, he readily agrees with Hertz in holding the existence of differential equations for the coordinates of masses more essential than the concept of force.[30] However, Ernst Mach is constantly at pains to deny that mathematical objects have any ontological connection with scientific facts. They constitute mere tools for increasing economy.

E. DECISIONS AND ONTOLOGY

But for Otto Neurath mathematics is based on formal logic, and it has accordingly become a study of logical structures and is not just calculation. But his step away from Mach in this context is actually only half-a-step. His classification of systems of hypotheses neglects what can be called: internal mathematical hierarchies. But I think that Neurath would have accepted this criticism provided that one does not draw extreme ontological consequences. His radical pragmatism requires an ontology that makes comparable all those things that are subject to decision. If any two alternatives can ultimately be resolved by means of an auxiliary motive, then they must share a common ontological framework, even if auxiliary motives often seem like a vague psychological force or procedure unable to accomplish the logical transition between distinct levels of a theoretical complex. It was Neurath's aim in his 1915 paper by comparing the wave and emission theories of light to foster ontological comparability of two theoretical complexes by identifying their common elements. Thus in effect, he considered fundamental principles and elementary experiences to be on an equal footing, and not just in terms of their empirical origin as Mach had done who was reluctant to believe current concepts to be as factual as experience. Thus, for Neurath, decisions between principles and elementary experiences could be made not just on the ground of economy as Mach had recommended but by *any* rational criterion, though ultimately through an auxiliary motive. The same holds for the comparison between already established systems of hypotheses or parts of them. Here the difference from Mach's concept of a theoretical system becomes evident. If a theory were only an economical arrangement of facts, there would be only one rational criterion deciding between two compelling theories: their respective economy.

Though unspecified, the auxiliary motive is a formal procedure that can be used in such a way that mathematical hierarchies are respected as long as they are not elevated into an ontology. But if the ontology is "smeared out"[31] throughout the system, then the comparability condition can be well met and human decisions made. Neurath's classification scheme treats elementary experiences and theoretical principles as equal. Do we therefore encounter some sort of "pragmatic realism" in him? Certainly, this holds in comparison to Mach, as Neurath only admits gradual distinctions between the ontologies of directly observable sensory elements and theoretical concepts. This allows him to skirt Mach's troubles with the existence of atoms [at a time when the theoretical work of Einstein and the empirical labor of Perrin had put the reality of atoms beyond serious question]. For Mach, atomism represented at most a useful hypothesis. Neurath's holistic ontology [of internal relations as if only a historical whole were fully real] is of course much more flexible in this respect, provided one is willing to eventually pay the price of a rather ubiquitous fallibilism. But this concession appears to be a rather strong gain, for one can still maintain a core element of Mach's thinking, namely, "that the true relation of the principles is a historical one."[32]

The criterion of sensory intuition (*Erschauen*) which Mach had appealed to has lost its justification for Otto Neurath since it is rooted in ontological preference for sensations over theories which his "smeared out" ontology rejects, even if his smearing does fall into fallibilism. Neurath attributes almost the same ontological status to sensory elements as to principles. This tendency is again reinforced by radical pragmatism because not only the definition of principles but also the selection of sense elements represents a decision. Ultimately, the construction can be traced back to its basic pieces which are later termed "protocol sentences".

And again, if the focus is changed, one arrives at the celebrated "Neurath Principle". If a protocol sentence were tocontradict a theoretical system, we may reject the protocol sentence or we may alter the system, either by amending a hypothesis or by entirely overthrowing it. But as this paper has attempted to show, decisions by means of an auxiliary motive are always defined with respect to rational systems of alternatives which they are designed to resolve. Having studied the genealogy of the "Neurath Principle" from its Machian roots,[33] it now becomes clearly reconcilable with the systematic attitude of Rudolf Carnap, [who would become the leading theoretician of the Vienna Circle]. In effect, the principle's core, the permanent necessity of decision, firmly urges us toward the development of a suitably structured theory.

2. SIMILARITY AND ANALOGY AS A LEADING FEATURE OF INQUIRY[1]

by Ernst Mach

1. Similarity is partial identity: The characteristics of similar objects are part identical and part different. Not a single observable mark of one object need coincide with a mark of the other, and yet even so the marks of the one may be interrelated in exactly the same way as the other. Jevons[2] calls analogy a more deep-seated similarity. Analogy may in some circumstances remain quite concealed from direct sense observations and reveal itself only through comparisons of conceptual interconnections between the marks of one object with the corresponding connections in the other. Maxwell[3] not only defines analogy but also underlines those features of it that are most important for scientific inquiry, when he describes analogy as partial similarity between the laws in one field and those in another, so that each illustrates the other. However, we shall see that Maxwell's approach is not different from ours. Hoppe[4] in his book regards the concept of analogy as superfluous, since as with similarity in general it is merely a matter of conceptual agreement of certain marks in the objects between which analogy is found. Although this is correct there are good grounds for taking analogy as a special case of similarity and distinguishing it from the general concept. Above all it is the inquirer into nature who is most driven to this view, since taking notice of analogies greatly furthers his work. Furthermore, it is clear that analogy or sameness of relation between elements may well occur in objects whose similarity is immediately observable by the senses, and this may seem so obvious that the analogy is overlooked.

2. Similarity which is perceived by the senses already determines an unconscious and unpremeditated similarity in behavior and in motor reactions towards similar objects; when the intellect becomes conscious it will do likewise, as Stern[5] has shown at length as regards popular thought. In addition Tyler's works[6] can offer ample evidence. As our conceptual thought becomes stronger, deliberate and purposive human endeavor to free oneself from practical or intellectual unease will likewise be guided by similarities and soon by more deep-seated analogies as well.

3. I have elsewhere[7] defined analogy as a relation which is between systems of concepts, in which we become clearly conscious both that corresponding elements are different and that corresponding connections between elements are the same.

It seems that mathematics, where things are indeed simplest, was the first field in which analogy first clearly revealed its clarifying, simplifying, and heuristic role. At any rate, Aristotle, in so far as he has spoken of it, relates analogy to quantitative relations of proportionality. Some simple analogies must have struck inquirers even in antiquity. Thus Euclid (Book 7, definition 16) calls the product of two numbers "area" and the factors "sides", a factor of three factors "body" with the factors as "sides" (definition 17), and also a product of two or three equal factors "square" and "cube" respectively (definitions 18, 19).[8] Where Plato has touched upon geometrical notions he uses similar terms. The invention of algebra rests on seeing the analogy of arithmetic operations whatever the different numbers: What is conceptually identical here will then be dealt with at one go and once and for all. Where magnitudes enter analogously into a calculation, then from one result we can obtain all of the others by simple analogous interchange of symbols. Descartes' geometry makes extensive use of the analogy between algebra and geometry, Grassmann's mechanics or vector theory of that between lines and forces, areas and torques, and so on. Every physical application of mathematics rest on taking note of analogies between facts and mathematical operations.

4. The great value of analogy for cognition was clearly recognized by Kepler.[9] In dealing with conic sections as to their optical properties, he says:

> The one focus of a circle is A, namely at the center; in the ellipse there are two foci, *A* and *B*, equidistant from the center of the figure in the more pointed part. In a parabola, one focus *D* is inside the section, the other must be set either outside or inside on the axis infinitely distant from the first, so that the line *HG* or *IG* drawn from this blind focus to any point *G* of the section is parallel to the axis. In the hyperbola, the external focus *F* is the nearer to the internal one *E*, the more obtuse the curve; and the one outside either of the opposite branches is inside the other and conversely. It follows by analogy that in a straight line either focus (we speak of straight lines not so much in the ordinary way but rather to complete analogy) falls on the line: there is but one, as in the circle. In the circle, the focus is thus at the center, as far from the circumference as possible, in the ellipse already less and in the parabola much less, and finally in the straight line it is at minimal distance, that is, it falls on the line. Thus, in the extreme cases of the circle and the straight line the foci coalesce, in the former as far as can be from the curve, in the latter falling fairly on it. In intermediate cases, the two foci are infinitely far apart in the parabola and finitely distant from the sides of a hyperbola and ellipse; in the ellipse the second is inside, in the hyperbola outside, whence the ratios are of opposite sign.... For the geometrical voice of analogy ought to be at our service; indeed, I very dearly love analogies, my faithful masters, aware of all nature's secrets: we must particularly look up to them in geometry, with infinitely many cases placed in between extreme ones, whenever however much we infer the intermediate case by absurd phrases, the analogies put any object's total essence clearly before our eyes.

5. Here Kepler not only emphasizes the value of analogy but also and rightly the principle of continuity which alone could lead him to the degree of abstraction that allowed him to grasp such deep-seated analogies. As to the workshop of ancient science we know rather little, we barely have their most important results of inquiry. However, the form of presentation is often such as to conceal the actual paths of inquiry, as was strikingly shown by the example of Euclid. Unfortunately, the ancient example has often been copied in more recent times, for the sake of a rigor that is overrated and to the detriment of the true interests of science. A thought is most completely and rigorously founded

when all the paths and motives that led to and confirmed it are clearly set out. Of this foundation the logical link with previous more familiar and uncontested thoughts is only one part. A thought, the motives for whose generation are completely clarified, can never be lost so long as those motives hold, while it can be given up immediately when they are recognized as decrepit.

6. The study of the classics from the renaissance of natural science is so enjoyable and rich in permanent and irreplaceable lessons precisely because those great and naive men tell us all the details of what they found and how, in the friendly joy of search and discovery, without any professional and leaned secrecy-mongering.

Copernicus, Stevin, Galileo, Gilbert, and Kepler furnish us examples of the greatest successes of scientific research that teach us without pomp what were the leading motives of human inquiry; thus the methods of physical and thought experiments,[10] the principles of simplicity and continuity and so on become familiar to us in the simplest way.

7. Besides the cosmopolitan trait of openness the science of that period is distinguished by an unusual growth in abstraction. It is from individual human experience that science develops, and it is at the level of the individual that ancient science usually sticks. However, if one starts by inheriting a rich fund, one is in a more favorable position to cast frequent, varied, and rapid glances at the whole range of special findings with a view to comparison, thereby discovering common features even in what is far apart, where the original finder or novice was still put off by differences. It is especially when a change takes place continuously or in small steps, that the infinity of far distant members of a series becomes noticeable, making one aware of what, in spite of change, has remained the same. Thus, a pair of intersecting lines may appear as a hyperbola, one straight line as two collapsed hyperbolic branches, a segment of a line as an ellipse and so forth. Parallels and intersecting lines differ for Kepler only by the distance of their point of intersection. To his younger contemporary Desargues[11] a line is a circle whose center lies at infinity, a tangent a secant with coincident points of intersection, an asymptotic a tangent at an infinitely distant point and so on. All these steps that are by now obvious to us offered insuperable difficulties to ancient geometers. With the help of the principle of continuity, we reach a higher level of abstraction, and so acquire ability to grasp analogies. In our geometrical intuition, analogies of continuously variable magnitudes lead to infinitesimal calculus both in its Newtonian and also in its Leibnizian forms, comparing algebraic symbolism with ordinary language gives Leibniz the idea of a universal characteristic or conceptual script and so leads to logical discoveries that are only just coming back to life.[12] Lagrange, at a high level of abstraction, was able to see the analogy between small increments as due to changes of independent variables and small increments as due to variation of functional form, which leads to the creation of the calculus of variation.

8. If an object *M* has marks *a, b, c, d, e* and another *N* agrees with it as regards *a, b, c,* one is inclined to expect that it will agree with *d, e* as well, an expectation not logically justified. For logic only guaranties agreement with what has been fixed, which cannot be contradicted as long as it has been retained. However our expectations depend on our physiological and psychological organization. Inferences from similarity and analogy are not strictly matters of logic, at least not of formal logic, but only of psychology. If *a, b, c, d, e* above are directly observable, we speak of similarity, if they are conceptual relations between marks, analogy is closer to normal usage. If the object *M* is familiar, then looking at *N* will by association remind us of *d, e* alongside the observed *a, b, c* which ends the process provided *d, e* are indifferent. Not so when *d, e* have strong biological

interest because of some useful or noxious property, or are specially valuable for some applied or purely scientific and intellectual purpose. In that case we feel compelled to search for *d, e*, the outcome being awaited with close attention. This will be reached whether by simple sense observation or by means of complex technical or scientific conceptual reactions. Whatever the outcome, we shall still have extended our knowledge of *N*, by obtaining a new agreement or difference with regard to *M*. Both cases are equally important and contain a discovery, but with agreement we have the further significant feature of an economic extension of a uniform conception to a larger field; that is why we are especially keen about looking for an agreement. This amounts to a simple biological and epistemological account of why we tend to value inferences from similarity and analogy.

9. Considerations of similarity and analogy are in several respects fruitful motives for extending knowledge. A still rather unfamiliar area of facts N may show some analogy with another *M* that is more familiar and more accessible to direct intuition: we feel at once impelled to look in thought, observation, and experiment for what in *N* corresponds to the known features or relations of features of *M*, and usually this will reveal hitherto unknown facts about *N* which are thus discovered. Yet even if our expectations are disappointed and we find unexpected differences between *N* and *M*, we have not worked in vain. We have come to know *N* better and thereby enriched our conceptual grasp of it. What initiates our use of hypotheses is just this attraction of similarity and analogy: a hypothesis enlivens our intuition and fantasy and thereby stimulates physical reactions. Besides, the function of hypotheses is partly to be reinforced and even sharpened and partly to be destroyed, which in either case amplifies our knowledge.[13]

10. Several equally known areas M, N, O. P may enter into analogy, in groups of two or more. Of course these areas have differences as well as agreements, otherwise they would be identical. Hence in analogizing we may prefer to start from now the one and now the other, and this will reveal different analogies each justified in its setting: clearly this process will show what is adventitious and arbitrary in our conceptions, and which of them are most widely and uniformly applicable and therefore most conformable with the ideal of science.

11. There is no lack of examples for the importance of analogy; it can indeed hardly be overrated in natural science. Even in ancient times water-waves directly visible clarified and made intelligible the process of sound propagation.[14] As to light the appropriate ideas were developed from the case of sound.[15] Galileo's discovery of the moons of Jupiter was more powerful than any other arguments in supporting the Copernican system: we have here a small scale model of the solar system. Huygens greatly valued this support.

12. The rotation of the plane of the polarization of light by electric currents, demonstrated by faraday in 1845, is one of the most remarkable examples of a great discovery through analogy. Twenty years earlier, J.F.W. Herschel had already surmised this relation between light and electricity, and was guided in his experiments by the right idea, even though the results were negative because the forces applied were just too small (letter to Faraday, November 9, 1845).[16] The rotation of the plane of polarization as rays proceed in certain rigid and liquid media gave Herschel the impression of a screw. He therefore looked for a helicoidal dissymmetry in quartz, and indeed discovered it: optical helicoidal dyssemmetry thus corresponds to the same feature in the medium. Now clearly a straight

electric current shows this screw structure in its associated magnetic field, so that John Herschel expected it to influence polarized light rather as quartz does: to test this he passed a ray of light first through a current-carrying coil (as Faraday later did), and then, in another experiment, parallel to and between two parallel wires with opposite currents, without positive result.

13. Another example, where analogy operates between known areas, has been Fourier's theory of heat currents, which was apparently developed by analogy with water currents. In turn, from his theory of heat conduction other theories were developed by analogy, for instance, that of electric currents and diffusion currents. Independently and alongside these, there developed a corresponding theory of attraction by forces at a distance. Comparing these theories that give a comprehensive account of vast areas of fact, many analogies emerge. William Thomson[17] (Lord Kelvin) compared the theories of heat conduction and attraction and found that the formulae of the first pass over into those of the second if we also replace the concepts of temperature and temperature gradient by those of potential and force respectively, a striking relationship seeing that the original fields seem so different since heat conduction was taken as based on contiguous action and attraction at action at a distance. These thoughts surely must have inspired Maxwell, since in this same way he recognized that Faraday's theories of electricity and magnetism, based on contiguous action, were just as valid as the theories of action at a distance, then alone recognized by mathematical physicists; Thus he came to turn his attention to the great advantages of the former.[18] In similar vein, by recognizing the analogy between the equations of light propagation and electromagnetic oscillations, he came to found the electromagnetic theory of light,[19] with subsequent experimental confirmation by Hertz.[20]

14. Clerk Maxwell[21] consciously developed the use of analogy into a very perspicuous physical method: he thinks that phenomena are too much lost from sight if we represent our results only in mathematical formulae. If, however, we use a hypothesis, then we look through colored spectacles, and explaining things from a one-sided point of view make us blind to facts. To him, the phenomena of electrostatics, magnetism, electric currents, and so on revealed common traits reminiscent of the flow of a fluid. To complete the analogy, he would idealize the fluid: it has no inertia (hence, no mass) and is incompressible and taken to flow through a resistive medium whose own resistance is proportional to the speed of the current. The picture is of course imaginary to us and based on analogy but nevertheless intuitive: we do not take it as real and we know precisely how it coincides conceptually with the facts which are to be represented. The pressure of the fluid clearly corresponds to the various potentials, the direction of the current to force and current, the pressure gradient to forces and so on. Without giving up intuition Maxwell thus succeeds in preserving his open mind and conceptual purity, combining the advantages of hypothesis and mathematical formulation.[22] His picture is such for us that its mental consequences are pictures of the consequences of facts, to adapt a phrase of Heinrich Hertz. Thus Maxwell comes close to an ideal method of scientific inquiry: hence his uncommon success.

15. To conclude, we must emphasize again that incomplete analogies too can promote inquiry, by revealing differences in the fields being compared. Thus, a theory of energy that merely rested on the conformity of energies would have remained within the bounds of the first law of thermodynamics whereas it is precisely by taking notices of differences that we reach the important recognition of dissipation.[23] A very instructive and historically important example of prematurely dropping a fruitful analogy occurs in

Newton's *Optics*, where (question 28) he is discussing Descartes' pressure theory and Huygens' wave theory[24]: Having dismissed the first, he then pronounces against the second because he finds no refraction of light into a shadow; although he knows that water-waves are refracted more strongly than sound, his experiments were such that even the slightest refraction of light into shadows could easily be missed, while only the opposite kind was noticed, and this last he prefers to attribute to a deviating force emanating from the body grazed. This assumption precludes any understanding of Huygens' work, and so Newton clings to his corpuscular projectile theory, trying to explain everything from the innate and immutable properties of rays, a task he regards as quite difficult enough as it is.

C. ANALOGIES BETWEEN OPTICS AND ACOUSTICS[1]

by Ernst Mach

Color has frequently been compared with musical notes, for example by Ptolemy, Descartes, and Newton. Grimaldi thought he was entitled to assume that an efflux of light was periodic in its properties and even to give expression to the principle that light added to light might produce darkness. This principle is still valid although it no longer follows as a result of the experiments of Grimaldi. The whole of Huygens' optics is based upon the analogy of light with water-waves and sound . . . .

As a result of his study of the eye and his Göttingen doctoral dissertation on the human voice, Thomas Young, a member of the medical profession, was led to take up the study of optics and acoustics, and recognized the numerous instances of similarity between light and sound. His thought was incited by the observation of beats, the separation of a vibrating string into equal aliquot parts, and the bending of sound round obstacles.[2] After an extensive comparison of optical and acoustical phenomena, together with a careful investigation of Newton's objections to the vibration theory, with a reduction of them to their proper perspective, he then came to the conclusion that the comprehension of *all* the known optical phenomena was facilitated by the wave hypothesis, and that some of them were explicable *solely* by this hypothesis.[3] Soon he was able to submit some original optical experiments which his theory showed to be completely explicable.[4] With the skillful repetition of Grimaldi's experiment on the diffraction of light by a wire, the *inner* fringes showed themselves to be produced by light from both edges of the wire deflected *into* the shadow, since they vanished when the light from one of the sides was screened off. Thomas Young in a cautious philosophical and general manner, lays down the interference principle somewhat in this manner[5]:

> As in the case of sound (a musical note), so also in a (colored) beam of light there then follows a succession at regular intervals of alternately opposite properties (states) which are mutually capable of destroying (neutralize) each other.

With this careful expression of the principle, we are no longer concerned merely with a hypothesis but with a fact fully confirmed by experimental analysis. The form of the interference experiment, which involves two neighboring slits illuminated by another *single* slit, is first given by Young in his *Lectures* published in 1807, Lecture 23, Plate 20. figure 267, and in Lecture 39, Plate 30, Figure 442.[6] The experiment is also referred to in Young's letter of October 19, 1819 to Fresnel.[7]

4. CORRESPONDENCE

(1)

Otto Neurath to Ernst Mach — Military Field Service, undated [1915?]

Please allow me to send my lecture: *"Zur klassification von Hypothesen-systemen"* by separate mail. I have tried in this work to build on ideas which long ago you had placed in my mind. It would make me happy should you consider one or another parts of it in a friendly way. In this lecture I have drawn broader philosophical consequences. I am investigating further some of the problems mentioned here.

With Sincere Respect,
Your Servant

Dr. Otto Neurath

(2)

Otto Neurath to Ernst Mach* — Military Field Service, undated [1915?]

I thank you courteously for your friendly lines. Let me take the freedom of forwarding my lecture on the classification of systems of hypotheses in which I seek to extend my ideas on optics a little further.

Let me proceed with a hesitant request. I would be sincerely thankful if you would tell me - provided that this places no burden on you - if, how, and where the attempt was made to bring the development of optics into close connection with that of acoustics, I naturally mean in a penetrating way. I have not previously had an occasion to tackle this very interesting problem about the alleged exchange: "Light is something like sound" and "Sound is something like light". The fact that theory of interference arose about the same time for both has often made me think that I should analyse the matter more exactly. I have indeed been able to find very good historical data on the interference theory of light, but apart from the original material which I already had in hand, I have ill-succeeded in my effort to learn the history of acoustical interference from an analytical and critical perspective, though I have already often turned my attention to this object in so far as my quite restricted scientific knowledge allows. But should you be willing to share a few words and perhaps a suggestion on this matter, I would be sincerely thankful. But I make this request only under the assumption that it requires no discomforting effort on your part.

I have often wished to look you up in Vienna, but have always refrained through fear of disturbing you or seeming immodest. I have had so much to ask you, so much, which apart from yourself few others could answer. Even letters are inadequate and would also place an immense burden on your time. I have heard with great interest about the latest developments in relativity theory which can be traced to your conception that gravitation as a function depends on the total distribution of mass and remains constant toward certain transformations (for example, rotation). It was this idea in your *Mechanics* which has never left me since my first reading, and has influenced my own intellectual development and by indirect paths even economics itself. It was your tendency to derive the meaning of particulars from the whole rather than the meaning of the whole from a summation of particulars, which has been so important. It is in value theory in particular that these impulses have benefitted me through associated indirect paths. I have always and evermore preserved a deep feeling of thanks toward your work. But it was not only these benefits from the field of physics, but even more what I learned in another direction from your process of thinking about physical matters. I feel myself forced to use this occasion to heartily and sincerely thank you as a human being, that is, so to say, *in concreto*.

With Sincere Respect,
Your Servant

Dr. Otto Neurath

NOTES

* This is a shortened version of a manuscript: "The Auxiliary Motive in the Forest and in Ethics", by Dr. Stöltzner. We have added what Neurath wanted to see from Mach's *Optik* (1921).

1 Otto Neurath, "Die Verirrten des Cartesius und das Auxiliarmotiv (Zur Psychologie des Entschlusses)", which was given as a speech before the *Philosophische Gesellschaft an der Universität zu Wien* on January 27, 1913 with a followup discussion on February 8, 1913. It was published by the Society in their yearbook for 1913 and later by Rudolf Haller and H. Rutte in Otto Neurath - *Gesammelte philosophische und methodologische Schridten*, 2 vols., Hölder-Pichler-Tempsky, Wien, 1981, pages 45-59. For this work in English see: R.S. Cohen and M. Neurath, *Otto Neurath - Philosophical Writings 1913-1945*, (Vienna Circle Collection **16**), Reidel: Dordrecht, 1983, pp. 1-12.

2 R. Haller & H. Rutte, "Geschichte und wissenschaftliches System bei Otto Neurath" in Ibid (1981), pages 125-134 and Rudolf Haller, "Das Neurath-Prinzip - Grundlagen und Folgerungen", in: *Fragen zu Wittgenstein und Aufsätze zur österreichischen Philosophie*, Rodopi: Amsterdam, 1986, pp. 108-124.

3 O. Neurath, "Zur Klassifikation von Hypothesensystemen (mit besonderer Berücksich-tigung der Optik)" was given as a lecture before the *Philosophisches Gesellschaft an der Universität zu Wien* on March 2, 1914 and was published in the Society's yearbook for 1914/15, pp. 39-63 reproduced in Neurath, Ibid (1981), pp. 85-101, translated as "On the Classification of Systems of Hypotheses" in Neurath , op. cit. (1983), pp. 13-31. See also:
Otto Neurath, "Prinzipielles zur Geschichte der Optik", *Archiv für Geschichte der Naturwissenschaften und Technik*, **5** (1915) 371-389. Also published in Neurath (1981) 71-84 and in translation: "On the Foundations of the History of Optics" in M. Neurath & Robert S. Cohen (eds.). *Empiricism and Sociology*, trans. by P. Foulkes, (Vienna Circle Collec-tion, vol. **1**), Reidel: Dordrecht, pp. 101-112.

4 J. Blackmore & K. Hentschel (eds.), *Ernst Mach als Aussenseiter: Machs Briefwechsel über Philosophie und Relativitätstheorie mit Persönlichkeiten seiner Zeit*, Braumüller: Wien, 1985, p. 151.

5 O. Neurath, op. cit. (See footnote #3.)

6 O. Neurath, op. cit. (See footnote #1.)

7 Ernst Mach, *Die Mechanik in ihrer Entwicklung*, edited by Renate Wahsner & Horst-Heino von Borzeszkowski, Akademie Verlag: Berlin, 1988, page 480. (*The Science of Mechanics*, trans. by Thomas J. McCormack, Open Court: La Salle, Ill., 1989, p. 559.)

8 Ibid., p. 479 (p. 559).

9 Neurath op. cit, 1981 (1913), p. 64. (See footnote #1.)

10 Ibid., p. 60.

11 Ibid., p. 59.

12 O. Neurath, op. cit. 1981 (1914), p. 60 & (1915), p. 83. (See footnote #3.)

* Mach unlike most physicists today seems to have identified facts entirely with sensory appearances, a reduction which made it *easier* for different theories (such as wave and particle theories in optics) to seem equivalent to him. In fact, he seems to have regarded all theories as transient and even equally supported by evidence. He thought of theories as a way to relate appearances, not appearances as "evidence" for theories - editor.

13 O. Neurath, op. cit 1981 (1915) p. 87. (See footnote # 3.)

14 Otto Neurath, "Die neue Enzyklopädie" in *Zur Enzyklopädie der Einheitswissenschaft*, edited by Otto Neurath et al, (Einheitswissenschaft, Band **6**), Van Stockum & Zoon: Den Haag, 1938, reproduced in Neurath, op. cit, 1981, p. 865f. (See footnote #1.)

15 See Robert Reininger, *50 Jahre Philosophische Gesellschaft an der Universität Wien 1888-1938*, Verlag der Philosophischen Gesellschaft: Wien, 1938, pages 33-34. For a discussion in English on the Society under Alois Höfler see: John Blackmore, "Late Appendix on the Philosophical Society of the University of Vienna 1888-1922" in John Blackmore, *Ludwig Boltzmann - His Later Life and Philosophy 1900-1906*, Book Two, Kluwer: Dordrecht, 1995, pp. 267-315.

16 The society normally published about half of its own lectures. They appeared in *Wissenschaftliche Beilage zum Jahresbericht* at least from 1902 to 1907 which was replaced by the *Jahrbuch der philosophischen Gesellschaft an der Universität zu Wien* until at least 1914-15. Starting again in 1926 and lasting perhaps until 1937 was a later series on much inferior paper which included excerpts from most lectures plus in later years a few complete ones.

17 Otto Neurath, "Das Begriffsgebäude der Wissenschaftslehre und seine Grundlagen", op. cit. 1981, p. 114. (This article appeared in *Zeitschrift für die gesamte Staatswissenschaft*, **73** (1917) 484-520.) Neurath also gave a lecture six years later with almost the identical title: "Das Begriffsgebäude der Wissenschaftslehre. Eine logische Betrachtung." before the University of Vienna Philosophical Society on February 26, 1923, but it does not seem to have been published.

18 Rudolf Haller, "Poetische Phantasie und Sparsamkeit - Ernst Mach als Wissenschaftstheoretiker" in R. Haller & F. Stadler, (eds.), *Ernst Mach - Werk und Wirkung*, Hölder-Pickler-Tempsky: Wien, 1988, p. 351. For an English version of this article see: Rudolf Haller, "Poetic Imagination and Economy - Ernst Mach as Theorist of Science" in *Ernst Mach - A Deeper Look*, edited by John Blackmore, (Boston Studies in the Philosophy of Science #143), Kluwer: Dordrecht, 1992, pp. 201-214.

19 Ernst Mach, *Erkenntnis und Irrtum*, Darmstadt, 1991 (1906), p. 220. For an English version of this book see: Ernst Mach, *Knowledge and Error*, with an introduction by Erwin N. Hiebert, Reidel: Dordrecht, 1976.

20 Ernst Mach, "Das Prinzip der Vergleichung in der Physik" in *Populär-wissenschaftliche Vorlesungen*, Böhlau: Wien, 1987, p. 274. For the article in English see: Ernst Mach, "On the Principle of Comparison in English", *Popular Scientific Lectures*, La Salle, Ill., 1943, pp. 236-258.

21 Ernst Mach, *Die Prinzipien der physikalischen Optik*, Leipzig, 1921, p. 372. For the English version of the book see: *The Principles of Physical Optics*, London, 1926.

22 Mach, op. cit. 1988, p. 406 & op. cit., 1989, p. 471. (See footnote #7.)

23 Mach, op. cit. 1987, p. 280. (See footnote #20.)

24 Neurath, op. cit, 1981, p. 96 (See footnote #3.)

25 Mach, op. cit., 1988, p. 501 and op. cit., 1989, p. 586. (See footnote # 7.)

26 *Nature*, No. 2807, **112**, August 18, 1923, p. 253 has translated remarks by Einstein originally appearing in the *Bulletin de la Société francais de philosophie* for July 1922 about events of April 6, 1922. They are also quoted in John Blackmore (ed.) *Ernst Mach - A Deeper Look*, Kluwer: Dordrecht, 1992, p. 162.

27 Paul Feyerabend, "Mach's Theory of Research and its Relation to Einstein", in *Studies in the History and Philosophy of Science*, **15** (1984) 1 - 22.

* Should scientific description be idealized or non-idealized? If it is idealized, especially in mathematical terms, then one's goal is often a system; if non-idealized, it is commonly a catalogue or encyclopedia. Both goals, those of Mach and Einstein, are almost surely legitimate, but by "non-idealized" Mach meant a rigorous description of appearances, whereas if one identifies the physical world with what exists *beyond* sensory appearances, then "non-idealized" is closer to the careful geographical or historical description which one normally finds in encyclopedias. But this should not be confused with Neurath's later interest in encyclopedias. - editor]

28 Neurath, op. cit. 1981, p. 76. (See footnote #1.)

29 Neurath, op. cit. 1988, p. 499 and op. cit. 1989, p. 583. (See footnote #7.)

30 Ibid., p. 285 (p. 321).

31 Here I import terminology from field theory to indicate that the step between levels is made smooth.

32 Neurath, op. cit. 1988, p. 99 and op. cit. 1989, p. 94. (See footnote #7.)

33 Our study of this genealogy would be more complete if Neurath's lecture: *Ernst Machs Geschichte der Optik und die Wissenschaftslehre* which was given before the University of Vienna Philosophical Society on January 21, 1922 had been published, but from about 1916 until 1926 it seems the Society did not issue its *Jahrbuch* or publish any lectures.

2

1 Ernst Mach, *Knowledge and Error*, translated by Thomas J. McCormack & Paul Foulkes, Reidel: 1976, pp. 162-170.

2 William Stanley Jevons, *The Principles of Science*, London, 1892, p. 627.

3 James Clerk Maxwell, X*Transactions of the Cambridge Philosophical Society*, **X** (1855) p. 27.

4 J. Hoppe, *Die Analogie*, Berlin, 1873.

5 W. Stern, *Die Analogie im volkstümlichen Denken*, Berlin 1893.

6 Edward B. Tylor, *Die Anfänge der Kultur*, German translation, Leipzig, 1873.

7 Ernst Mach, *Die Analyse der Empfindungen*, 4. Auflage, G. Fischer: Jena, 1903, p. 3.

8 Euclid's *Elements*, (quoted in the German from the edition by J.F. Lorenz, Halle, 1798.)

9 Johann Kepler, *Opera*, ed. Frisch, Vol. II, p. 186. The relevant diagrams are obvious and will be omitted.

10 Cf. Ch. XI.

11 *Ouvres de Desargues*, ed. Poudra, Paris, 1864.

12 Louis Couturat, *La Logique de Leibniz*, Paris, 1901.

13 Ernst Mach, "Bemerkungen über die historische Entwicklung der Optik", *Zeitschrift für den physikalischen und chemischen Unterricht* (Berlin), **11** (1897-98) 3-8.

14 Marcus Vitruvius Pollio, *De architectura*, V, Cap. III, 6.

15 Christiaan Huygens, *Traité de la lumière*, Leiden, 1690.

16 Bence Jones, *The Life of Faraday*, London, 1870, Vol. **II**, p. 205.

17 William Thomson, *Cambridge Mathematical Journal*, **III**, February 1842.

18 James Clerk Maxwell, *A Treatise on Electricity and Magnetism*, Oxford, 1873, Vol. **I**, p. 99.

19 James Clerk Maxwell, "Dynamic Theory of the Electromagnetic Field", *London Philosophical Transactions*, 1865.

20 Heinrich Hertz, *Untersuchungen über die Ausbreitung der elektrischen Kraft*, Leipzig, 1892.

21 James Clerk Maxwell, *Transactions of the Cambridge Philosophical Society*, **X** (1855) page 27. When I myself mentioned these analogies in a similar way, in the Prague periodical Lotos (February 1871) and in Erhaltung der Arbeit (Prague 1872), the work of Thomson and Maxwell

was still unknown and inaccessible to me. It seems that S. Carnot was the first consciously to have adopted this mode of thought.

22 Mach, 1897-98, op. cit, pp. 3-8. (See footnote #13 above.)

23 Ernst Mach, *Die Principien der Wärmelehre*, 2. Auflage, Barth: Leipzig, 1900.

24 : Sir Isaac Newton, *Optice*, ed. Clarke, London, 1719, p. 136.

3

1 Ernst Mach, *The Principles of Physical Oprics - An Historical and Philosophical Treatment*, translated by J.S. Anderson and A.F.A. Young, Dover Publications: New York, 1953 (1926), pp. 273-275.

2 Thomas Young, "Outlines of Experiments and Inquiries respecting Sound and Light", *Philosophical transactions of the Royal Society*, 1800.

3 "On the Theory of Light and Colours", Ibid., 1802.

4 "An Account of Some New Cases of the Production of Colours not Hitherto Described", Ibid.

5 "Experiments and Calculations Relative to physical Optics", Bakerian Lecture, Ibid., 1804.

6 Thomas Young, "A Course of Lectures on Natural Philosophy and mechanical Arts", 1807.

7 Thomas Young, *Miscellaneous Works*, edited by G. Peacock, Volume I, 1855, p. 393.

4

* The original letters are now in the Deutsches Museum in Munich, Germany. Published versions in the original German are in Joachim Thiele, *Wissenschaftliche Kommunikation*, A. Henn Verlag: Kastellaun, 1978, pp. 99-101.

REFERENCES

Anon. "Einstein and the Philosophies of Kant and Mach". *Nature*, No. 2807. vol. 112. August 18, 1923. p. 253.

Blackmore, John, and Klaus Hentschel (Hrsg.). *Ernst Mach als Aussenseiter: Machs Briefwechsel über Philosophie und Relativitätstheorie mit Persönlichkeiten seiner Zeit.* Wien: Braumüller, 1985, p. 151.

Couterat, Louis. *La Logique de Leibniz*. Paris: 1901.

Desargues, Gérard. *Ouvres de Desargues*. edited by Poudra. Paris: 1864.

Euclid's *Elemente*, edited by J.F. Lorenz. Halle: 1798.

Feyerabend, Paul. "Mach's Theory of Research and its Relation to Einstein". *Studies in the History and Philosophy of Science*, 15 (1984) 1-22.

Haller, Rupert, and H. Rutte (Hrsg.). "Geschichte und wissenschaftliches System bei Otto Neurath". *Otto Neurath - Gesammelte philosophische und methodologische Schriften*. two volumes. edited by R. Haller and H. Rutte. Wien: Hölder-Pichler-Tempsky, 1981, 125-134.

Haller, Rudolf. "Das Neurath-Prinzip -Grundlagen und Folgerungen". *Fragen zu Wittgenstein und Aufsätze. zur österreichischen Philosophie*. Amsterdam: Rodopi, 1986, pp. 128-134.

Haller, Rudolf. "Poetische Phantasie und Sparsamkeit - Ernst Mach als Wissenschaftstheoretiker". *Ernst Mach - Werk und Wirkung*. edited by R.Haller and F. Stadler. Wien: Hölder-Pichler-Tempsky, 1988, pp. 342-355.

Hertz, Heinrich. *Untersuchung über die Ausbreitung der elektrischen Kraft*. Leipzig: 1892.

Hoppe, J. *Die Analogie*, Berlin: 1873.

Huygens, Christiaan. *Traité de la lumiere*. Leiden: 1690.

Jevons, William Stanley. *The Principles of Science*. London: 1892.

Jones, Bence. *The Life of Faraday*. two volumes. London: 1870.

Kepler, Johannes. *Opera*. 8 volumes. edited by Charles Frisch. 1858-1871.

Mach, Ernst. "Eine Bemerkung über den zweiten Hauptsatz der mechanischen Wärmetheorie".

Lotos. 21 (1871) 17-18, 139-147.
Mach, Ernst. *Die Geschichte und die Wurzel des Satzes von der Erhaltung der Arbeit*. Prag: Calve, 1872.
Mach, Ernst. "Bemerkungen über die historische Entwicklung der Optik". *Zeitschrift für den physikalischen und chemischen Unterricht* (Berlin). 11 (1897-98) 3-8.
Mach, Ernst. *Die Principien der Wärmelehre*. 2. Auflage. Leipzig: Barth, 1900. [1896].
Mach, Ernst. *Die Analyse der Empfindungen*. 4. Auflage. Jena: G. Fischer, 1903 [1886].
Mach, Ernst. *Die Prinzipien der physikalischen Optik*. Leipzig: J.A. Barth, 1921.
Mach, Ernst. *The Principles of Physical Optics*. translated by J.S. Anderson & A.F.A. Young. New York: Dover. 1953 [1926].
Mach, Ernst. *Knowledge and Error*. Pranslated by Thomas J. McCormack and Paul Foulkes. Dordrecht: Reidel, 1976, pp. 162-170.
Mach, Ernst. "Über das Prinzip der Vergleichung in der Physik". *Populär-wissenschaftliche Vorlesungen*. Wien: Böhlau, 1987, pp. 266-289.
Mach, Ernst. *Die Mechanik in ihrer Entwicklung*. edited by Renate Wahsner and Horst-Heino von Borzeszkowski. Berlin: Akademie Verlag, 1988.
Mach, Ernst. *The Science of Mechanics*. trans. by Thomas J. McCormack. La Salle: Open Court, 1989.
Mach, Ernst. *Erkenntnis und Irrtum*. Darmstadt: 1991.
Maxwell, James Clerk. *Transactions of the Cambridge Philosophical Society*. X (1855) 27. [included in Maxwell's *The Scientific Papers*. Volume I. Cambridge: 1890.]
Maxwell, James Clerk. "A Dynamical Theory of the Electromagnetic Field". *London Philosophical Transactions*. 1864-65. [included in *The Scientific Papers*. Volume I. Cambridge: 1890.]
Maxwell, James Clerk. *A Treatise on Electricity and Magnetism*. Oxford: 1873.
Neurath, Otto. "Principielles zur Geschichte der Optik". *Archiv für Geschichte der Naturwissenschaften und Technik*. 5 (1915) 371-389.
Neurath, Otto. "Das Begriffsgebäude der Wissenschaftslehre und seine Grundlagen". *Zeitschrift für die gesamte Staatswissenschaft*, 73 (1917) 484-520. [also in Neurath 1981.]
Neurath, Otto. "Die neue Enzyklopädie". *Zur Encyklopädie der Einheitswissenschaft. edited by Otto Neurath et al.* (Einheitswissenschaft, Band 6.) Den Haag: Van Stokum and Zoom, 1938.
Neurath, Otto. "Die Verirrten des Cartesius und das Auxiliarmotiv (Zur Psychologie des Entschlusses)". *Otto Neurath - Gesammelte philosophische und methodologische Schriften*. two volumes. Hrsg. R. Haller & H. Rutte. Wien: Hölder-Pichler-Tempsky, 1981, pp. 45-59.
Neurath, Otto. "On the Classification of Systems of Hypotheses". *Otto Neurath - Philosophical Writings 1913-1945*. edited by R.S. Cohen and M. Neurath. (Vienna Circle Collection 16). Dordrecht: Reidel, 1983, pp. 1-12.
Reininger, Robert. *50 Jahre Philosophische Gesellschaft an der Universität Wien 1888-1938*. Wien: Selbstverlag, 1938.
Stern, Wilhelm. *Die Analogie in volkstümlichen Denken*. Berlin: 1893.
Stöltzner, Michael. "The Auxiliary Motive in the Forest and in Ethics". manuscript, 1997.
Thiele, Joachim. *Wissenschaftliche Kommunikation*. Kastellaun: A. Henn Verlag, 1978,
Thomson, Williams. [Lord Kelvin]. *Cambridge Mathematical Journal*. Volume III, February 1842.
Tylor, Edward B. *Die Anfänge der Kultur*. Leipzig: 1873.
Vitruvius Pollio, Marcus. *De Architectura*. Volume V, Chap. III, Sect. 6.
Young, Thomas. "Outlines of Experiments and Inquiries respecting Sound and Light". *Philosophical Transactions of the Royal Society*. 1800. [See also 1802, 1804, and 1807.]
Young, Thomas. *Miscellaneous Works*. edited by G. Peacock. Volume I, 1855. p. 393.

Chapter Six

VON HAYEK, BERGMANN, AND MAYERHÖFER

by Themselves

Section Guide
1. "Ernst Mach and Viennese Social Science"
2. "To Otto Neurath about the Vienna Circle"
3. "Mach and My Student Days in Vienna"

1. ERNST MACH & VIENNA SOCIAL SCIENCE

by Friedrich von Hayek[1]

My task is exclusively to bear witness to how penetrating Ernst Mach's influence already was before the year 1922 when Moritz Schlick formed the so-called "Vienna Circle". It so happens that I studied in my own hometown Vienna precisely three years from 1918 until 1921, and that what passed for philosophical discussion revolved essentially around Mach's ideas. Vienna as a whole was already extraordinarily inclined toward philosophy of natural science.[2] Apart from Heinrich Gomperz,[3] Adolf Stöhr[4] also taught in Vienna and both men followed a similar line, and Robert Reininger[5] was at the very least friendly toward such an outlook.

I can no longer recall exactly how I came almost at once to turn to Mach when I returned from the field in November 1918. Unfortunately my [extant] reading list only starts with early 1919, but it already includes the remark "and now also *Erkenntnis und Irrtum*", which assumes that I had already become acquainted with other philosophical works by Mach in my academic study during the previous four months. I know that I became quite absorbed in his *Populäre-wissenschaftlichen Vorlesungen*, *Die Mechanik in ihrer Entwicklung*, and especially his *Analyse der Empfindungen*, which probably is related to the fact that while officially I was a student of law during those three years that my time was actually shared with a lot of economics and psychology and only in passing with law.

It is not so easy to say what the immediate reason was why we were so deeply occupied with Mach's philosophy. Already in the years before the war it must have been the same. One can point to the work of Schumpeter[6] as typical, who in his very first work in 1908 was so manifestly shot through with Mach's way of thinking that Friedrich von Wieser[7] in his review of the book concentrated on the question of the applicability of Mach's thoughts to the social sciences. And I also know that his contemporary, the philosopher Ludwig Wittgenstein, a distant cousin of mine, had already grappled with Mach at the same time.[8]

Immediately after the war when I came to the university, a remarkable situation had arisen which affected the social sciences. Lenin had previously attacked Mach's philosophy [1909], but Friedrich Adler,[9] who if I remember correctly was one of the most prominent figures of the time in politics, had shot Minister Stürgkh [1916], and while in

J. Blackmore, R. Itagaki, and S. Tanaka (eds.). Ernst Mach's Vienna 1895-1930. 123-138.

prison [1918] had published a defence of Mach, which led to lively discussions over this matter within the extreme left of socialism between communists and left-wing socialists. But the controversy also spilled over onto us, that is, on the non-socialists, and it became particularly relevant when Wieser's successor, a somewhat metaphysically-oriented economist, Othmar Spann,[10] took over and we looked for arguments against metaphysics, which we found in Mach, although it was not very easy for us to fully swallow Mach's positivism. We also found that it served in too outspoken a way as a foundation for a socialist conception which was very uncongenial to us, especially as advocated by Otto Neurath, who would later become one of the founders of the "Vienna Circle", who I believe wanted to change Mach's positivism into a particularly crude form of physicalism, or, as it was called at that time, scientism. But on the other side, we found in Mach almost the only arguments against a metaphysical and nebulous point of view, so that in those years we endeavored to take from positivism what was manifestly sound while bracketing out what was not applicable to the *Sozial und Geisteswissenschaften*, even though what we excluded was also clearly a great source of reliable information.

Mach's works led me personally to psychology and sense physiology. I even wrote a study at that time on these questions, which thirty years later even led to a book.[11] In reflecting on Mach's phenomenalism, I came to the conclusion that the conception of pure, simple sensations, which constitute the elements of all sensory perception did not appear to be fully plausible. During an experience which was very similar to the one which Mach himself describes[12] when he suddenly recognized that the "thing-in-itself" of Kantian philosophy was an unnecessary and empty concept which one could throw away I was suddenly enlightened by a similar idea that in Mach's psychology of the senses his "pure and simple sensation" was essentially a superfluous notion. Since Ernst Mach had explained so many of the connections between sensations as relations, I was finally driven to the conclusion that one could trace back the entire structure of the world of sensations to those relations, and that pure and simple sensation, which played such an important role with him, could be thrown out.[13] But this is only an example in passing of the great influence which Mach had exercised on our thought at that time.

One could say, that for young men with an interest in philosophy, that immediately after the war in 1918-19 for those who came to the University of Vienna and had no taste for orthodox philosophy, that Mach was the only alternative which was available. We also tried to read Avenarius,[14] but it was an attempt which we all gave up very soon. I don't know why. In any case, we found him very hard to understand. The path from Mach went rather to Helmholtz, to Poincaré, and to similar thinkers, and for those who sought something more systematic, like my friend Karl Popper, naturally to all of the great natural scientists and philosophers of the period.

With that I have basically exhausted what I wanted to say. I would only like to add that not only in the narrower reaches of the natural sciences, but most especially in those disciplines in which the methodological or scientific character of their theories were much less certain than in the natural sciences and therefore had yet to arrange a clear picture of what science was, that the influence of Ernst Mach was especially great. And that had little to do with the fact that in part Mach had become a political rallying point. Also, it seems to me to have been due to purely accidental circumstances that later in 1929, for example, at a time when I was already in the process of leaving Vienna, that the *Verein Ernst Mach* was established [by the Vienna Circle] and acquired a certain reputation for having [left-wing] political tendencies. It was indeed composed overwhelmingly of socialists, but that by itself does not mean that it was politically active, even if that composition did serve as a pretext in 1934 for [the right-wing Austrian dictator] Dollfuss[15] to suppress the *Verein Ernst Mach*.

2. "FROM A LETTER TO NEURATH ABOUT THE VIENNA CIRCLE (1939)"[16]

by Gustav Bergmann[17]

I was brought into the Circle through [Friedrich] Waismann,[18] to whom I, like many other younger guests of the Vienna Circle,[19] would be indebted for much stimulation and instruction. I had encountered Waismann when I was still a *Gymnasium* student and also attended the lectures on mathematics and on philosophy which he delivered at the *Vienna Volksheim*. At that time, I also became personally acquainted with him. I mention this because the fact that Waismann and [Edgar] Zilsel[20] had been very successful teachers for many years at this adult evening school (somewhat leftist-oriented) is characteristic of the social milieu from which, on another level, the Vienna movement emerged.

I would meet again [in the Vienna Circle] not only a number of the older students and younger doctors of the exact sciences, but also my teachers Hans Hahn[21] and Karl Menger,[22] and you, dear doctor, whom I had already known from the student movement. The ideological unity and the actual cooperation between the many different exponents of a progressive scientific world-view contributed not a little to the sympathy with which we young people followed the proceedings. This was perhaps in general characteristic of the mood of the more progressive portion of Viennese students at the time of the democratic republic in Austria. [before 1933?].[23]

Among the academic youth, this "progressive" group also included some of the younger students of [Hans] Kelsen[24] as well as several who were then interested in psychoanalysis. To be sure, that group was as a whole not very numerous and, not surprisingly, predominantly Jewish. If, however, up until the collapse in 1938, a relatively large segment of the Viennese intelligentsia regarded the "Vienna Circle" as their representative philosophy, then that is due - in addition to the "Ernst Mach Society"[25] and skillful popularizers like Edgar Zilsel - not least to several student members of the Boltzmanngasse circle (mostly young mathematicians and physicists), whose numbers fluctuated but who composed somewhat fewer than half of the listeners.

[Moritz] Schlick,[26] during the classical period of the Circle [1925-1930], always liked to emphasize the genetic relationship of the Vienna Circle with the modern development of mathematical and natural-scientific thought, the tradition of Ernst Mach, which was viewed as having been in opposition to the philological-historical philosophy of the old school.This followed since the mathematician Hans Hahn and his student Karl Menger belonged to the innermost core of the Boltzmanngasse group; mathematicians and physicists like [Philipp] Frank and [Richard] von Mises;[27] students trained by Mach and Boltzmann who were themselves from the Viennese milieu, were among the leaders of the movement who published the most. Accordingly, it was principally the young mathematicians and physicists who were welcome guests of the Circle. Schlick at that time placed little importance on professional contact with the other representatives of philosophy in Vienna . . . .

The group, which met two or three times a month on Thursday evenings in a room of the new Physics Institute on Boltzmanngasse - where the library of the Philosophy Department was then also housed - numbered on average twenty people . . . .

Schlick actually seldom spoke in the circle and even then mostly briefly and in his own rather hesitant and halting manner which I believe very much diminished his influence on the broad mass of lecture students.[28] he himself, moreover, felt that the tasks of lecturing and grading had become a burden and he would occasionally deplore the lack of interest and comprehension of his students. Toward the end of the longer winter semester he felt fairly tired and looked forward to his regular Easter trips to an Italian spring. In personal interaction he was quite informal, friendly, and of a natural good will, but at the same time reserved. His humor was on occasion not without sharpness. As a rule, however, as far as I could observe, he didn't change his customary reserved manner. Outside of the Circle, I hardly entered into any personal contact with him, and I believe that the same is true of most of the members of the Circle. Yet we all loved him very much and still preserve a luminous memory of his exceptionally attractive personality.

Leadership of the discussion, especially when the topic was technical, was carried out by Carnap and Hahn. These two, together with you, were the triarians who were engaged in a fight for maintaining the purification and consistent articulation of the views of the school for which it was so difficult to find a name. This was so difficult because against each available name that was approximately correct - such as "Positivism", "Empiricism", and the like - objections were raised mostly because of their historical associations; thus finally the factually neutral name "Vienna Circle" - which you proposed - became the only one that could be agreed upon, and in fact has prevailed . . .

Always a prepared leader in battle ... you and Carnap ... were the only two forces that stood behind all of the Circle's organizational events, like the publication of the monograph series and the holding of the first congress. It must not have been easy to win Schlick's cooperation with the "Ernst Mach Society", your favorite project, which was intended to be simultaneously the organizational center of the movement and a kind of adult education school.

I took part in the meetings in the Boltzmanngasse through approximately five years, from 1927 until 1931 - in the first years very regularly, but in the final years with certain interruptions. With the start of 1932 I stayed away from the meetings entirely, and whatever I have since learned from a few personal friends about the proceedings in the Circle and its development can hardly be considered a sufficient basis for a report. Nevertheless I might in this connection offer several points.

First: Unless I expressly indicate otherwise I will speak here only of that which *geographically* and in terms of *personnel* was the "Vienna Circle", that is, of the group whose meetings took place first in the private home of Schlick and then later in the Boltzmanngasse ... The group, however, would, as I now view it, already reach its highpoint in 1927/28, maintain momentum for several years and by 1931/32 already show clear signs of splintering and, as a consequence, would decline. The advocates of a consistently physicalistic theory of science, with "logical reconstruction of the world" and its radically rationalistic general attitude, would lose with Carnap's departure [in 1931] the personality which would have been appropriate to form the center of an academic colloquium . . . .

In the pre-Wittgensteinian period the most important topics were as far as I remember Brouwer's[29] ideas and Hilbert's[30] mathematics questions about Russell's theory of types (always a special interest of Hahn's), the conceptual representations of the probability calculus, and Carnap's reports on what was then his current work. Often, though, the discussion ran through the course of several evenings in a quite general vein and without a fixed program, and was usually the case when basic questions came up again from the viewpoint of a technical problem; such as the so-called problem of reality, the unified character of all scientific propositions, causality, the problem of other minds, and methodological questions, things about which the views of the Circle had already been largely clarified. You can imagine that these evenings were for us younger mathematicians and physicists the most interesting ones. On such occasions questions concerning the so-called humanistic sciences were discussed and the basic attitude of the Circle about the *Problematik* and the methodology of these disciplines sketched. In enumerating the participants, by the way, I cannot fail to mention the name of [Viktor] Kraft,[31] who next to [Felix] Kaufmann[32] was the only representative in the Circle from the so-called humanistic sciences. He was one of the most regular visitors at the Thursday evening meetings, but he almost never said a word . . . .

During the period in which I am to report, [Heinrich] Gomperz[33] spoke only once in the Circle, and the way the discussion went did not exactly help encourage further attempts in that direction. On the other hand, there also existed organizationally a not too loose connection through the "Ernst Mach Society" with the theoretical physicist [Hans] Thirring,[34] whose precision in epistemological matters and general questions earned him much sympathy in the Circle. Occasionally Schlick liked to invite one of Thirring's students in order to have him report in the Circle, for example, on the new approaches to quantum theory, which were then of general interest from the point of view of the philosophy of science . . . .

In the case of Hahn, the psychological situation may not have been quite so easy to fathom. As chairman of what was then the Vienna Association of Socialist University Teachers and as someone who had even earlier held clear political positions he had earned the respect of his students as a human being. But he conveyed little warmth, and his thinking at times was coldly abstract. Personally inhibited, somewhat ceremonious and professorial, his manner was curiously stylized .... His curiously stubborn interest in spiritism fits in well with this picture. I never could suppress the tawdry image of a "flight into the fourth dimension". And in fact Hahn actually did meet in spiritist salons with high society, the ladies of the old Austrian aristocracy .... Hans Thirring's interest in spiritism was of an entirely different sort. He had convinced himself that the examination of so-called "psychic phenomena", conducted by scholars who were invested with an outside authority, was at least indicated,[35] and he had thereupon decided to engage in the mentioned activities. For Hahn, however, spiritism was a well-known affair of the heart.

Of younger participants in the Circle I shall mention only [Kurt] Gödel[36] by name. Frequently, foreign guests, older students, and younger scholars who had come to Vienna to work for a time with Schlick would also attend. Corresponding to the significant role of Anglo-Saxon scholars in the rise of [what we considered to be] scientific philosophy, most of these philosophers were either American or British . . . .

Besides the Thursday evening meetings, there were also during my time, especially during the Christmas and Easter University vacations when Frank and Mises regularly came to Vienna [Philipp Frank and Rudolf Carnap were then at the German University of Prague], coffeehouse meetings in a smaller circle. To these meetings one or the other of us younger people were drawn in as listeners. I participated now and then in those evening meetings, just as I would then occasionally participate in private discussions between Carnap and Waismann . . . .

The Wittgensteinian esoterics, however, under whose influence Schlick himself increasingly fell, were not reluctant to see the decline of the former Circle. Their aversion to "beating an issue to death" and to discussions with people lacking the "capacity to grasp the essential, the necessary intuition", had likewise increased their alienation from those of us who maintained the traditional attitude of the Circle. This is by the way not merely a conjecture but rather is substantiated by repeated remarks made by Waismann about his own and Schlick's attitude toward the Circle at this time.

In any case, starting in 1930 Waismann, throughout almost two winters (the Circle followed the division of the academic year) expounded the theses of Wittgenstein with great enthusiasm and a commitment that was until then not customary in the Vienna Circle, and critical discussion, which had been the custom in the group, was met with a certain distaste. Schlick himself, the personal center of the Circle, in which Waismann had up to that point not really distinguished himself, followed Waismann's lead, although in a very courteous way. So meetings became more infrequent and it became very difficult to find genuine topics of discussion, which earlier had always been in plentiful supply. Later, so far as I know, the group renewed itself through new members, but it fell more and more under the influence of Wittgenstein and his pupil Waismann, who at the same time began to organize a fairly extensive teaching activity on the fringes of academia. That, however, had hardly anything to do with the :Vienna Circle", either in the sense of the title of the school or the group about which I am supposed to be telling you .... The attitude of the young people in the Circle was also not very favorable toward the "Ernst Mach Society", and I admit to you today I still have a certain sympathy for this restraint.

As far as Wittgenstein is concerned, first a factual statement: except for Schlick, Waismann, and - as one hears - with little success Carnap, hardly anyone from "the guild" ever laid eyes on him. The idea that he could appear in a congress or participate in a discussion was almost blasphemous. At the very first congress in 1929 in Prague the mathematician [Otto] Toeplitz[37] asked, not without an undertone of serious criticism about this peculiarity, whether there actually was a real person hidden behind the Wittgenstein Myth. But at any rate Wittgenstein is said to have sat in Schlick's apartment occasionally with his face turned to the wall in this most intimate of circles, reading aloud from Hamsun[38] . . . .

You will allow me here a clarifying remark. If you should already have gotten the impression that in these sketches too much space is given to my polemic against the Wittgensteinian school, that is correct and with only two restrictions. First, *historically* speaking, the major event in the Circle during the time of my participation in the meetings just had been in fact the debate involving Wittgenstein,[39] and I therefore ought to report especially about that. Second, *philosophically* speaking, so far as I know and understand the Wittgensteinian theses and so far as I, who have lived for some time at a distance from these matters, can really form a judgment about them, I believe that large parts of the Wittgensteinian perspective are either correct or at least point toward something of essential importance. In order for you to verify the extent to which my remarks are really of any importance to you, I will first try a brief formulation of what seems to me to be the fundamental thesis of the Wittgenstein philosophy, a thesis not altogether easy to grasp. It would go something like this:

Language, which mirrors the world and whose structure we must study if we want to become clear about philosophical-methodological questions (that is, more accurately, questions pertaining to the "logic of language"), is of a nature, that the complex of propositions representing the actually obtaining states-of-affairs is embedded in "logical space" somewhat like a mechanical model in its logical space, by the isolating and separating of what is correctly said about "reality" in contrast with "what it is possible to say", which does allow what we are able to comprehend of the structure of language and the nature of logical operations to "show itself".

If you are displeased by the little word "reality" in this formulation I will not hide from you something still worse. Where now I placed above the two words "structure" and "nature", I had in my rough draft the word "essence", and I only subsequently replaced it with the other two, which might sound somewhat less offensive to you.

What other direct information which I have acquired, I owe to long-standing, fairly intensive interaction with Waismann, which occurred in spite of a later estrangement that increased as Waismann came to accept his present views. I broke off completely with him only after the memorial speech he delivered to the University of Vienna Philosophical Society after the death of Schlick.

After Moritz Schlick's death in 1936, Edgar Zilsel, who had been only an infrequent guest in the Boltzmanngasse, regularly gathered a group together in his own apartment and he is said to have made an effort to continue the tradition of the classical period of the Circle . . . .

3. "MACH AND MY STUDENT DAYS IN VIENNA 1932-1938"[40]

by Josef Mayerhöfer[41]

In early 1932, therefore shortly before I passed my *Matura*, which let me into the University of Vienna, I read in the newspaper about a series of new lectures which would be given in the Physical Institute of the University before the Summer Sumester started and which carried the fascinating title "Crisis and Reconstruction in the Exact Sciences".[42] It is worth remarking that a not inconsiderable fee to attend was requested in order to help finance an elaborate gravestone-monument for the physicist, Ludwig Boltzmann. But in spite of the economic depression of the time the lecture hall was full.

This lecture series strengthened me to finally decide to study physics and mathematics at the University in Fall 1932. The lecturers in the series then became my teachers. I studied and did my doctoral dissertation with Hans Thirring .... He was *Ordinarius* for theoretical physics and had shared in the development of the theory of relativity, but was very reserved about Mach as a physicist .... Indeed, among physicists there was much dispute at the time about how to evaluate Mach. Felix Ehrenhaft,[43] who taught experimental physics and who knew about the contrast between Mach and most physicists, used to speak of "Mach and the physicists".But his own fate as experimental scientist was also linked to this formulation with its rather alienating aspect. He had barely missed receiving a Nobel prize about the electrical charge on elements. He would have received priority had his work been correct, but he took his own already submitted work back, because he believed that he had discovered "sub-electrons".[44] Millikan[45] then submitted his work and in 1923 received the Nobel prize. But Ehrenhaft persisted in his mistaken belief and thus originated the phrase "Ehrenhaft and the physicists".

I also remember Hans Hahn and the icy clarity with which he carried out his mathematical proofs and Karl Menger who gave brilliant and captivating lectures. He liked to improvise his proofs and thus did not draw back from making mistakes. He also brought the work of Kurt Gödel to our attention, who was already famous and was then still pursuing his studies among us.

The "Five [Cycles of] Vienna Lectures" occurred during the years which followed and under very similar titles: "Old Problems - New Solutions", "New Progress in the Exact Sciences", "New Paths to Exact Knowledge of Nature", and so forth.[46] Moritz Schlick, the philosopher, would be so to say the spiritual leader at many of these meetings, but mostly he remained in the background. There was also at this time a *Wiener Kreis* which included a lot of outstanding scientists, and on its perimeter there was an organization of a popular nature called the *Verein Ernst Mach*. This reminded me of my own teacher in *Gymnasium* who had imparted to us the theories of Ernst Mach, so I entered the *Verein* during my first semester in Fall of 1932 as member #197 as underwritten by Hans Hahn who was secretary. I immediately went to lectures by Heinrich Gomperz, Otto Neurath, Rudolf Carnap, and others. But unconnected with the Vienna Circle I was especially captured by the very stimulating lectures of Karl Bühler[47] who would also examine me for my *Philosophicum*. Ludwig von Bertalanffy[48] was also present at several of these lectures and would insert striking comments which would place some of the more extreme positivist opinions in question.

Unfortunately, this interesting constellation of philosophers of science who had been attracting us students with such enthusiasm soon came to an end. I remember that in February 1934 a kind of civil-war-like unrest began, and the consequence of this well-known outbreak was that four days later on February 16 the Social Democratic Party was outlawed and all of its several organizations and possessions were either disbanded or confiscated. About a month later on March 20th an order was given to Moritz Schlick, President of the *Verein Ernst Mach,* that the Society was dissolved. Three days later he wrote to the *Sicherheitskommissär des Bundes für Wien* and protested that the Society was not political[49] [even if most of the members but not Schlick himself were leftist-oriented] . . . .

But naturally his protest had no effect, but even so, he was permitted to continue his University lectures unhindered. I had a discussion seminar with Schlick, and it was he who gave me the topic which would be later used for my *Philosophicum (Rigorosum*].[47] The leading members of the Circle which under more circumscribed conditions still survived, had earlier in the year 1933 put out an interesting series of writings on unified science which were able to stay in print until 1935. But the number of major thinkers decreased inspite of occasional additions from Prague [Carnap and Frank]. Hans Hahn suddenly died on July 24, 1934 and Moritz Schlick himself was murdered in the University on June 22, 1936 . . . .

I received my doctor's degree in Fall 1937 with Thirring and became an officially approved assistant but with no salary. Thirring lost his position as professor in March 1938 when Adolf Hitler Marched into Austria. I was able to pass my teacher's exam in math and physics and have a year of training on the job [*Probejahr*] before I too was drawn into the military in 1940.

Meanwhile the Vienna Circle and the Ernst Mach Society were scattered like dust by the winds of time. But emigrants would take care that the name *Wiener Kreis* would soon become world famous. But in this paper I will not make any attempt at an overall judgment on the philosophy associated with the Vienna Circle. I shall merely attempt to communicate a few personal impressions in order to show how far the Circle had strayed from the original ideas of Ernst Mach.

Hans Hahn, for example, had tried to interpret historical assertions from the standpoint of unified science [*Einheitswissenschaft*].

> There is no difference in principle between historical and scientific truth. There is as little absolute historical truth as there is an absolute scientific truth. Testing is the criterion for both. A historical proposition has just as much to do with the facts as a scientific one, and is as much and as little hypothetical as what is scientific. And just as every natural law includes a term for an unconstituted character, namely, the grammatical form of the past, so does every historical proposition as well.[50]

Ernst Mach would scarcely have accepted such a view, since he almost always in his fundamental works took a historical approach, and he rejected Auguste Comte's Positivist, law-based idea that "To know is to be able to predict" (< *savoir pour prévoir* >). Hahn the mathematician manifestly had a weird way of understanding history, which was certainly not the case with Mach [since the one thing one *cannot* do in historiography is to consistently or infallibly predict the future].

Another example from the same series: Otto Neurath, the noted inventor of picture-statistics, held a lecture which has subsequently been translated into English (1989) as "Unified Science and Psychology" and which became published as Chapter One of the series *Unified Science*.[51] He attempted in that work to bring all of the different streams of psychology under one hat. But anyone who had actually heard the lectures of the psychologist Karl Bühler, who had the very largest University auditorium at his disposal at that time, would surely have rejected Neurath's imagined reconciliation of behaviorism, gestaltism, psychoanalysis, etc., as being unrealistic [in light of the radically different underlying philosophical assumptions].

When I think back on my student time 1932-1937, it certainly was Ernst Mach, but not the experimental scientist, who significantly influenced me as the harbinger of new methodological practices such as economy of thought, fitting thoughts to facts, and detecting similarities. It was Hans Thirring who interpreted and applied these principles for doing science, particularly in his four year cycle of courses on theoretical physics. He avoided judging all of our physical hypotheses in terms of possibility. and instead traced everything back to the functional dependency of mathematical quantities.

What made a particular impression upon me was Hans Thirring's interest in and understanding of the history of physics (and the history of culture in general) which would become more widely circulated through the books[52] that he would publish after the War . . . .

My thoughts went back to the lecture series: "Crisis and Reconstruction in the Exact Sciences" which we have mentioned before where money was raised for Ludwig Boltzmann's statue in the Vienna Central Cemetary, and indeed the title seems to have come from Hans Thirring. But the content of his lectures gave a deep insight into the crisis of science as a whole at that time, though perhaps it had not been his intention to probe so deeply. If there is already a crisis in physics, why at the same time should there also be one in biology? And even in Mathematics? In any case, I had the impression at once that it couldn't just be accidental. The crisis manifestly could not lie in the fields of science of the individual scientists, but in the people, the human beings who were living during that period of history.

At the end of my university studies, even though I had enjoyed lasting contact with many representatives of the Vienna Circle, I had come to the conclusion that scientific and cultural development did not run in straight lines of progress but in successive epochs which were partly opposed to each other and which were separated by crises.

After the War, I did not return to the Institute for Theoretical Physics but in 1947 turned to the Austrian National Library. I completed the prescribed library training, learned paleography, bibliography, medieval and modern book theory, etc. and I was also entrusted to work in the archives. It was this schooling which helped me pursue further the ideas I had already adopted before the War. Above all, I learned that in contrast to Austria a new branch of science was now flourishing abroad, namely, the history of science, which possessed a rich international literature, scientific societies and congresses, and professorships in West Germany and especially in the U.S.A.; and in particular George Sarton (a Belgian) had founded an important journal, *Isis*, which provided an introduction into the history of the sciences, and which moreover included several thousands of pages which included unbelievably exact data on bibliographical sources. This new direction of serious research corresponded exactly to what I was looking for . . . .

That the concept of historical epochs was not foreign to Mach himself is clear from the title of the lecture series he announced for Summer 1898 in the University of Vienna: "Main Epochs in the Development of Science treated in an epistemological-psychological manner"

Nor should we forget his emphasis on an historical method of presenting his ideas on science and philosophy as were represented by: *Die Mechanik in ihrer Entwicklung; historisch-kritisch dargestellt* (1883), *Die Principien der Wärmelehre. Historisch-kritisch entwickelt* (1896); and *Die Prinzipien der physikalischen Optik Historisch und erkenenntnis-psychologisch entwickelt* (1921) .... A new way of thinking developed in the 18th century, "historical thinking", and with this a reluctance to judge or criticize the past, but would try to clarify events in terms of the presuppositions current at the time of the events themselves. This means that Mach was not as extreme a Positivist as Comte [with his three stages of human progress and with the final, law-based "scientific" stage judged as superior to the earlier stages]. But Mach also fell short of finding a genuine historical way of thinking. For example, he did not show any appreciation at all for Pierre Duhem's distinction between broad and deep thinking minds and was only inclined to acknowledge the abstract theories of deep thinkers.

There was also another aspect of modern historical thinking which Mach did not value. In the foreword to his *Principien der Wärmelehre* he writes:

> Although a great many sources have been used, one should not look for the results of archive investigation; the question here is much more about the connection and growth of thoughts than about interesting curiosities.

This is naturally a very hard judgment about archive research. I must let the reader decide whether the results of my own work in the archives . . . are only "interesting curiosities" . . . .

An official of the University Library, Viktor Kraft, lectured after the War on the Vienna Circle, and after his retirement from the Library even became professor of philosophy,[53] but his efforts to breathe new life into the Vienna Circle proved to be vain. But I must emphasize again that the *Wiener Kreis* had already deviated far from Ernst Mach's views.

NOTES

1 Friedrich von Hayek (1899-1992) was an Austro-English economist, philosopher, and Nobel prizewinner. He is often considered a late representative of the Austrian School of Economics largely begun by Carl Menger. This is a translation of Hayek's "Diskussionsbemerkungen über Ernst Mach und das sozialwissenschaftliche Denken in Wien", *Symposium aus Anlass des 50. Todestages von Ernst Mach*, Ernst Mach Institut: Freiburg im Breisgau, 1966, pp. 41-44. The footnotes in this chapter have been added by the editors.

2 In addition to Mach's influence one should also recall that the physicist-philosopher Ludwig Boltzmann (1844-1906) had many students and his books were also widely read. Furthermore, the very popular University of Vienna Philosophical Society was also very much science-oriented under its longtime president, Alois Höfler, who died in 1922. See John Blackmore, "Late Appendix", *Ludwig Boltzmann - His Later Life and Philosophy 1900-1906, Book Two*, Kluwer: Dordrecht, 1995.

3 Heinrich Gomperz (1873-1942), son of the classical scholar Theodor Gomperz, was Professor of Philosophy in Vienna from 1924 to 1934 and later in Southern California. He was a longtime friend of Ernst Mach.

4 Adolf Stöhr (1855-1921) taught philosophy at the University of Vienna from 1901 to 1921, was much influenced by Mach, but as a loner never seems to have had important followers of his own. See William M. Johnston, *The Austrian Mind*, University of California Press: Berkeley & Los Angeles, 1972, pp. 199-201.

5 Robert Reininger (1869-1955) is known as a Neo-Kantian philosopher with strong ethical tendency toward Stoicism. He was Professor of Philosophy in Vienna from 1913 until the 2nd World War and quasi-officially or unofficially kept teaching for many years after that. He succeeded Alois Höfler (See footnote #2) as leader of the University of Vienna Philosophical Society in 1922.

6 Joseph Schumpeter (1883-1950) was an Austro-American economist who taught at various Austrian and German universities until he immigrated to America where he taught at Harvard from 1932 until 1950. While clearly preferring capitalism he advocated a mixed economy. Like Mach he was a strong pacifist during much of his life.

7 Friedrich von Wieser (1851-1926) helped systematize Menger's views on economics and thus became a very central figure in the Austrian School of Economics. He was a marginalist in economics but like Schumpeter tended to advocate a mixed economy.

8 See Henk Visser's article in this collection on Mach and Wittgenstein. The difficulty in plausibly understanding and hence in translating the phrase *sich...mit...auseinandersetzte* could conceal important factual information which Friedrich von Hayek may have known about Wittgenstein and his familiarity with Mach and his ideas.

9 Friedrich Adler (1879-1960) was the son of Viktor Adler, leader of the Austrian Social Democratic Party. The son was indeed a key figure in Austrian politics, especially after the death of his father in 1918. It is often alleged that Friedrich Adler was instrumental in 1919 in preventing Austria from going Communist as neighboring Hungary and Bavaria had briefly done. For his defence of Mach see his book: *Ernst Machs Überwindung des mechanischen Materialismus*, Vienna, 1918.

10 Othmar Spann (1878-1950) was a Catholic social theorist who employed holistic ways of thinking which reminded some critics of Hegel. It is not clear what von Hayek meant by "metaphysics" in Spann, but either his Catholicism or holism might deserve some uses of that unfortunate expression. Johnston defends Spann against several of his critics. See Johnston, op. cit., pp. 311-315 (See footnote #4 above.)

11 *The Sensory Order*, London & Chicago, 1952.

12 Ernst Mach, *Die Analyse der Empfindungen*, (chapter #1, section 13), Darmstadt, 9th edition, 1987, p. 24, footnote #1.

13 See M. Katsumori's article in a subsequent volume on Mach. He has added to the work of M. Santone (See Blackmore 1992) in clarifying the philosophy of Wataru Hiromatsu which also reduces Mach's theory of elements to a theory of relations.

14 Richard Avenarius (1843-1896) taught in Zürich from 1877 until his death where his philosophy which except for difficult terminology was very similar to that of Ernst Mach influenced many Russian students which in turn raised the ire of Plechanov and Lenin. His main work *Kritik der reinen Erfahrung*, two volumes, was first published in 1888 and republished in 1908, that is, on the eve of Lenin's sharp reply in 1909.

15 Engelbert Dollfuss (1892-1934) was Austrian Prime Minister from 1932 to 1934, an extremely difficult period. Militant Social Democrats on the left and armed supporters of Hitler on the right forced him to seek the aid of Mussolini, whose price of support was to abolish all Austrian political parties and form a corporate, semi-fascist state under a so-called Fatherland Front. The Nazi's retaliated by assassinating him.

16 See Gustav Bergmann, "Memories of the Vienna Circle", in *Scientific Philosophy: Origins and Developments*. The article was translated by W. Heald & G. Zoeller; the editor of the book is Friedrich Stadler, Kluwer: Dordrecht, 1993, pp. 194-208. Neurath requested that Bergmann write a report on what he had remembered about the Vienna Circle. Bergmann did this on a boat while fleeing across the Atlantic to America six months after Hitler had marched into Austria. On the

Statendam he met the Austrian novelist Hermann Broch who gives the date of their arrival in New York as October 9, 1938. See Paul M. Lützeler, *Hermann Broch*, Quartet Books: London, 1987, p. 174. Bergmann's account has been somewhat shortened.

17 Gustav Bergmann (1905-) who from 1938 would teach at the University of Iowa seems to have passed through several stages in his philosophy from phenomenalism, to phenomenology, to a kind of "realism" which would incorporate the ideal languages of linguistic philosophy with a revived concern with epistemology and ontology as if a more or less reasonable synthesis were possible. See *The Metaphysics of Logical Positivism* (1954) and *Meaning and Existence* (1960).

18 Friedrich Waismann (1896-1959) was assistant to Moritz Schlick in Vienna and later advocated his own interpretation of Wittgenstein's views. For his later philosophy see: *Principles of Linguistic Philosophy* (1965) and *How I see philosophy* (1968). Both books were edited and prepared for posthumous publication by Rom Harré.

19 Beginning shortly after Moritz Schlick's arrival in Vienna in 1922 as a professor of philosophy of science he held small conversation sessions in his home with Hans Hahn and other mathematicians and physicists. Herbert Feigl (1902-1988) has said that he and Waismann requested in 1924 that Schlick form a discussion group and that he consented. (Meetings before then were subsequently called "pre-historic".) Rudolf Carnap arrived temporarily in 1925 and on a more permanent basis the next year. As the Circle grew it moved to what is now the mathematics library on the first floor in the physics building on Boltzmanngasse 5. Otto Neurath who had long been a major contributor to the University of Vienna Philosophical Society seems to have switched to Schlick's Circle some months after the discussion sessions in the Society were stopped. These four men were the main leaders, though many young mathematicians and physicists increasingly attended. The Vienna Circle declined amidst the political turmoil in Austria in the 1930's.

20 Edgar Zilsel (1891-1944) was a left-wing Austrian socialist with interests in sociology, the Vienna Circle, and his own circle in Vienna. His philosophical views have recently attracted attention. See Edgar Zilsel, *Wissenschaft und Weltanschauung*, ed. G. Mozetic, Wien, 1992.

21 Hans Hahn (1879-1934) was the mathematician who invited Schlick to Vienna in 1922. He was an active figure in the Vienna Circle, but his cold personality and unsophisticated positivism rarely attracted praise. His wife Lili (or Lily), however, was the life of every party and became famous for jokes and humor about her old teacher, Ludwig Boltzmann.
Hans Hahn's sister married Otto Neurath.

22 Karl Menger (1902-1988) emigrated to America in 1937, and while never as famous as his father Carl Menger (1840-1921) the economist, he would still become a prominent mathematician and positivist philosopher in his own right, first at Notre Dame (1937-1946) and then at Illinois Institute of Technology.

23 The First Republic of Austria technically endured until 1938, but Engelbert Dollfuss became strongman in 1932, parliament was suspended in 1933 and the Social Democratic Party outlawed in 1934, such that personal and political rights were being increasingly whittled down through most of the 1930's until Hitler completed the job.

24 Hans Kelson (1881-1973) drafted the Constitution for the First Austrian Republic and then developed a "pure theory of law" free from value judgment but very dependent on positivist notions, especialy an identification of value with feeling. See his book: *The Doctrine of Pure Law* (1934).

25 This is an English translation of the *Verein Ernst Mach*. For history of it. See Friedrich Stadler, "The 'Verein Ernst Mach' What Was It really?" in *Ernst Mach - A Deeper Look*, ed. J. Blackmore, Kluwer: Dordrecht, 1992, pp. 363-377.

26 Moritz Schlick (1882-1936) had been an assistant of Max Planck and a direct realist in epistemology before coming to Vienna in 1922. It appears Wittgenstein's *Tractatus* and discussion in his Circle persuaded him to adopt a linguistic approach as if epistemological and ontological differences were "merely linguistic". Like most Austrian and German thinkers he regarded indirect and representational realism as "metaphysical" and not as epistemological positions at all, a lapse

or omission which made linguistic reduction of philosophy seem plausible. His wife was American, and he often taught summer school in the U.S.A. He was murdered by a disgruntled student in 1936.

27 Richard von Mises (1888-1953) became respected for his contributions to probability theory which he worked out with Hans Reichenbach (both men taught in Istanbul during part of the 30's) and which was partly adopted by Sir Karl Popper. He also wrote a *Heft* for the *Verein Ernst Mach* titled *Ernst Mach und die empirische Wissenschaftsauffassung* (1938) and the following year began to teach at Harvard.

28 But Karl Menger who attended sessions of the Vienna Circle starting in Fall 1927 has described Schlick's role in somewhat more exact and positive terms: "We would stand in little groups just talking until Schlick would clap his hands and then we would be seated. Sometimes Schlick would begin by reading to us a letter that he had received dealing with problems that we had discussed or had planned to consider. His correspondents included Einstein and Russell. He would open other sessions by reading announcements of new publications (especially British ones) and would promise to report about some or ask volunteers to review them. Occasionally Schlick would introduce a guest passing through Vienna. Then there would begin a discussion on the topic which had been proposed in the preceding meeting or in someone's report about work in progress." (See Karl Menger, "Memories of Moritz Schlick" in *Rationality and Science*, ed. E. Gadol, Springer Verlag:
New York, 1982, p. 86.)

29 Luitzen E.J. Brouwer (1881-1966) was a Dutch mathematician who is known as the founder of the intuitionist school of modern mathematics.

30 David Hilbert (1862-1943) was a German mathematician who taught in Göttingen. He was a formalist who emphasized the importance of axioms.

31 Viktor Kraft (1880-1975) had his own circle, was critical of the Schlick Circle's bent toward phenomenalism, and was conceivably aware that the "linguistic move" to employ a "physical object language" did not really abolish underlying presentist epistemology. He eventually became the Vienna Circle's first major historian. He had a lot to be silent about during the reign of Wittgenstein and Carnap.

32 Felix Kaufmann (1895-1949) a lecturer in jurisprudence developed phenomenological views influenced by Edmund Husserl, but in his book *Methodology of the Social Sciences* (1944) his approach was close to logical positivism.

33 Heinrich Gomperz (1873-1942) like his father Theodor Gomperz was a personal friend of Mach and like both men also became a Professor of Philosophy. His realization that both phenomenalism and direct realism were presentist in epistemology helped lay the groundwork for the Circle's false allegation that all epistemological differences were merely linguistic. Gomperz was a leading figure in Viktor Kraft's Circle in Vienna, he is often alleged to have had his own discussion group, and often appeared to be somewhat sceptically-minded about philosophy itself.

34 Hans Thirring (1888-1976) had carried out experiments on what Albert Einstein called "Mach's Principle", but the results have been interpreted in different ways. He became a strong moralist and opposed Nietzsche, Spengler, and of course Adolf Hitler. After the war he also attended pacifist conferences.

35 During the early and middle 1930's there was considerable interest in Vienna in using scientific tests to help determine if psychic phenomena were legitimate. Karl Przybram, a physicist, was a leader in examining PSI. Some tests suggested fraud and others seemed inconclusive.

36 Kurt Gödel (1906-1978) was a famous logician who demonstrated that "the axiomatic method of mathematics cannot yield information concerning both the completeness and the consistency of the system axiomatized". His approach encouraged philosophers to accept the distinction between first-order and meta-languages, even though the distinction distorts intended meaning and communication not just by objectifying what is actually subjective as all formal logic does, but also by trying to detach meaning from reference and assumptions which often has the effect of

trivializing the importance of both of them, and especially the conditional factors within the assumptions. These steps are commonly followed to give the appearance of objectivity and certainty to "first-order" formal logic and mathematics. But while historians rightly distrust idealization, most mathematicians and logicians seem to think that they would be utterly lost without it, indeed, as if it were a defining characteristic of science itself! The Vienna Circle tended to encourage this sort of illusion. Observational, experimental, historical, and applied science can and normally should try to avoid most or all idealization.

37 Otto Toeplitz (1881-1940) joked about "The Wittgenstein Myth", but in the absence of readily accessible biographical information, he may have been virtually a myth himself.

38 Knut Hamsun (1859-1952) was a rather grim Norwegian novelist who emphasized the harsh psychological condition of people under pressure from hunger or work. His simple style was influenced by Nietzsche and Strindberg. He won a Nobel Prize for literature in 1920.

39 Gustav Bergmann seems to have been unaware that Hahn, Schlick, and later Carnap, under the prodding of Kurt Reidemeister had already studied Wittgenstein's *Tractatus* in 1924 or 1925 and once again with Carnap included in 1926, which possibly influenced the latter's vain attempt to outlaw epistemology and ontology by reducing them to matters of language. As for Schlick, he seems to have been so deeply affected by "Ludi" that he apparently regarded what Bergmann considered to be the high point of the Circle, which corresponded closely with Carnap's partial domination from 1926 to 1931, to be merely a brief interim before Wittgenstein's intellectual dominance could be resumed. See Herbert Feigl, "The Wiener Kreis in America", *The Intellectual Migration*, ed.D. Fleming & B Bailyn, Harvard University press: Cambridge, 1969, pp. 630-673.

40 Our short and partly rearranged selection is from Josef Mayerhöfer, "Ernst Mach in Wien" , *Mitteilungen der österreichischen Gesellschaft für Geschichte der Naturwissenschaften*, **9** (1989) 19-42.

41 Josef Mayerhöfer was born in 1914 and entered grade school in Vienna in 1920. He describes much of his life in the course of this article so we will not repeat it here. He has also published other articles on Mach including: "Ernst Machs Berufung an die Wiener Universität 1895", *Clio Medica*, **2** (1967) 47-55.

42 The series consisted of five lectures.

43 Felix Ehrenhaft (1879-1952), who Boltzmann so often confused with Paul Ehrenfest, was a Viennese physicist and brilliant lecturer. His alleged discovery of "sub-electrons" and tenacious defence of this view hurt his reputation. See Gerald Holton, "Subelectrons, Presuppositions, and the Millikan-Ehrenhaft Dispute", in *The Scientific Imagination - Case Studies*, Cambridge University Press: Cambridge, 1978, pp. 25-84.

44 Ernst Mach's only daughter, Carolina Lederer (1873-1965), has claimed in an unpublished manuscript that Mach and Ehrenhaft were friends, and a letter from Anton Lampa to Mach dated May 1, 1910 mentions Ehrenhaft's work and praises it, but there is no extant correspondence between Mach and Ehrenhaft and no mention of him in the indices of Mach's books. In short, the common suspicion that Mach somehow influenced Ehrenhaft to deny the existence of electrons as particles by discovering fractional charges without end lacks significant supporting evidence at this stage.

45 Robert Andrews Millikan (1868-1953) was long a prominent American physicist. His autobiography mentions several American supporters of Mach's ideas and discusses the controversy over the reality of atoms which occured at the St. Louis World's Fair in 1904.

46 Dr. Mayerhöfer has listed the lecture cycles & their dates of publication. Fünf Wiener Vorträge: (Erster Zyklus) Krise und Neuaufbau in den exakten Wissenschaften (1933); Zweiter Zyklus: Alte Probleme - neue Lösungen in den exakten Wissenschaften (1934); Dritter Zyklus: Neue Fortschritte in den exakten Wissenschaften (1936); Vierter Zyklus: Neue Wege exakter Naturerkenntnis (1939).

47 Karl Bühler (1879-1963) made a major contribution to the theory of imageless thought at Würzburg in 1907, a success in experimental psychology which helped refute George Berkeley's claim that "only an idea can be like another idea" i.e. only a sensation or image can be like another sensation or image, a view which had been used by phenomenalists, idealists, and many positivists

to restrict the scope of reference in science. Bühler's wife was also a prominent psychologist. Bühler taught in Vienna from 1922 to 1938 when he emigrated first to Norway and then to America. (A *Philosophicum* is normally a brief oral exam in philosophy taken after about two years, but if it is equated with a *Rigorosum* as above, then perhaps at Vienna it was an oral examination for the Doctor's degree.)

48 Ludwig von Bertalanffy (1901-1972) was an idealist philosopher who taught in Vienna 1934-1948 and after that in Ottawa, Canada. Like Ludwig Boltzmann he imagined that Kant's a priori categories of the mind might be subject to evolution, thus becoming one of several forerunners of a recent movement called: "Evolutionary Epistemology".

49 See Stadler, op. cit., pp. 375-377 & Mayerhöfer (1989), pp. 34-35.

50 Hans Hahn, "Logik, Mathematik, und Naturkennen" (1933) as *Heft #2* in the series *Einheitswissenschaft*. It can be found in English translation as pages 24-45 in *Unified Science*, edited by Brian McGuinness and translated by Hans Kaal, (in the Vienna Circle Collection, Vol. **19**), Kluwer: Dordrecht & Boston, 1989.

51 Ibid. pp. 1-23. The original German-language article was "Einheitswissenschaft und Psychologie" and was published as Heft #1 in 1933 as part of the *Einheitswissenschaft* series.

52 See Hans Thrring *Der Weltfriede als Psychologisches Problem* (1946), *Anti-Nietsche, Anti-Spengler* (1947), and *Atomkrieg und Weltpolitik* (1948).

53 For some late remarks by Viktor Kraft on Ernst Mach see "Nachwort", in K.D. Heller, *Ernst Mach*, Springer Verlag: Wien - New York, 1964, pp. 168-169.

REFERENCES

Adler, Friedrich. *Ernst Machs Überwindung des mechanischen Materialismus*. Wien: 1918.

Avenarius, Richard. *Kritik der reinen Erfahrung*. 2 vols. Leipzig: 1888-1890.

Bergmann, Gustav. *The Metaphysics of Logical Positivism*. New York: Longmans, 1954.

Bergmann, Gustav. Meaning and Existence. 1960.

Bergmann, Gustav. "Memories of the Vienna Circle". in *Scientific Philosophy: Origins and Development*. edited by F. Stadler. Dordrecht: Kluwer, 1993, pp. 194-208.

Blackmore, John. (ed.). *Ernst Mach - A Deeper Look - Documents and New Perspectives*. Dordrecht: Kluwer, 1992.

Blackmore, John. "Late Appendix". *Ludwig Boltzmann - His Later Life and Philosophy 1900-1906. Book Two: The Philosopher*. Dordrecht: Kluwer, 1995, pp. 269-315.

Hahn, Hans. "Logik, Mathematik, und Naturkennen", in *Unified Science*, edited by B. McGuinness. Translated into English by Hans Kaal. Dordrecht: Kluwer, 1989, pp. 24-45.

Hayek, Friedrich von. *The Sensory Order*. London & Chicago: 1952.

Hayek, Friedrich von. "Diskussionsbemerkungen über Ernst Mach und das sozialwissenschaftliche Denken in Wien". *Symposium aus Anlass des 50. Todestages von Ernst Mach*. Freiburg in Breisgau: Ernst Mach Institut Selbstverlag, 1966, pp. 41-44.

Holton, Gerald. "Sub-electrons, Presuppositions, and the Millikan-Ehrenhaft Dispute". in *The Scientific Imagination - Case Studies*. New York: Cambridge University Press, 1978, pp. 25-84.

Johnston, William M. *The Austrian Mind*. Berkeley & Los Angeles: University of California Press, 1972.

Kaufmann, Felix. *Methodology of the Social Sciences*. 1944.

Kelson, Hans. *The Doctrine of Pure Law*. 1934.

Kraft, Viktor. "Nachwort". in K.D. Heller, *Ernst Mach*. Wien: Sprenger Verlag, 1964, 168-169.

Mach, Ernst. *Die Analyse der Empfindungen*. 9th edition. Darmstadt: Wissenschaftliche Buch-Gesellschaft, 1987 [1886].

Mayerhöfer, Josef. "Ernst Machs Berufung an die Wiener Universität 1895". *Clio Medica*. 2 (1967) 47-55.

Mayerhöfer, Josef. "Ernst Mach in Wien". *Mitteilungen der österreichischen Gesellschaft für*

Geschichte der Naturwissenschaften 9 (1989) 19-42.
Menger, Karl. "Memories of Moritz Schlick". *Rationality and Science,* edited by E. Gadol. New York: Springer Verlag, 1982.
Mises, Richard von. *Ernst Mach und die empirisiche Wissensschaftauffassung*. included in *Unified Science,* edited by B. McGuinness. Dordrecht: Kluwer, 1989 [1938], pp. 130-165.
Stadler, Friedrich. "The 'Verein Ernst Mach'. What Was It Really?". in *Ernst Mach - A Deeper Look.* edited by J. Blackmore. Dordrecht: Kluwer, 1992, pp. 363-377.
Thirring, Hans. *Der Weltfriede als psychologisches Problem.* Wien: Deuticke, 1946.
Thirring, Hans. *Anti-Nietzsche, Anti-Spengler*. Wien: A. Sexl, 1947.
Thirring, Hans. *Atomkriedg und Weltpolitik.* Wien: Danubier-Verlag, 1948.
Waismann, Friedrich. *Priciples of Linguistic Philosophy.* edited by Rom Harré. 1965.
Waismann, Friedrich. *How I see Philosophy.* edited by Rom Harré. 1968.
Zilsel, Edgar. *Wissenschaft und Weltanschauung*. edited by G. Mozetic. Wien: 1992.

Chapter Seven

WITTGENSTEIN'S MACHIST SOURCES

by Henk Visser

Section Guide

1. INTRODUCTION

Most readers find Wittgenstein and his writings very enigmatic. He is brief, takes much for granted, and it is normally difficult to relate his ideas to anything familiar.[1] Indeed, his first book, which is often simply called the *Tractatus* is so abbreviated as to become an almost inexhaustible source of diverse interpretation. Indeed, anyone who is bold enough to allege that he "knows" what Wittgenstein means has almost surely jumped to conclusions on insufficient evidence or is presuming a lot. His posthumously published *Philosophical Investigations* [2] may seem to represent an improvement in that at least several parts of it seem virtually comprehensible, but even here there are heated debates about individual passages. One of the many reasons why readers often feel so frustrated is Ludwig Wittgenstein's comparative lack of references to thinkers with whom he agreed or opposed.[3] Names can be vital signposts to help us see the points he wanted to make, or even the problems he wanted to solve, but they are generally absent.

There are of course a few exceptions. In the *Tractatus* there are a handful of names: Frege, Hertz, Mauthner, Russell, and Whitehead,[4] but dearth about *what* he learned from them. He does explicitly mention "the great works of Frege and the writings of my friend Bertrand Russell",[5] but again nothing about *why* he thinks they are great, *what* in them deserves to be considered great, and *what* in particular influenced him, greatly, or even at all. Maybe he didn't believe in influence, a concept closely connected with causes.

But once again the *Philosophical Investigations* seem a little less Spartan in style. More authors are touched on, including Plato, Augustine, Luther, Faraday, Carroll, James, Frege, Russell & Lewis.[6] And in at least two cases he even appears to allow the possibility of some influence, acknowledging that Frank Ramsey's criticism and that of P. Sraffa had "helped" him in Cambridge. But in truth, since there is an absence of detail on *how* they "helped" him or *what* it was they helped him with, one still remains in the dark. Advertently or inadvertently, absence of stated sources for one's ideas can give the impression of great originality, which many readers, especially non-historians, place above both clarity and soundness. This is not to deny that Wittgenstein was original, but that more information about sources and what was learned from them can help both communication and judging more accurately the real nature and extent of his originality.

On the other hand, maybe he had forgotten many key sources, especially a lot he had

J. Blackmore, R. Itagaki, and S. Tanaka (eds.). Ernst Mach's Vienna 1895-1930. 139-158.

read or heard when he was a child or *Gymnasium* pupil in Austria. Or to be more exact, perhaps in concentrating on problems it rarely occurred to him to think about likely sources or parallels, especially early ones. This reticence has naturally stimulated research into his childhood, and there are now many Austrian candidates who he casually mentioned or who are said to have influenced him. Boltzmann, Kraus, Loos, Wahle, and Weininger are among the leading Viennese possibilities. From the larger German-speaking area one also hears names like Frege, Hertz, Schopenhauer, and Spengler, but the ideas of Spengler like Moore, Russell, Ramsey, and Sraffa if they struck Wittgenstein as true or important did so afterwards. The name of Fritz Mauthner must also be added as a candidate.

Given his general sparsity of references to sources and especially to what if anything he learned from them, it might seem that the least likely origin for many of his ideas would be precisely the writings of someone so obscure that it seems Wittgenstein almost never commented on him, and only then to express sharp criticism: "Mach writes in such a horrid style that it makes me nearly sick to read him; however, I am glad that you [Bertrand Russell] think so much of a countryman of mine."[7]

But that Wittgenstein did in fact read Mach extensively is affirmed by his contemporary and fellow Austrian, the Nobel prize winner in economics, Friedrich von Hayek: "Others with whose ideas he [Wittgenstein] was very familiar with, though he disliked their manner, were Ernst Mach and the psychologist Otto Weininger"[8] and "And I also know that his [Friedrich von Wieser's] contemporary, the philosopher Ludwig Wittgenstein, a distant cousin of mine, had already grappled with Mach at the same time."[9]

Many other writers have also taken for granted or speculated that inspite of Wittgenstein's own announced aversion to his style that Ernst Mach was in fact the source for many of Wittgenstein's ideas. These writers include, Philip P. Wiener,[10] Stephen F. Mason,[11] Nicola Abbagnano,[12] one of the editors of this volume, John Blackmore,[13] and recently John W. Cook.[14] But in my article I will avoid most speculation and taking things for granted and specify particular passages which indicate Mach's influence on Wittgenstein. I will also argue that Mach's contribution was neither "philosophical" nor scientific in the sense that Mach would have presented him with substantial claims on the structure of logic, language, or the world. Mach was important to Ludwig Wittgenstein for giving him, first, just the results of an analysis of scientific statements he could use, and second, a method for the treatment of many philosophical troubles in the form of thought experiments of a kind. This means that Mach's influence was different from that of the "names" we have already mentioned as well as the additional Viennese we will discuss in the next paragraph. Thus, both views, first, that Ludwig Wittgenstein was not influenced by Mach at all, and second that Wittgenstein in the *Tractatus* had adopted Mach's "neutral monism" or "phenomenalism", are repudiated. What is kept, is that Wittgenstein read Mach in different periods of his life, and may have learned more from him than he has admitted.

But if Mach is not the answer to most of the deepest "origin questions", then who is? First, we don't know. Maybe he really was original. But second, there are four candidates with promising credentials for the early period: Boltzmann, Kraus, Wahle, and Mauthner. Wittgnstein's anti-philosophy and effort to reduce philosophical questions to linguistic ones can be discovered almost completely intact in the very late writings of Ludwig Boltzmann (1844-1906) from 1903 to 1905, especially in his lecture on Schopenhauer in his *Populäre Schriften* (1905).[15] Richard Wahle (1857-1935) a contemporary in Vienna held equally negative views and may even have first expressed them before Boltzmann did. Both men also had intellectual debts to Mach.[16]

Austria was flooded with linguistic philosophers at the time who very likely had some impact on Ludwig Wittgenstein concerning the importance of words and expression. One thinks of Fritz Mauthner (1849-1923)[17] and of Anton Marty (1847-1914) who both corresponded with and were influenced by Mach, and especially Karl Kraus (1874-1936)[18] who was the most famous and caustic of all the Viennese critics, and who treated language and "correct expression" as virtually sacred.

But our main concern is Ernst Mach and his influence on Wittgenstein. Most readers already suspect that Wittgenstein's heavy reliance on "thought experiments" plus his *Tractatus* analogy of "throwing away" one's ladder after employing it *may* have come from Mach[19] or at least were suggested by him, and his ideas were widely circulated in Vienna, but we are looking for more certain influence from Mach, not indirectly through other thinkers and the *Zeitgeist* - and there was probably a lot of that given the range and bulk of Mach's impact in Vienna during Wittgenstein's childhood - but directly from Ernst Mach's own books *as can be pinned down chapter and page*. In practice, this means finding passages original to Mach but later found with unmistakable similarity in the writings of Wittgenstein. This is our primary but by no means our exclusive focus.

2. EARLY PRIORITY OF LOGIC AND FORM

But fairness to Ludwig Wittgenstein demands that we explicate his work at least to some extent before we trace influence and pontificate about it. Let us start with the *Tractatus*, the only major philosophical work by him which was published during his lifetime. The *Tractatus* has been widely seen as a sort of claim that Wittgenstein had solved "the problem of philosophy" by eliminating philosophy. But the radical nature of this claim is partly hidden by an extremely abbreviated and elliptical way of writing, which at its very clearest can still seem cryptic to many readers.

Let us begin with an effort to capture Wittgenstein's apparent perspective as he saw it. In order to do this well, we will need to augment the *Tractatus* with information from his notebooks plus correspondence, since evidence from them will be necessary in order to navigate through many parts of the *Tractatus* itself without foundering.

In so far as Ludwig Wittgenstein seems to hold that philosophy has been replaced by properly understood logic and mathematics, then he would appear to be taking a page out of Ludwig Boltzmann's anti-philosophy, in particular, from his second and third lectures in nature philosophy given in Vienna in late October 1903, but excluded from his *Populäre Schriften* of 1905.[20] Shifting to England, it is rather clear from Wittgenstein's letters to Russell written in 1913 that the problems which Wittgenstein wanted to solve were derived from difficulties in Volume I of the *Principia Mathematica*, notably those connected with the three non-evident axioms which Whitehead and Russell had added to their logical theory in order to give a logical reconstruction of mathematics: the axiom of infinity, the reducibility axiom, and the axiom of choice or the multiplicative axiom. As soon as one in line with Wittgenstein refuses to regard these axioms as logical axioms, and considers the axiom of infinity as a statement of physics, then the question arises how to distinguish logical propositions from propositions of physics, or, more generally from propositions about "the world". For it is clear, according to Wittgenstein - after having presented a thought experiment in which it is imagined that "we live in a world in which nothing more exists than $\aleph_0$ things" - "that whether or not the world in which we live is really such or not, is not a matter for logic to decide".[21]

Wittgenstein's answer to this question, which he regarded as "definitive", was that "all the propositions of logic are generalizations of tautologies and all generalizations of tautologies are propositions of logic". However, at that time he could "not yet say quite clearly what tautologies really are", but so much seemed certain that "it is the peculiar (and *most* important) mark of *non*-logical propositions that one is *not* able to recognize their truth from the propositional sign alone". "But the propositions of logic - and only they - have the property that their truth or falsity is already expressed in their symbolizations".[22] (In his letter of December 15, 1913 Ludwig Wittgenstein called the explanation of the nature of tautology "the fundamental question of all logic".[23]) Eventually, he chose to repeat the insight in the *Tractatus*: "The propositions of logic are tautologies" (6.1). "It is the characteristic mark of logical propositions that one can recognize from the symbol alone, that they are true; and this fact contains in itself the whole philosophy of logic. And so also it is one of the most important facts that the truth or falsehood of non-logical propositions can *not* be recognized from the propositions alone." (6.113)

Another way of setting logic apart from say, physics, is given in the first remark of the surviving notebooks: "Logic must take care of itself".[24] This would return again in the *Tractatus* (5.473). But this remark led almost immediately to a critical note about "the task of philosophy", as we shall see with far-reaching consequences: "How is it reconcilable with the task of philosophy that logic should take care of itself?" The question is that logical symbolisms are used in order to describe (part of) the world we live in. So despite the fact that logic must take care of itself, it has some connection with the world. But what is this connection?

Already in the notes dictated to G.E. Moore in Norway, Wittgenstein attributed logical properties to the world: "Logical so-called propositions *shew* logical properties of language and therefore of the universe, but *say* nothing!"[25] He then also reflected on "the relation of propositions to reality". Given an ordinary proposition such as "Moore is good", he distinguished between spatial properties, for example, that the word "Moore" is to the left of the word "Good". and *logical* properties, for example, that this proposition has the subject-predicate form.[26] His insight was that both kinds of properties are *shewn* by the proposition, but that only the former can be said by another proposition, as we already did. As to the relation of the proposition "Moore is good" to reality, he was prepared to say that the spatial property already postulates a sort of correspondence between the proposition and reality, but since the spatial property is only conventional or arbitrary, it is the logical property of the proposition that makes the proposition a "logical picture" of a situation. The last formulation cannot yet be found in the notes to Moore, but arises in the course of 1914 in the notebooks. Then he also took a decisive step: a situation has a logical form too, which cannot be described either. (It becomes later a main subject of the *Tractatus*.) Hereafter, Wittgenstein sees himself confronted with a philosophical problem that arises from the above sketched view of "the logical structure of the world", but it is very similar to many traditional philosophical questions about "the apriori" though without epistemological connotations. "The great problem round which everything I write turns is: Is there, a priori, an order in the world, and if so, what does it consist in?"[27]

The first problem which Wittgenstein hoped to solve with his "doctrine of the logical form" was the question of the a priori character of logic. His solution, formulated in the first "note" dictated to G.E. Moore had set the standard for all of the other problems concerning the a priori: "Logical so-called propositions *shew* logical properties of language and therefore of the universe, but *say* nothing."[28]

That is to say: the problem is reformulated as the question of the right explanation of *propositions* (*Sätze*), the approach takes account of both language and "the world", and eventually the solution is primarily organized with reference to *forms*. The following quotations from Wittgenstein's notes for Moore may serve as an illustration:

> Logical propositions *are forms of proof:* they *shew* that one or more propositions follow from one (or more).[29]

> From the fact that I see one spot is to the left of another, or that one colour is darker than another, it seems to follow that it *is* so; and if so, this can only be if there is an *internal* relation between the two; and we might express this by saying that the *form* of the latter is part of the *form* of the former. We might thus give a sense to the assertion that logical laws are *forms* of thought and that space and time are *forms* of intuition.[30]

Such conclusions seem to be natural consequences from Wittgenstein's "doctrine of the logical form of (scientific) propositions" - as he already characterized philosophy in a manuscript of 1913.[31] However, no reader of Wittgenstein's planned book would agree with them without some further elucidations and support. In the next section it will be argued by us that Wittgenstein got this support, if not the idea itself from a reading of Mach's *Popular Scientific Lectures*, which also provided him with a clue to Kantian problems about space, time, and causality.

3. SOURCE: MACH'S *POPULAR SCIENTIFIC LECTURES*

Much of what Wittgenstein wrote in the *Tractatus* can be considered to be an attempt at explication of the fundamental conceptions within the previous account: language, propositions, logical forms, and "the world". Whether or not he was able to succeed in a satisfactory manner is a matter of dispute, but so much is sure, that Wittgenstein struggled hard with the problem of making clear to himself what it was he wanted to affirm. His notebooks bear witness of this, especially in the form of examples, for Wittgenstein seemed to prefer not to give detailed logical analyses of specific propositions. Thus, on December 23, 1914, he made the following note, after a row of remarks on such issues as "the nature of negation":

> A characteristic example for my theory of the significance of descriptions in physics (*physikalischen Naturbeschreibung*): the two theories of heat, heat conceived at one time as a stuff, at another time as a movement.[32]

This is a very peculiar remark, appearing out of the blue; is it possible that Wittgenstein had just read something about those old theories of heat, which encouraged him to think of them as examples for his theory? If so, then there was one thinker contemporary with young Wittgenstein who did discuss - not heat as feeling and movement, a common contrast between mind and matter - but *stuff* and *movement*. Who was it? It was Ernst Mach, notably through his discussion of the two conceptions of heat in Chapter 12 of his *Popular Scientific Lectures*, the chapter titled: "On the Principle of the Conservation of Energy".[33] Mach's purpose had been to distinguish between essential and non-essential conceptions of the phenomena of heat. In the description of what happens when two masses of water having a different temperature are brought together, a pioneer like Joseph Black thought of different quantities of heat, thus distinguishing between quantity of heat and temperature, but as if heat were a stuff or liquid which is generally called the caloric theory of heat. Most later scientists, however, starting with Benjamin Rumford tended to think of heat as a movement (thus returning to the view of Galileo and Francis Bacon).

Ernst Mach, however, in his opposition to the reality of atoms, which followed naturally from his phenomenalistic world view, opposed heat as movement, presumably because it was generally presupposed that what one meant was the movement of very small particles such as atoms. His response was to allege that the conception of heat as a movement was just as non-essential as the conception of heat as stuff or liquid had previously been.[34]

But while most physicists rejected Mach's position on heat or would soon do so, Wittgenstein was apparently impressed by Mach's distinction between what was essential and non-essential which later reappeared in Wittgenstein as a sharp distinction between what was essential and what was arbitrary. He applied this to his understanding of "the nature of the proposition" or "the nature of all facts whose picture the proposition is". For it is rather clear that the arbitrary, non-essential properties of propositions do not shew properties of the Universe. Ernst Mach's implication that neither stuff nor movement was essential to the Universe also seems to have reinforced Wittgenstein's belief that logic or form was primary.

Ludwig Wittgenstein's favorite example of different divisions of a plane for describing black spots was apparently used as an analogue for "different systems of describing the world".[35] (In this example, a system of divisions is a purely geometrical system, i.e. all of its properties can be stated a priori.)[36]

On March 29, 1915, Wittgenstein reflected again on the problem of a priori "knowledge" (as Russell would say), this time neither about stuff and motion as alleged forms of heat or even as spatial and temporal phenomena, nor even about space and time themselves, but with respect to causality. His logico-philosophical solution is almost predictable. If spatial and temporal phenomena such as stuff and motion are non-essential, that is, arbitrary, then causality, if essential, must belong elsewhere. Thus he maintains: "the law of causality is not a law, but the form of a law", that is, as if only form can be essential. Yet, there is room for comment:

> "Law of causality" is a class-name. And just as there are let us say minimum laws in mechanics, e.g. that of least action, there is *a* law of causality; a law of the causality *form* in physics.
>
> Just as men also had an inkling of the fact that there must be *a* "law of least action", [so it must have had logical form] before they knew precisely how it would run. (Here as so often, the a priori turns out to be something purely logical.)[37]

Mach appears to have formulated the same view in his interpretation of the first law of thermodynamics as if it were identical with the principle of excluded perpetual motion. Once again this opinion can be discovered in his *Popular Scientific Lectures*. he also insists that the principle is "certainly not a new discovery" as if many other people had used or run across it before he had done so. The conviction of the correctness of the principle must have existed, according to Mach, already long before the elaboration of mechanics as part of science.[38] This view not only tallies with Wittgenstein's emphasis on the a priori as something logical in form but accessible to human intuition as what can be shewn or at least suspected, but appears most likely to be a position derived from reading Mach's *Popular Scientific Lectures*.

Again, the evidence for this is largely based on the peculiarity of Mach's opinion and partly on the strong opposition to it by other scientists. Neither Mach's allegation that "heat as the motion of particles was non-essential in physics" nor his opinion that "the first law merely meant the impossibility of perpetual motion and had been anticipated many times before" found favor with many other scientists or writers. The first was rejected by all atomists and the second was opposed by Ludwig Boltzmann and Max Planck, two major authorities on thermodynamics and statistical mechanics. Boltzmann insisted that the first law was fully empirical and that it always remained theoretically possible in the future to refute it with contrary data based on empirical observation.[39] Planck insisted that Mach had failed to distinguish between two basic meanings of the first law.[40] In short, Wittgenstein would have had great difficulty in discovering any other source which would have presented just these singular theories of Mach *in positive terms*.

That the particular source from Mach was very probably his *Popular Scientific Lectures*, which went through five German editions from 1900 to 1911 and three more from 1918 to 1922, is strengthened further by several other passages. In the *Tractatus*, Wittgenstein's discussion of the problem of induction and the alleged unassailableness of the laws of nature is preceded by a somewhat more elaborated treatment of space and time, including Kant's famous problem of the right and left hand. Here the similarities between his and Mach's views become even closer: Wittgenstein's example of the one-dimensional version of the problem was already given by Ernst Mach, who credited Möbius for it. As for Kant's argument that congruous figures with different orientation cannot be made to cover one another, Mach remarked: "One could do it Möbius thinks, if one had the disposition of a fourth dimension."[41] Wittgenstein wrote: "A right-hand glove could be put on a left hand if it could be turned round in four-dimensional space."[42] In other words, both Mach and Wittgenstein appeal to four dimensions here. If anything represents Mach's influence on Wittgenstein as probable fact, don't these two quotations do it?

It is generally known that Wittgenstein regarded "most propositions and questions that have been written about philosophical matters" as senseless. They result, according to Ludwig Wittgenstein, "from the fact that we do not understand the logic of our language." Let us compare two sets of passages by Wittgenstein and Mach. *Primo*, Wittgenstein: "And so it is not to be wondered at that the deepest problems are really *no* problems."[43] "The *riddle* [proposed by Du Bois Reymond's *Ignorabimus*] does not exist."[44] But Mach somewhat earlier: [Du Bois Reymond's] "*ignorabimus*" is only correctly applied vis `a vis "wrongly posed problems, which really are *no* problems".[45] And *secundo*, Wittgenstein: "If a question can be put at all, then it can also be answered."[46] And Mach, again somewhat earlier: "Every *true* problem can and will with sufficient time be solved ..."[47] Surely, these quotations are so similar that no reasonable person can doubt the fact of Mach's influence on Wittgenstein.

But there is more. Wittgenstein characterized the process of induction (1) as the process of assuming the simplest law that can be made to harmonize with our experience, and (2) by considering it as a process with no logical, but only a psychological foundation.[48] What was Mach's earlier view? First, in his famous theory of economy he emphasized that science in general and laws in particular should be the simplest descriptions of the appearances, and second, he also wrote that laws of nature were "a result of our psychological need to orientate ourselves in nature".[49] Is it possible to distinguish between the positions of Wittgenstein and Ernst Mach on induction and the nature of scientific laws? Not as this author can determine. The similarities are surely too conspicuous to deny either the fact or the nature of Mach's influence.

4. SOURCE: MACH'S *ANALYSIS OF SENSATIONS*

Some readers may be willing to admit that Mach influenced Wittgenstein on a number of issues but still feel that there is something incommensurable about the phenomenalist Ernst Mach with his emphasis on sensations having a significant influence on Ludwig Wittgenstein with respect to logical issues, especially when Mach thought that logic itself could be reduced to matters of sensation. But another school finds nothing amiss in Mach's impact, largely because they seem to believe that both Mach and Wittgenstein were neutral monists, that is, phenomenalists. I will try to use historical analysis and some more quotations, not so much to resolve the unresolvable as to follow a path between these two extreme positions.

Wittgenstein did formulate two problems which are somehow related to sensory experience, to whit, the problem of the material *a priori* - why it is impossible that two colours can be at the same place in the visual field?[50] and the question "to what extent is solipsism true"?[51] The latter problem was discussed by Bertrand Russell in Chapter Three of his book *The Problems of Philosophy*, in connection with the fact that "other people are represented to me by certain sense data", whereas the former problem about colours arose in Moore's lectures, in connection with possible properties of sense data. In Moore's famous example of the envelope, the patch of a whitish colour and its size and shape, which he actually did see, are sense-data things given or presented by the senses; but later he called *only* the patch, *not* its colour, size or shape, a *sense-datum*, that is, its *form*.[52]

But Russell, who took over the term "sense data" from Moore (who in turn had derived it from William James), looked at the matter differently. He considered the knowledge let us say the resemblance between two shades of green is greater than the resemblance between a shade of red and a shade of green to be "immediate" and "indubitable".[53] Indeed, Wittgenstein just *might* have had this view in mind when he commented on a "fact" such as that "this blue colour and that stand in the internal relation of brighter and darker *eo ipso*": "it is unthinkable that these two objects should not stand in this relation."[54] So it looks as if Ludwig Wittgenstein was, after all, interested in the structure of sensations, and this would make Mach's contribution to the analysis of sensations a possible source of inspiration for Wittgenstein.

As a matter of fact, in addition to the half-dozen critics we mentioned before (See footnotes #6 to #13) who believed that Mach had influenced Ludwig Wittgenstein, there is yet another Wittgenstein scholar, who is called David Favrholdt, who reached almost the same opinion in his *An Interpretation and Critique of Wittgenstein's Tractatus*, Munksgaard: Copenhagen, 1964. He has chosen to write on page 147 that "Wittgenstein's empiricism must be a sort of neutral monism which resembles the views of Russell, James, (In *Does Consciousness Exist?*) and Mach (in *Analyse der Empfindungen*)." Thus, many people seem to feel that there was influence.

But before studying the matter further, it is necessary to point out that "neutral monism" has different epistemological and ontological meanings. Bertrand Russell's *ontology* as if both mind and matter consisted of *sensibilia* was combined with an indirect epistemology as if physical sensibilia could not be directly sensed. Ernst Mach's neutral monism on the other hand was *epistemological* in that one could sense sensations in one set of relations as physical and in another as mental with Mach normally denying that he had an ontology or was even a philosopher. James's neutral monism seems to have been much closer to that of Mach than to Russell. Most critics identify the neutral monism of both Mach and James with phenomenalism or a kind of conscious experience monism (inspite of James's own alleged "pluralism" and denial that consciousness exists), but withhold both classifications from Russell's ontological neutral monism.

Most recently (1994), John W. Cook has compared Mach's view "that the world consists only of our sensations" with Wittgenstein's remark that "all experience is world and does not need the subject": "This is", for Cook, "an explicit affirmation of neutral monism, although Ludwig Wittgenstein in his notebooks, calls it 'realism' (NB, P. 185) and in the *Tractatus* 'pure realism' (TLP, 5.64)".[55]

Indeed, for some critics the very fact that young Wittgenstein reverses the normal definitions of the terms "realism" and "idealism" in the manner in which many followers of Berkeley, Hume, and Kant[56] have done "proves" that he was a phenomenalist or even a subjective idealist, but this criticism is not fair since many philosophers, especially in the German-speaking realm, think that Kant-inspired counter-definitions are the "normal" ones and not the anti-Berkeley definitions that one generally finds in most dictionaries and encyclopedias of philosophy to whit: "Realism is the position that there is a physical world and it exists independently of all consciousness of it".

But at the risk of giving too much space to critics who want to link Wittgenstein too closely to Mach's ideas, especially his neutral monism, let us emphasize that there are strong objections against any assimilation of Wittgenstein's alleged epistemology to neutral monism. Nowhere in any of the *Notebooks* does Wittgenstein's interest in the logical structure of sensory phenomena turn into a theory of perception or indeed into any conception of how perception takes place. In fact, as David Pears once remarked: Ludwig Wittgenstein attempted to be "careful not to commit himself to the view that acquaintance with the objects named in elementary propositions would have to be sensory, or to the view that they could not be sensory".[57] Nor was this position unusual for any young man exposed to the extreme distrust which Viennese philosophers shewed toward what they called "metaphysics", a term sometimes defined as the study of alleged trans-sensory objects and sometimes as any inquiry into "reality", whether sensory or trans-sensory.

But whether Pears was right about Ludwig Wittgenstein's neutrality concerning epistemology and ontology is not easy to answer. Calling these branches of philosophy as well as all philosophy "linguistic" or even "metaphysical" may not exclude the possibility that Wittgenstein presupposed an epistemological or ontological position, but even if he did, if the presuppositions were fully unwitting and unconscious to him, and more important, incompatible with what he *thought* was his philosophy or antiphilosophy (and Mach's reduction of logic to sensations surely was incompatible), then surely there would be a measure of injustice in identifying his logical position with what to his own conscious understanding was fully unknown. He may well have picked up phenomenalist assumptions during his Viennese childhood, but in the *Tractatus* the main focus is so much on logic and forms, that any lingering empirical residue from that childhood can hardly deserve to be treated as philosophically primary, especially if as may well be the case he *wanted* to be neutral about epistemology and ontology. Nor was he fully indifferent to the habit of physicists to distinguish between sensory and physical objects:

In a notorious remark, Ludwig Wittgenstein on the one hand assimilates the objects of non-scientific appearance to physical knowledge as studied by physics, such as colours and sounds, but on the other hand, he also speaks of reduction of what is non-scientific to physics:

> A point cannot be red and green at the same time: at first sight there seems to be no need for this to be a logical impossibility. But the language of physics [*die physikalische Ausdrucksweise*] reduces it to a

> kinetic impossibility. We see that there is a difference of structure between red and green.
>
> And then physics arranges them in a series. And then we see how here the true structure of the objects is brought to light.[58]

In fact, if the first quotation above is clarified in terms of Wittgenstein's probable intent and assumptions we get a standard physical account as if sensory objects and physical objects are always different, which is certainly not either neutral monism or phenomenalism:

> A point cannot be red and green at the same time: at first sight there seems no need for this to be a logical impossibility [since it might just be a factual impossibility.] But the language of physics reduces it to a kinetic impossibility [where the different colors have different wave lengths] We see that there is a difference of structure [that is, in the electromagnetic spectrum] between red and green [as understood in physical as opposed to sensory terms].

But if we interpret these passages as if Wittgenstein were aware of how physicists normally think in distinguishing between sensory and physical objects, then we risk allowing that he understood mind-matter dualism and indirect epistemology, that is, as if there were real philosophical differences, which in turn compromises his antiphilosophy and his attempt to reduce philosophical to linguistic problems. In short, in order to successfully defend Wittgenstein against the charge of neutral monism or phenomenalism (which may not be possible), we risk compromising his linguistic approach. It may be the case that only phenomenalists unable to understand anything outside of phenomenalism are most inclined toward linguistic reductionism as if it were an antiphilosophy, but this does not prove that any particular person was a phenomenalist, and in Wittgenstein's case we have already shewn that he was aware of other epistemological and ontological persuasions.

In fact, there is no denying that he was capable of distinguishing between sensory and physical objects when discussing physics and that he sometimes did so. For example, he writes: "the propositions of physics usually deal with other objects than those of our ordinary language."[59] The joker in this claim is what Wittgenstein means by "ordinary language", since most mind-matter dualists who hold an indirect epistemology are likely to identify most of the objects of physics with the objects referred to in what they call "ordinary language" whereas this is not the case with Machian-type neutral monists.

In fact, Wittgenstein's conclusion about the difference between red and green by being considered a logical one already means that he is referring to what physicists mean by red and green, not what phenomenalists mean. The fact that Mach himself was a physicist notwithstanding, since his position regardless of how popular outside of physics, or even within some physical circles in France and Germany during the 1880's and 1890's, was always a minority one within physics *as a whole*. Also, when Mach discussed say the ordering of tones, or the transition from pure red into pure yellow - by the addition of yellow - he was explicitly concerned with the purely sensory aspects of the basic theories of Helmholtz and Hering,[60] and this approach was foreign to Wittgenstein.

Up until now, we have resisted the likelihood that Ludwig Wittgenstein was significantly influenced by Mach's book *The Analysis of Sensations* with its blatant phenomenalism and talk about sensory elements. He simply did not use Mach's favorite vocabulary from that book, and as we have noticed previously, he was apparently very averse to Mach's literary style, but the American philosopher William James did rely heavily on *The Analysis of Sensations*.

An astute look into James's famous *Principles of Psychology* (in two volumes, 1890) will not disappoint the reader who searches for parallels between Wittgenstein and James, and since a version of James's discussion in fig. 76 in the second volume of a plane diagram which can be conceived as a cube and even in two different ways, can also be found in the *Tractatus*, but starts in James with examples from Mach's *Contributions to the Analysis of Sensations* (first published in German in 1886 with the primary title later shortened to *The Analysis of Sensations*), the indirect connection between Wittgenstein and Mach's book is there. Thus, we can read William James in order to find a number of debts which Wittgenstein probably owed Mach. The young Wittgenstein with his sensitive stylistic tastes may not have read much in Mach's *Analysis* directly, but significant parts of the content surely did exercise influence.[61] Let us quote from the *Tractatus* about the above:

> 5.5423 To perceive a [sensory] complex means to perceive that its constituents are combined in such a way. This also explains, I think, that the figure can be seen in two ways as a cube; and all similar phenomena. For we really see two different facts. (If I look first at the corners *a* and only cursorily at *b*, then *a* will appear in front, and vice versa.)[62]

The importance of Mach, from whom James took over the example of a folded visiting-card - which by the way also occurs in slightly altered form in *Popular Scientific Lectures* - can be seen to consist in his attempt to eliminate the supposition of a somehow "enigmatic change of orientation" [*rätselhafte Orientierungsänderung*].[63] To that purpose, Ernst Mach drew auxiliary lines, but Wittgenstein showed that this is not necessary in the case of the picture of one of the other cube. Though he did not suggest any law", Wittgenstein did contribute to Mach's research question in his example, and this may make his connection with Mach's *Analysis of Sensations* stronger than only indirect. Yet the above arguments do not support the hypothesis that Wittgenstein had read Mach's book when he worked on the *Tractatus*. But it still remains to be investigated whether Wittgenstein's treatment of the problem of solipsism might.

John Cook explicitly declared that Wittgenstein followed Mach's lead to solipsism. The statement that "the thinking, presenting subject does not exist"[64] could well be dependent on Mach's discussion which ended with the memorable statement that the I must be given up. [*Das Ich ist unrettbar.*][65] But it was not necessary to read Mach in order to hear about solipsism or the superfluous nature of the ego. Vienna had no shortage of subjectivists and gloom and doom philosophers. One need only mention Karl Kraus and Otto Weininger, who were it seems both favorites of Ludwig Wittgenstein, and as with William James, "carriers" of some of Mach's more extreme views.

It is also possible to argue that Wittgenstein's remark on the way he had traveled - from idealism to realism[66] - has also some similarity to Mach's road which started with the idealism of Kant and Berkeley[67] and ended with the "realist" Hume. But while the influence might be real, there are two major problems with Wittgenstein's claim. First, it violates Wittgenstein's own linguistic reduction and his firm stand against philosophy by allowing the legitimacy of epistemological and ontological categories. And second, it is based on Wittgenstein's reversal of traditional philosophical definitions. For most historians David Hume was not a "realist" but a phenomenalist or even a subjective idealist. He by no means accepted that there was a real physical world and that it existed independently of consciousness, at least not in his philosophy books. More exactly, Wittgenstein apparently identified idealism with belief in Kant's *Ding-an-sich,* such that rejecting the thing-in-itself or noumenal world made one a "realist", a curious German counter-definition which as mentioned before is fully at odds with virtually all definitions of "realism" and "idealism" to be found in a majority of reference books and in the usage of most traditional historians of philosophy. On the other hand, since Mach also seems to have adopted the counter-definition, perhaps this raises the likelihood that Wittgenstein did absorb it directly or indirectly from him.

While opponents of Ernst Mach naturally allege that phenomenalism by denying the existence of a trans-conscious physical world automatically dooms one to solipsism, one should recall that Mach himself believed that by denying the ego or self as well, one avoids solipsism by denying the solip, the existence of the self who would be alone. But whether this maneuver truly avoids solipsism or merely makes it worse by denying the existence of people as well as the external world, it is clear that Wittgenstein followed Mach with little or no suspicion of how most non-phenomenalists would reject Mach's approach. But Wittgenstein's counter-definitions and identification of the world with appearances seemed to have a dynamism of their own and he appeared to plunge ever further into the abyss, at least as outsiders might see it.

> This is the way I have traveled: idealism singles men out from the world as unique, solipsism singles me alone out, and at last I see that I too belong to the rest of the world; so on the one side *nothing* is left over, on the other side, as unique, *the world.*[68]
>
> The key to the question is to what extent solipsism is a truth: There really is only one world soul, which I for preference call *my* soul and as which alone I conceive what I call the souls of others.[69]
>
> The I makes its appearance in philosophy through the fact that the world is my world.[70]

The conclusion lies at hand that either young Wittgenstein was a solipsist in much of his thinking or that he adopted Mach's figleaf that by abolishing the self as distinct from other appearances ("the world") that he was also denying solipsism , since if there was no self how could he be alone in "the world"? But for us, the important point is that even if Wittgenstein had not yet studied Mach's *Analysis of Sensations* he obviously reached a conclusion very similar to it, which may well bespeak indirect influence.

5. LATER WITTGENSTEIN AND CONCLUSION

When Wittgenstein "broke his philosophical silence" in the years 1929 and 1930 and resumed "anti-philosophical philosophizing", he was it seems struck by the need to be more consistent in his reduction of philosophy to language. Nor was this easy to do, since like Ludwig Boltzmann he loved to philosophize against philosophy, but whereas Boltzmann largely under the influence of Franz Brentano appears to have abandoned his attempt to reduce everything philosophical to matters of language, Wittgenstein chose to take the opposite tack: try harder to reduce philosophy to language or at least to mere semantic clarification. Nor did he linger long with his old approach to language. He turned toward grammar, and in time, thought more in terms of linguistic conventions within small groups of speakers as if the context were essential and could vary a great deal from group to group.

But one occasional casualty of his linguistic crusade was Ludwig Wittgenstein's identification of his own philosophical obligation to Ernst Mach as no longer being philosophical in nature but merely linguistic. And lo and behold, he actually mentions Mach by name, as well as a method Wittgenstein had earlier used while philosophizing against philosophy, that is, what Mach had christened as "thought experiments", though he still seems reluctant to credit the source of that method to Mach. But Wittgenstein's point is that from now on thought experiments will be understood merely as an aspect of language.

> What Ernst Mach has called a thought-experiment, is of course not an experiment at all. It is [merely] a grammatical consideration (*eine grammatische Betrachtung*).[71]

This is more than a cursory remark on Mach's method. From now on, Wittgenstein tries to satisfy his purpose, namely the clarification, "solution", if not the full elimination of philosophical problems, by thought experiments of a new kind, that is, by descriptions of imaginary situations which throw light on "grammatical" features of crucial concepts. Thus, Wittgenstein modified his original view that most questions and statements of philosophy can be attributed to the inability to understand the logic of our language (*unsere Sprachlogik*).[72] For then it seems that a critique of language would consist of logic, whereas Wittgenstein's new view emphasizes the need for grammar - logic being only concerned with a *part* of a system of language, not language in extenso. And here Mach's method offers Wittgenstein an excellent means for further investigations about old and new subjects.

Among the old issues, we encounter again the problem of the alleged material a priori character of colour statements and the problem of the "I". A new theme - outside of the philosophy of mathematics - is the distinction between two languages: the phenomenalistic and the physicalistic. And in all of these subjects it is Ernst Mach who can be seen to figure in Wittgenstein's treatment, nor is this surprising, given the interest of the Vienna Circle in both Mach and the distinction itself and the fact of Wittgenstein's contact with selected members shortly before his return to Cambridge.

The tendency of the Vienna Circle to prefer physicalistic expression as a kind of frosting on a phenomenalistic cake apparently upset Wittgenstein. He wrote in 1929: "the worst philosophical errors always arise when we want to apply what is our ordinary - physicalistic - language in the sphere of the immediately given.[73] Also:

> One of the clearest examples of the confusion between physical and phenomenological [phenomenalistic] language is a picture that Mach designed of his visual field, and in which the so-called morbidity of the things (*Gebilde*) towards the edge of the visual field is represented by a turbidity (in a wholly different sense) of the drawing. No, a visible picture of the visual field cannot be made.[74]

But Wittgenstein thought Mach had misapplied physicalistic language elsewhere in his treatment of immediate phenomena. For example, in *The Analysis of Sensations* Mach had argued that the distinction between left and right had to do with a human motor asymmetry with respect to operations of the left and right hand, while the distinction between above and below was linked with the direction of gravity. Wittgenstein rejected this as follows:

> It can be said that visual space is . . . an oriented [*gerichteter*] space, a space in which there is an above and a below, and a right and a left, but this above and below, right and left has nothing to do with gravity or the right and left hand . . . [75]

If this quotation does not yet give sufficient evidence that Wittgenstein was seriously occupied with Mach's work, then still another critical remark by Ludwig Wittgenstein can be given, this time directed against the view that physical time measurements could be used for more exact determination of successive thoughts. Again it was Mach, who formulated such a view, in spite of his warning that the time of the physicist does not coincide with the system of time sensations. Mach writes:

> When the physicist wishes to determine a period of time, he applies processes which are assumed to be identical to serve as his standard of measurement.[76]

Wittgenstein on the contrary preferred to consider the concept of time in the domain of the given - or the world of data [*die Welt der Daten*] - as something radically different from the time-concept of physics. One can also observe his stronger emphasis on grammar:

> We cannot . . . apply the concept of time, i.e. the rules of syntax, as they hold for physical substantives, to the world of presentations, that is, where we use a radically different way of expression.[77]

So far it seems that Wittgenstein used statements which he found in *The Analysis of Sensations* as targets of abuse, perhaps partly to articulate more fully his own apparent views that the Vienna Circle following some ideas in Mach may have been misusing his distinction between a phenomenological [phenomenalistic][78] and a physical language. He may have been especially bothered by the apparently blithe attitude of Carnap, Neurath, and others who appeared to grossly underestimate how difficult it could be to reconcile physicalistic language with the study of "the immediately given".

To be sure, one of the seeming advantages of a linguistic philosophy is precisely that it can assign different languages to what had previously been considered different epistemological and ontological positions, and thus appear to do without philosophy, not to mention metaphysics, but in the opinion of Wittgenstein the cavalier attitude of the Vienna Circle toward the difficulties in actually carrying this out in practice to the whole range of human experience, largely went back to Ernst Mach, who had made it a practice to simplify his philosophical views while looking for rough parallels in the works of other thinkers while skirting numerous potential and actual difficulties. But Wittgenstein's dismay at this practice which he also observed in propaganda emanating from the Vienna Circle and its adult educational organ the Ernst-Mach-Verein apparently did turn his attention back to Mach such that he likely read or re-read his *Analysis of Sensations*, and as mentioned before, began to refer to Ernst Mach by name, perhaps as a kind of *bête noire*. His earlier comment that he disliked Mach's style also begins to make increased sense. What he disliked was simplifying complex problems à la Mach and in the process distorting them. But since one of the main goals of much theoretical physics is precisely simplification and idealization, one can also notice a possible reason for Wittgenstein's growing disenchantment with science itself.

But it is conceivable that Ludwig Wittgenstein was blaming Mach too much for the shortcomings of the Vienna Circle. Even though Ernst Mach had been a physicist and generally preferred to express himself in physicalistic terms, he was in fact as interested as Wittgenstein in "the given - or the world of data", and his reservations about applying physical measurements of time to direct awareness of sensory phenomena were almost as strong.

What is new after Wittgenstein started thinking about philosophy again circa 1929 was his emphasis on "grammar". This is such a central concept of the *Philosophical Remarks* that one is tempted to speak of a "grammatical turn". How very important grammatical rules are in particular follows from Wittgenstein's new treatment of the internal properties of colours - well-know from the *Tractatus* - and in general his position on phenomenalistic language.[79] Already in the beginning of the remarks written down in 1929, what he designates as *Phänomenologie* is called "the grammar of those facts, on which physics constructs its theories"[80] This point which was made in the context of the distinction between "explaining" and "describing" could evoke some reminiscences of Ernst Mach's article "Description and Explanation", or his lecture "On the Principle of Comparison in Physics". Yet Wittgenstein was less interested in the question of two views on *physics*, than in the solution of philosophical problems by an investigation of forms of language. Indeed, it is amazing how much attention Wittgenstein paid such things as "the space of colours" even though he was really trying to reduce them to grammatical forms. If grammar had replaced logic as his primary focus at least he still spoke about forms, but now grammatical forms, but the real constant in his investigations was the *study* of conscious appearance within which he hoped to find those forms. But he never became as uncritical as most followers of Edmund Husserl or as obsessed with universals. His own phenomenology remained closer to phenomenalism, that is, to Mach.

As for possible weaknesses in linguistic philosophy the main one has surely centered on its difficulty in handling causes and causal explanation. Each different language may well seem to imply a different conception of causes and causal explanation which would be much better handled if one were to link different causal positions to different world views, that is, to different epistemologies and ontologies. Wittgenstein by restricting himself to narrow "grammatical" questions seemed to decline to elaborate a plausible conception of causes and causal explanation, but a majority of linguistic philosophies, including logical positivism and analytic philosophy, cannot reasonably take the same course. Nor can simplistic "covering law" theories satisfy the practical world.

Unlike more recent thinkers such as Hempel and Popper, both Ernst Mach and Wittgenstein held that scientific laws merely describe and do not explain. Mach appealed to his theory of economy, psychology, and mathematical and volitional functions as tools of explanation, but observers have not been slow to point out that Mach had not fully reconciled his phenomenalism with his theory of economy or his appeal to functional explanation, as if perhaps no causal theory of any kind could be totally compatible with phenomenalism, except maybe David Hume's allegation that mere constant conjunctions between appearances was the best that one could do, an approach which surely falls far short of being sufficient in practical life.

To the extent Wittgenstein could actually develop a linguistic philosophy which could exclude philosophical problems, then one might argue that even if he tended to presuppose a neutral monist or phenomenalistic world view it was in the process of being superseded to the best of his ability, and part of that progress, it could be maintained, came at Ernst Mach's expense, either by Wittgenstein increasingly rejecting Mach's views or by reducing them to linguistic phenomena. In short, refuting Mach's phenomenalism or reducing its scope by linguistic reduction became Wittgenstein's way of *purging himself of his own phenomenalistic residue*. But whether he actually made significant progress or not remains a matter of dispute. He is rightly known as a major pioneer in linguistic philosophy or anti-philosophy, but even the most able pioneers may fail to reach their goal. Sometimes there are just too many obstacles, both internal and external.

NOTES

1 The first edition of Ludwig Wittgenstein's "Logisch-Philosophische Abhandlung", *Annalen der Naturphilosophie*, vierzehnter Band, drittes und viertes Heft, Unesma: Leipzig, 1921, is not wholly reliable, hence quotations will be taken from the bilingual edition *Tractatus Logico-Philosophicus*, second impression, Routledge & Kegan Paul: London, 1933.

2 The first issue of *Philosophische Untersuchungen / Philosophical Investigations* came out in 1953. The edition which is used is the second, corrected edition, Basil Blackwell: Oxford, 1958. References are given in the form of the numberings of the editors.

3 In the *Tractatus*, Ludwig Wittgenstein wrote that he gave no sources, because it was "indifferent" to him whether his thoughts had already been thought before him (Preface).

4 *Vermischte Bemerkungen; eine Auswahl aus dem Nachlass*, Suhrkamp: Frankfurt 1977.

5 *Tractatus* (Preface).

6 *Philosophical Investigations*, pp. 111 & 133.

7 Ludwig Wittgenstein to Bertrand Russell, January 16, 1913.

8 Friedrich von Hayek, "unfinished draft of a sketch of a biography of Wittgenstein" (1953), p. 3.

9 Friedrich.von Hayek, "Diskussionsbemerkungen über Ernst Mach und das sozialwissenschaftliche Denken in Wien", in *Symposium aus Anlass des 50. Todestages von Ernst Mach*, edited by W. Merzkirch, Freiburg im Breisgau, 1967, p. 42.

10 Philip P. Wiener, "Method in Russell's Work on Leibniz", *The Philosophy of Bertrand Russell*, Paul Arthur Schilpp ed., Chicago, 1944, p. 274.

11 Stephen F. Mason, *A History of the Sciences*, New York, 1960, p. 501.

12 Nicola Abbagnano, "Positivism", in *The Encyclopedia of Philosophy*, Vol. VI, New York, 1967, p. 223.

13 John Blackmore, *Ernst Mach - His Life, Work, and Influence*, Berkeley, 1972, p. 185.

14 John W. Cook, *Wittgenstein's Metaphysics*, Cambridge University Press: Cambridge, 1994.

15 In addition to the Schopenhauer Lecture (pp. 385-402), one can find Boltzmann's linguistic philosophy in Ilse M. Fasol-Boltzmann (ed.), *Ludwig Boltzmann Principien der Naturfilosofi - Lectures on Natural Philosophy 1903-1906*, Springer Verlag: 1990, Berlin and in John Blackmore, *Ludwig Boltzmann - His Later Life and Philosophy 1900-1906*, Volumes I & II, Kluwer: Dordrecht, 1995.

16 Richard Wahle's book: *Das Ganze der Philosophie und ihr Ende*, Vienna and Leipzig, 1895, (2nd edition 1896) is an attempt "to clean phenomenal occurences" from elements which can give rise to metaphysical theories (cf. op. cit. p. 82). Some possible parallels with Wittgenstein's earlier work appear to lie in Wahle's treatment of such different subjects as, for example, the Necker cube (op. cit., p. 240) and punishment and reward (op. cit. p. 439), which were dealt with in a similar way to Wittgenstein (5.5423 and 6.422) Also in *Die Tragikömedie der Weisheit*, Braumüller: Vienna & Leipzig, 1915. Wahle's aim was the increase of clarity, for he thought the disappearence of philosophical unclarities would have a purifying effect (cf. op. cit.. p. 18). His main conclusion as to philosophy that "it is best to be silent" (cf. op. cit. p. 4) was shared by Wittgenstein in the last statement (#7) of the *Tractatus*. See also: William M. Johnston, *The Austrian Mind*, Berkeley,

1972, pp. 201-203 for mention of Mach's influence on Wahle.

17 Wittgenstein's remark that "all philosophy is critique of language" - but not at all in Mauthner's sense - is generally interpreted as a statement in favor of philosophy as *logical* analysis instead of Mauthner's *historical* approach. But this is not to say that Wittgenstein would not have found some inspiration in Mauthner's *Beiträge zur Kritik der Sprache*, as Mach did (*Erkenntnis und Irrtum*, p. 82). Yet as to antiphilosophical views, there are more similarities between Mauthner and the *later* Wittgenstein than with the author of the tractatus. For a linguistic predecessor of Fritz Mauthner see O.F. Gruppe and Arthur Trebitsch's article: "O.F. Gruppe. Ein vergessener Philosoph" in *Deutscher Bibliophilen - Kalender für das Jahr 1914*, Moritz Perles: Vienna, 1914.

18 Please See William M. Johnston, in "Karl Kraus's Idolatry of Language: The Curse of a Photographic Memory", *The Austrian Mind*, Berkeley, 1972, pp. 203-207.

19 Martin Gardner once pointed out that this ladder argument goes back at least to the second century A.D.: It was Sextus Empiricus who used it at the end of his second book "against the logicians". Nevertheless, it seems reasonable to assume that Wittgenstein's actual source was probably Ernst Mach, though of course Wittgenstein may have read Sextus Empiricus directly and found it, or even in the writings of O.F. Gruppe, the little known linguistic philosopher mentioned in footnote #17.

20 Fasol-Boltzmann, op. cit., pp. 80-82.

21 Ludwig Wittgenstein, *Letters to Russell, Keynes, and Moore*, Basil Blackwell: Oxford, 1974, page 39.

22 Ibid.

23 Ibid., p. 45. Earlier he had formulated "the fundamental question of logic" as "the big question how a system of signs must be constituted in order to make every tautology recognizable *as such in one and the same way*" (op. cit., p. 41).

24 Ludwig Wittgenstein, *Notebooks 1914-1916*. Edited by G.H. von Wright and G.E.M. Anscombe, Basil Blackwell: Oxford, 1961, p. 2.

25 Ibid., p. 107.

26 Ibid., p. 110.

27 Ibid., p. 53.

28 Ibid., p. 107.

29 Ibid., p. 108.

30 Ibid., p. 117.

31 Ibid., p. 93; cf. Ludwig Wittgenstein, *Notebooks 1914-1916*, second edition, Basil Blackwell: Oxford, 1979, p. 108.

32 Notebooks 1914-1916, p. 37.

33 Ernst Mach, *Populär-wissenschaftliche Vorlesungen*, 4. vermehrte und durchgesehene Auflage, J.A. Barth: Leipzig, pp. 167-216.

34 Ibid., pp. 198-199.

35 *Notebooks 1914-1916*, p. 35; *Tractatus*, 6.341.

36 Ibid., 38; Ibid., 6.35.

37 Ibid., 41; Ibid., 6.32, 6.321, 6.3211.

38 Mach, op. cit., p. 187.

39 Ludwig Boltzmann to Ernst Mach, Munich, October 1, 1893. See *Ernst Mach als Aussenseiter*, edited by J. Blackmore & K. Hentschel, Braumüller: Vienna, 1985, p.p. 10-11. But three letters later Ludwig Boltzmann writes that he is happy they have reached an agreement about the first law, but since Mach's letters to Boltzmann have disappeared it is not clear just what their agreement was.

40 See Max Planck, *Physikalische Zeitschrift*, **11** (1910), pp. 1180-1190.For an English translation see "On Mach's Theory of Physical Knowledge - A Reply", in Ernst Mach - a Deeper Look, Kluwer: Dordrecht, 1992, p. 143:

> He [Mach[speaks very often in his book [*Prinzipien der Wärmelehre* (1896)] about perpetual motion, but he does not attach any clear-cut meaning to this expression. He continually confuses perpetual motion of the first kind (production of work from nothing) with perpetual motion of the second kind (uncompensated production of work from heat).

41 Mach, op. cit., p. 304.

42 *Tractatus*, 6.36111.

43 *Tractatus*, 4.003.

44 Ibid., 6.5. See also Emil Du-Bois Reymond, *Ueber die Grenzen des Naturerkennens. Die sieben Welträthsel.* Dritte Auflage, Veit: Leipzig, 1891. Du-Bois Reymond argued that there were seven problems in science which he doubted would ever be solved. Let us add that a hundred years later they have still not been solved.

45 Mach, op. cit., p. 410.

46 *Tractatus*, 6.5.

47 Mach, op. cit., p. 410.

48 *Tractatus*, 6.36, 6.3631, and 6.36311.

49 Ernst Mach, *Erkenntnis und Irrtum. Skizzen zur Psychologie der Forschung*. J.A. Barth: Leipzig, 1905, pp. 453-454.

50 *Tractatus*, 6.3751.

51 Ibid., 5.62.

52 G.E. Moore, *Some Main Problems of Philosophy*, George Allen & Unwin: London, 1953, p. 30.

53 Berrtrand Russell, *The Problem of Philosophy*, Henry Holt and Company: New York, 1912, page 161.

54 *Tractatus*, 4.123; cf. the quotation from the notes dictated to G.E. Moore at the end of the second section.

55 John W. Cook, *Wittgenstein's Metaphysics*, Cambridge University Press: Cambridge, 1994, p.15.

56 A.C. Fraser became well-known in the 19th century for calling Berkeley's philosophy "realism" as Barry Stroud has in the 20th for calling Hume's philosophy "realism". But the reversal has its source in Kant who did not want his philosophy associated with that of Berkeley, who had been called an idealist because he identified the physical world with sensations which he called ideas.

57 David Pears, *Wittgenstein*, Fontana/Collins: London, 1971, p. 71.

58 The *Notebooks*, p. 81.

59 Ibid., p. 68.

60 Ernst Mach, *Die Analyse der Empfindungen*, siebente Auflage, Gustav Fischer: Jena, 1918, pp. 54-57.

61 William James, *The Principles of Psycholgy*, Volume **II**, Henry Holt & Co.: New York, 1890, p. 182, and Wittgenstein, *Notebooks*, p. 87.

62 Wittgenstein's use of the word "perceive" is how phenomenalists employ it. Most of the people who accept the representative theory of perception reject Wittgenstein's usage and would call what he terms "perceive" as to "sense" or "consciously apprehend".

63 Mach, *Analyse der Empfindungen*, op. cit., p. 183.

64 *Tractatus*, 5.631.

65 Cook, *Wittgenstein's Metaphysics*, op. cit., p. 16. Mach's expression which literally means that the ego or self cannot be saved became a popular catch-phrase in turn-of-the-century Vienna, but

seems not to have penetrated far into Anglo-American usage.

66 *Notebooks*, p. 85.

67 Mach, *Analyse der Empfindungen*, op. cit., p. 295.

68 *Notebooks*, p. 85.

69 Ibid., p. 49.

70 Ibid., p. 80; *Tractatus*, 5.641

71 Ludwig Wittgenstein, *Philosophische Bemerkungen. Aus dem Nachlass herausgegeben*, edited by Rush Rhees, Basil Blackwell: Oxford, 1964, p. 52

72 *Tractatus*, 4.003.

73 *Philosophische Bemerkungen*, p. 88.

74 Ibid., pp. 260 & 267. It can be asked whether Mach's picture does indeed represent turbidity, or whether it was Ernst Mach's intention to do such. For in his original drawing, nothing of the kind can be seen.

75 Ibid., p. 255.

76 *Die Analyse der Empfindungen*, op. cit., p. 285.

77 *Philosophische Bemerkungen*, p. 820; cf. p. 83.

78 Though both Mach and Wittgenstein used the term "phenomenological" what they meant comes closer to the term phenomenalistic as used today in philosophy, especially since the labels "phenomenological" and "phenomenology" have more or less become the firm property of Edmund Husserl and his followers who favor an emphasis on universals which seems foreign to both Mach and Wittgenstein.

On the other hand, some psychologists and many physicists seem to retain the term "phenomenological" in its pre-Husserlian sense of phenomenalistic, which complicates matters. But we historians of philosophy have no choice. We must distinguish between the terms and associate Mach with phenomenalism and Wittgenstein with a preference for phenomenalistic language, at least when focusing on what is given in conscious attention.

79 Ludwig Wittgenstein, *Bemerkungen über die Farben / Remarks on Colour*, edited by G.E.M. Anscombe. Trans. by Linda L. McAlister and Margaret Schättle, Basil Blackwell: Oxford, 1977, p. 2; Here Wittgenstein distinguishes two meanings of propositions of the form "X is lighter than Y", first as an external relation, and then, as an internal relation.

80 *Philosophische Bemerkungen*, p. 51.

REFERENCES

Abbagnano, Nicola. "Positivism". in *The Encyclopedia of Philosophy*. Vol. VI. Macmillan, 1967.

Blackmore, John. *Ernst Mach - His Life, Work, and Influence*. Berkeley: University of California Press, 1972.

Blackmore, John, & Klaus Hentschel (Hrsg.). *Ernst Mach als Aussenseiter*. Wien: Braumüller, 1985.

Blackmore, John (ed.). *Ernst Mach - A Deeper Look*. Dordrecht: Kluwer, 1992.

Blackmore, John. *Ludwig Boltzmann - His Later Life and Philosophy 1900-1906*. Volumes One and Two. Dordrecht: Kluwer, 1995.

Cook, John W. *Wittgenstein's Metaphysics*. Cambridge: Cambridge University Press, 1994.

Du-Bois Reymond, Emil. *Ueber die Grenzen des Naturerkennens. Die sieben Welträtsel*. Dritte Auflage, Leipzig: Veit, 1891.

Fasol-Boltzmann, Ilse M. *Ludwig Boltzmann Principien der Naturphilosofi - Lectures on Natural Philosophy 1903-1906*. Berlin: Springer Verlag, 1990.

Gruppe, O.F. & Arthur Trebitsch. "O.F. Gruppe. Ein vergessener Philosoph". *Deutscher Bibliophilen - Kalender für das Jahr 1914*. Wien: Moritz Perles, 1914.

Hayek, Friedrich von. "Unfinished draft of a sketch of a biography of Wittgenstein" (1953). A copy was given to John Blackmore either at UCLA or Universität Freiburg during the 1960's or early 1970's.
Hayek, Friedrich von. "Diskussionsbemerkungen über Ernst Mach and das sozialwissenschaftliche Denken in Wien". in *Symposium aus Anlass des 50. Todestages von Ernst Mach*, edited by W. Merzkirch , though his name is not listed as the editor. Freiburg im Breisgau: Ernst-Mach-Institut Selbstverlag, 1967, pp. 41-44.
James, William. *The Principles of Psychology*. Volume II. New York: Henry Holt & Company, 1890.
Johnston, William M. *The Austrian Mind*. Berkeley: University of California Press, 1972.
Mach, Ernst. *Principien der Wärmelehre*. Leipzig: J.A. Barth, 1896.
Mach, Ernst. *Erkenntnis und Irrtum. Skizzen zur Psychologie der Forschung*. Leipzig: J.A. Barth, 1905.
Mach, Ernst. *Populär-wissenschaftliche Vorlesungen*. 4. vermehrte und durch gesehene Auflage. Leipzig: J.A. Barth, 1910.
Mach, Ernst. *Die Analyse der Empfindungen*. siebente Auflage. Jena: Gustav Fischer, 1918.
Mason, Stephen F. *A History of the Sciences*. New York: Collier, 1962.
Mauthner, Fritz. *Beiträge zu einer Kritik der Sprache*. 3 vols. Stuttgart: Cotta, 1901-1902.
Moore, G.E., *Some Main Problems of Philosophy*. London: George Allen & Unwin, 1953.
Pears, David. *Wittgenstein*. London: Fonatana/Collins, 1971.
Planck, Max. "Zur Machschen Theorie der physikalischen Erkenntnis - Eine Erwiderung". *Physikalische Zeitschrift*. II (1910) 1186-1190.
Russell, Bertrand. *The Problem of Philosophy*. New York: Henry Holt and Company, 1912.
Wahle, Richard. *Das Ganze der Philosophie und ihr Ende*. second edition. Wien und Leipzig: 1896.
Wiener, Philip P. "Method in Russell's Work on Leibniz". in *The Philosophy of Bertrand Russell*. edited by P.A. Schilpp. Chicago: Open Court, 1944.
Wittgenstein, Ludwig. "Logisch-Philosophische Abhandlungen". *Annalen der Naturphilosophie*. vierzehnter Band. drittes und viertes Heft. Leipzig: Unesma, 1921.
Wittgenstein, Ludwig. *Tractatus Logico-Philosophicus*. second impression. London: Routledge & Kegan Paul, 1933.
Wittgenstein, Ludwig. *Philosophische Untersuchungen / Philosophical Investigations*. Oxford: Basil Blackwell, 1958 [1953].
Wittgenstein, Ludwig. *Notebooks 1914-1916*. edited by G.H. von Wright & G.E.M. Anscombe. Oxford: Basil Blackwell, 1961.
Wittgenstein, Ludwig. *Philosophische Bemerkungen. Aus dem Nachlass herausgegeben*. edited by Rush Rhees. Oxford: Basil Blackwell, 1964.
Wittgenstein, Ludwig. Letters to *Russell, Keynes, and Moore*. Oxford: Basil Blackwell, 1974.
Wittgenstein, Ludwig. *Vermischte Bemerkungen; eine Auswahl aus dem Nachlass*. Frankfurt: Suhrkamp, 1977.
Wittgenstein, Ludwig. *Bemerkungen über die Farben / Remarks on Colour*. edited by G.E.M. Anscombe. translated by Linda L. McAlister and Margaret Schattle. Oxford: Basic Blackwell, 1977.

Chapter Eight

CARNAP'S MACHIST "PHASE"

by himself

Section Guide
1. Mach's Influence on Carnap
2. "Metaphysical Reality" (Carnap)
3. Inclusive and Fair Classification
4. Carnap's Basic Epistemology

1. MACH'S INFLUENCE ON CARNAP

Rudolf Carnap was born in 1891 in Ronsdorf near Barmen in Northwest Germany. He was the 14th and last child of Johann Sebulon Carnap (1826-1898). The father was a self-made man in the weaving business who became wealthy late in life, which after the death of his first two wives allowed him to marry (1887) into a well-educated family[1] which included a prominent educator and archaeologist.[2] After his death, Anna née Dörpfeld, his third wife, a poet, teacher, and writer, tutored young Rudolf for three years. In Gymnasium his favorite subjects seem to have been Latin and mathematics.[3] He studied at the universities of Jena and Freiburg from 1910 to 1914.[4] He remembered that Bruno Bauch at Jena had taught him Kant's philosophy and Hermann Nohl had discussed Hegel's views, but it was Gottlob Frege (1848-1925) who impresed him most.[5] He told Carnap about a new kind of formal logic which "could serve for the construction of the whole of mathematics".[6]

Frege would also become rightly or wrongly famous for the notion that "constructs should replace inference", an approach which suggested that all "real" science was idealized, an opinion which no historian or lover of non-idealized science is likely to accept. But the popularity of theoretical physics following the theories of Einstein on special and general relativity early in the century and quantum mechanics later also helped popularize starting from axioms and laws and then deducing the resulting idealized science, especially abstract, mathematical science.

During the nineteenth century the experimental discoveries of Faraday, Pasteur, and many others had given the impression to numerous people that discoveries came first and theories and mathematics afterwards,[7] but Einstein seemed to reverse that as if theories and mathematics came first in science and observation along with most experiments were primarily ways to "test theories". Ernst Mach had grown up during the earlier period and as a keen experimental physicist he naturally looked at theories and mathematics as secondary. Theories were merely temporary[8] and mathematical idealization was primarily an economic tool to simplify and save time and effort. Carnap as a child of the age of Einstein but who was also familiar with Mach's ideas sought some kind of compromise in methodology of science which would retain the strongest points of both theoretical and

J. Blackmore, R. Itagaki, and S. Tanaka (eds.). Ernst Mach's Vienna 1895-1930. 159-186.

and idealized side of science, which the new quantum theory of the 1920's would clearly strengthen.

Rudolf Carnap would serve in the First World War,[9] finish his studies in physics and philosophy afterwards, and arrive in Vienna in 1925 and more permanently in 1926, but it can be said that the background above already disposed him toward the career and publications which would later make him well-known, especially in the United States of America. Like many mathematicians and physicists he had the ideal of doing science without being bothered by what Kant and later Mach had called "metaphysics", an abusive term which sometimes meant the study of all reality, sometimes the study of alleged trans-sensory or trans-conscious reality, and sometimes simply anything which seemed to be "unscientific".[10]

Three maneuvers to avoid the bugaboo of "metaphysics" were in vogue at that time. First, one could extend the term to include all philosophy and everything outside of science, an approach which Boltzmann sometimes had seemed to favor.[11] Second, one could objectify meaning as logicians and mathematicians often did as if language were primary and meaning could be determined by conventions or dependent on or part of language. Having done that one could then attempt to conflate positions within philosophy, especially in epistemology and ontology, as if unwelcome epistemology were "metaphysical". and later "merely linguistic" or "external to science".[12] And third, one could lean toward the views of Mach and Richard Avenarius that methodological phenomenalism[13] and *Denkökonomie*[14] - the former being what William James and Bertrand Russell would later call "neutral monism"[15] - formed useful guidelines about how to conduct science, since it could be interpreted to make virtually all philosophy or at least "non-neutral" philosophy seem unnecessary. Though he appears to have been as a logician and physicist sympathetic with the first two approaches, Carnap during most of the 1920's attempted to write a book which incorporated at least part of Mach's point of view.[16] But in addition to working on his book, Carnap also wrote many articles during the 1920's where the influence of Mach rose and fell more than once.

Let us turn to Rudolf Carnap's early writings but start in a rather modest manner by quoting from Edmund Runggaldier's account of Carnap's earliest manuscripts and short publications in so far as they touch on Ernst Mach or phenomenalism..

> Carnap is influenced not only by conventionalist philosophers like Hugo Dingler and Poincaré, who had strongly criticized the notion of absolute space, but also by many German positivists. Among the latter Mach and Avenarius had developed the idea of "neutral monism" which Carnap adopts in his account of the basis for the constructionist system in the *Aufbau.* (page ii)
>
> However, Carnap's attempt to reduce the "secondary world" of common sense to the phenomenal "primary world" was not successful. (page ii)

Several of Carnap's early works including his doctoral dissertation *Der Raum* (1921) published a year later[17] and "*Dreidimensionalität des Raumes und Kausalität*" (1924)[18] reflect an intellectual struggle between Carnap's attraction toward conventionalism and Mach's phenomenalism with the former gradually winning out, a fact indicated by the title of Runggaldier's book *Carnap's Early Conventionalism* (1984).[19] But the acute reader must not imagine that the end of Mach's conspicuous phenomenalist influence meant the end of all his influence. Carnap's basic presentism would survive and it is possible that Mach as well as Kant and many other thinkers contributed to that. Indeed, as mentioned before it was endemic in much German philosophy. Here are selected quotations from Runggaldier

> It is worth noting here that Rudolf Carnap began to be interested in the construction of an axiomatic system for a physical theory of space (and time) as soon as he passed his [oral?] examinations in Jena. He thought that he might develop such a system into a doctoral dissertation. But the outline he wrote was rejected by Professor [Max] Wien, head of the department of physics at the University of Jena, as well as by Professor Bauch [his old teacher of Kant before the war]. The topic of the outline seemed to fit neither into physics nor into philosophy. But in the end he came to an agreement with Bauch that he should switch the project to the philosophical foundations of space.[20]
>
> Thus, because a philosophy professor, but not the physics professor who was his first choice, accepted his plan for a dissertation Carnap would find himself headed toward a career in professional philosophy rather than one in physics. From a Kantian perspective, Professor Bauch might naturally have hoped that Carnap's "philosophical foundations of physics" might somehow evolve into an a priori world of mental categories resembling that of "the master", but Carnap's focus in his doctoral dissertation turned out to be much narrower than that. Using Mach's distinctions between metric, physical, and physiological [intuitive] spaces as a starting point, Rudolf Carnap perhaps with some familiarity with Husserl's work on different kinds of spaces as well attempted to think through the idea of topological space. Runggaldier has added in a sympathetic note that "Carnap's concentrating on topological relations as having a different cognitive role from metrical relations is basic to any theory of knowledge."[21]

Runggaldier, however, is rather critical of some aspects of Carnap's dissertation:

> Unfortunately, there are unsolved problems which stem from Rudolf Carnap's account of topological relations. One can ask, e.g. is T 'phenomenalistic' (at least in the sense of his methodological solipsism) . . . or, why is Carnap's topological physical space a three dimensional one (it is not at all clear why it should be a three-dimensional one); or, does Carnap's topology include any neighborhood relations, and if so, where do they come from; how does time affect topological relations; and so on.[22]

But for all of the apparent flaws which Runggaldier says he discovered in Carnap's doctoral dissertation, it was accepted by the Universiy of Jena presumably in 1921 and published the next year. But Carnap was still eager to pursue his rather narrow interests. His focus on logic, mathematics and related philosophy naturally inclined him to prefer the lectures and seminars of those philosophers who also had a strong attachment to mathematics, especially Mach's mathematically-minded friend Hugo Dingler in Munich, and in the years 1924-25 Mach's sharp opponent Edmund Husserl in Freiburg.[23]

But Dingler was not only a theorist in geometry and a philosopher of theoretical science but a friend of Ernst Mach and was preparing a book on his ideas which would appear in the year 1924.[24] Carnap has alleged that he reacted against Dingler's excessive conventionalism, presumably in favor of a more modified conventionalism. In short, Dingler may have had the effect of briefly strengthening the empirical or Machist side of Carnap's approach but without significantly shaking Carnap's primary interest in logic and ideal constructions. Nevertheless, the dualism remained and the phenomenalist aspect which the Vienna Circle would also emphasize would not be as easily reduced or subordinated as Carnap himself would shortly come to wish.

From 1921 until 1924 Carnap worked on a number of articles as well as his future book *Der Aufbau*[25] such that because the dates of publication do not necessarily reflect the dates or order of composition it can be very difficult to determine the historical sequence or the true development of his philosophical ideas. Runggaldier, however, believes that Rudolf Carnap's dissertation was followed by a rapid repudiation of much of Bauch's Kantian influence and his own extreme phenomenalism in favor of "Carnap's early conventionalism" and a more modest Machism. But this picture makes it a bit awkward to explain the strong Machist influence on Rudolf Carnap's 1924 article "*Dreidimensionalität des Raumes und Kausalität* and *Der Aufbau* (1928). Nevertheless, it is true that his very first article "*Über die Aufgabe der Physik* . . ." (1923)[26] does favor conventionalism while placing it in a partly Einsteinian context which went well beyond his previous three-dimensional topological analysis in his doctoral dissertation. But since it is possible that one can reconcile greater *interest* in Mach's phenomenalism on Carnap's part with abandonment of an *extreme* Machist position, it may well be that for all of our reservations that Edmund Runggaldier does have the chain of events in Rudolf Carnap's intellectual development *generally* correct, though it may be useful to remember that Mach's influence seems to have become stronger in *Der Aufbau* (1928). But even though Carnap alleges that the latter work was basically written from 1921 to 1925, the renewed or "never removed" early focus on Ernst Mach's phenomenalism and theory of elements may have been partly influenced by either Mach' s friend Hugo Dingler or the largely Machist orientation of the Vienna Circle circa 1925 when he first arrived in Vienna. But because the matter is still so unclear and Runggaldier gives us the most detailed and largely plausible account let us once again quote from him:

> There are many similarities between his [Mach's] and Carnap's early phenomenalism. But Ernst Mach's general approach to "the given" is different from Carnap's; this will be especially clear as soon as one looks at his "biological" interpretation of "tolerance". (page 21)
>
> In his introductory remarks to his first article "*Über die Aufgabe der Physik* ... " Carnap maintains that the question about the source of physical knowledge is not purely phenomenal or empirical; the construction of physics is not solely the result of empirical observation and experimental investigation but also of non-experimental principles, including conventions. Carnap adopts Henri Poincaré's and Hugo Dingler's conventionalist thesis that, to build up physics, conventions ("*Festsetzungen*") which depend on free choice are necessary.[27]

Conventionalism was fashionable among Machists. There seems to have been no great difficulty among some positivistically sympathetic physicists to combine Machist phenomenalism with Henri Poincaré's conventionalism or the more extreme form favored by Dingler. After all, both Poincaré and Dingler acknowledged debts to Ernst Mach's epistemology,[28] Philipp Frank, another physicist, was a rabid Machian but also an extreme conventionalist at least circa 1910, even arguing that causes were merely conventions.[29] Furthermore, another member of the Vienna Circle, indeed a professor who helped bring Moritz Schlick to Vienna in 1922, namely Hans Hahn, was a fervent Machist[30] and also a mathematician who was fully capable of using and defending abstract constructs and conventions.

Under the influence of Frege and Husserl they seemed to reject Ernst Mach's "psychologism" about the nature and origin of formal logic and while they defended the importance of observation, this in no way weakened their trust in mathematical abstractions and in formal conventions. In other words, one may question how seriously Carnap was attracted to Mach's phenomenalism even while writing his *Aufbau.*

First, let us admit that what is called "conventionalism" is a very natural position for logicians and mathematicians to hold. They are familiar with many abstract problems where more than one answer seems appropriate, and where one cannot say that one is "more true" than another. And second, even when one crosses over the line between mathematics and physics, there are scientific laws which seem valid even though they are not literally true. This situation occurs because while the law may imply or be implied by other laws, because it takes fewer variables and constants into account than may actually exist in particular situations, it is both idealized and literally false. Even the claim that it is true under idealized or counterfactual conditions, i.e. if the universe only had the constants and variables specified by the law, such may still be false since in fact an indefinite number of unexpressed variables and constants may be presupposed by the scientific law or their understanding of it.

Nevertheless, most physicists themselves seem to treat most laws as "approximately true", a convenient touch of hopeful thinking which assumes that the constants and variables which are not stated or assumed by laws or are not part of our understanding of the "initial conditions" cannot be important or influential enough to significantly affect relevant situations. But naturally people who allow the possibility that some unanticipated variables may be powerful and quite disruptive (such as human intervention) cannot blindly accept that all scientific laws are always "approximately true" in real situations. Nor is it easy for historians to accept the notion that scientific laws and other idealizations are more real than either mind or matter.

Some of the mathematicians familiar with these problems, led by Henri Poincaré in the late 19th century, proposed that law-focused, mathematical science must aim at something else than truth as correspondence with reality, whether exact or "approximate", since *deterministic* laws could not qualify as being true in correspondence terms, and neither *probabilistic* nor so-called *empirical* laws allowed enough implication and deduction of other laws to permit the creation of a mathematically necessary inclusive system in physics, especially of the kind which Einstein would increasingly come to favor.

> Carnap's article "dreidimensionalität des Raumes und Kausalität eine Untersuchung über den logischen Zusammenhang zweier Fiktionen" [1924] is another attempt at a synthesis between phenomenalism and conventionalism . . . . Carnap denies that there is causality in the phenomenal world which he calls, here, the 'primary world'. Causality is given only in the constructed 'secondary world'.[31]

> But Carnap, following the new trend of "positivism" in the German-speaking world, attaches "reality-value" only to the "primary world", the "secondary" being " . . . merely an arbitrary transformation of the former carried out on economical grounds. For Carnap the "secondary world", viz. the world of "causality" is a "fiction" in a metaphysically negative sense. (page 38)

Rudolf Carnap in "*Dreidimensionalität ...*" clearly sought to extend this argument by pointing out that if Einstein's special and general theories of relativity are truly correct, such that non-Euclidean or multi-dimensional assumptions about space must be introduced into theoretical physics, even though reality has no intrinsic dimensional character and traditional physical causes can only be understood in terms of three spatial dimensions, then the whole notion of correspondence with reality and of causes as forces in a three-dimensional and temporal context may be compromised.

If in light of Einstein's multi-dimensional theories there is no coherent notion of either reality or causes, especially "in themselves", then a rather conventionalist approach in theoretical science may appear to be one of the more reasonable abstract options left. Laws are thus viewed as no more than internally consistent and "necessary" conventions but which are neither true nor false in terms of correspondence with reality. But since it is normally convenient to regard them as true, valid, or at least "approximately true", the argument arose from several sources that one might appeal to "relative truth", either on pragmatic, coherence, or on Vaihinger's "as if" grounds,[32] the latter being a perspective which was quite popular during the 1920's, with his main book going through at least ten editions by the end of 1927.[33]

It is also interesting that *Erkenntnis*, the Logical Positivist house-organ of the 1930's is said to have been taken over from the "As If" people even before Vaihinger's death in 1933.[34] In fairness, we all tend at times to treat what seems highly probable "as if" it were true, real, right, or certain, but Vaihinger's and Carnap's fictionalism went well beyond that stage, namely, to the position that much that *seems* improbable, neither true nor false, or even certainly false should in some or even many particular scientific contexts be treated "as if" it were indeed true, real, right, certain, or at least valid, a position which Ludwig Boltzmann had also held during the late 19th and early 20th centuries in his defence of Newtonian physics and his own kinetic theory of gases. But it is possible that the Vienna Circle and most Logical Positivists were less than enthusiastic about Vaihinger's tendency to call his fictionalist philosophy: "Idealist Positivism", an expression which might remind some readers of Vladimir Lenin's attempt in 1909 to associate the positivists of his day (Mach, Avenarius, and especially Bogdanov) with the Idealist epistemology of Bishop George Berkeley.

Even though many scientists and philosophers since Boyle and John Locke have generally been considered "empiricists", this did not mean they accepted a phenomenalist or presentist epistemology about matter or the external world, but Rudolf Carnap even from his earliest work seems to have thought of himself as an "empiricist" or "empirical constructionist" as if that *excluded* the indirect epistemologies of Boyle and Locke. Carnap seems to have been cavalier with many other definitions as well, often dismissing views he didn't like as "metaphysical" even if held by successful scientists.

From the perspective of most serious research historians: labels, definitions, and classification should not be used for rhetorical purposes, but as instruments to help accurately place one's point of view within the larger constellation of philosophical positions, hence it is difficult for historians to be fully neutral toward thinkers who violate this point of honor, such as Philipp Frank, Edmund Husserl, and especially Rudolf Carnap. All three men and Logical Positivists generally often narrowed the definition of terms in order to exclude opposition views from fair consideration. But during the 1920's Carnap was still new at this practice, which makes it a littke easier to describe his ideas without undue personal discomfort. Nevertheless, when possible we will rely on quotations from Edmund Runggaldier who seems to have been less bothered by "doctored definitions". In order to be fair to Carnap, however, we will in Section II let him speak for himself without interruption.

Carnap of course with his emphasis on idealization on the one hand and presentist assumptions on the other found that truth as correspondence with reality was a hard concept to absorb into construction-focused methodology of science. Still, he continued to seek aspects of compromise particularly in his "*Dreidimensionalität* . . ." article in which he distinguished between "primary science" and "secondary science",[35] but since his conception of primary science was still empirical in a sense with phenomenalist or largely Machist overtones most scientists would probably have rejected Carnap's approach as Planck and Einstein would do.

Carnap at some point during the 1920's, perhaps under the influence of Wittgenstein, began to seriously pursue the goal of developing a plausible methodology of science fully free from epistemology and ontology while also including a greater emphasis on theories and idealization, a goal which could not be accomplished by incorporating Mach's phenomenalism. Frege had suggested objectifying meaning as if language were primary as allegedly in logic and mathematics and of course in Vienna when Schlick's Circle read or rather re-read Wittgenstein's *Tractatus* about 1926, then "better" avenues to "free" modern physics and science from its own deeper epistemological and ontological assumptions began to seem more possible. By the end of the 1920's Carnap was well on his way to adopting convention-modified phenomenalism as expressed in terms of a physical object language. Rungaldier, however, has summed the matter up as follows:

Der Aufbau (first version) 1925

> In the *Aufbau* the application of set theory enabled him [Rudolf Carnap] to take the phenomenalist doctrine of Mach, Avenarius, and Russell seriously. (page 45)
>
> Carnap defends the free choice of method viz. "language", for semantic analysis - on the grounds of a presumed or alleged absence of any ontological commitment. (page 46)
>
> A physicalistic instead of a phenomenalistic language is not merely another mode of speech, but it implies a formulation of a different philosophical position as well. (page 47)
>
> Toward the end of the article Rudolf Carnap claims ontological neutrality for his constructional system. (page 52)

Physikalische Begriffsbildung (1926)

It is not the case, Carnap points out, that any physical concept has meaning independently of the rules of correlation. (page 54)

Carnap already has a strong aversion against investigations into what entities like "concepts" are .... Thus to refer to a concept means referring to a uniform and regular use of a sign or symbol ... Carnap's account of concept formation given in *Physikalische Begriffsbildung* has little to do with phenomenalism. (page 57)

In briefly addressing the contents of Carnap's long-delayed work, his *Aufbau*, which in English became called *The Logical Structure of the World*, his attempted homage to Ernst Mach, whose contents he had already partly rejected by the time it appeared (1928), we do feel obligated, nonetheless, to mention his many references to Mach and his ideas.

Carnap on Mach

The system which is formulated in this book takes as basic elements the elementary experiences ... something similar to Mach's elements, e.g. concrete sense data.[36]

He has also added several additional references to Mach while discussing the book in his autobiographical remarks (1963):

At first I made the analysis in the customary way, proceeding from complexes to smaller and smaller components, e.g. first from material bodies to instantaneous visual fields, then to color patches, and finally to single positions in the visual field. Thus the analysis led to what Ernst Mach called the elements. My use of this method was probably influenced by Mach and other phenomenalist philosophers. But it seemed to me that I was the first who took the doctrine of these philosophers seriously. I was not content with their customary general statements like "A material body is [only] a complex of visual, tactile, and other sensations", but tried actually to construct these complexes in order to show their structure.[37]

The choice of a phenomenalistic basis was influenced by some radical empiricist or positivist German philosophers at the end of the last century whom I had studied with interest, in the first place Ernst Mach, and further Richard Avenarius, Richard von Schubert-Soldern, and Wilhelm Schuppe. For the construction of scientific concepts on the phenomenal basis I found fruitful suggestions in the works of Mach and Avenarius, and above all, in the logical constructions which were made by Russell.[38]

> For the construction of the world of physics on the basis of the temporal sequence of sensory experiences, I [Carnap] used the following method. A system of ordered quadruples of real numbers serves as the system of coordinates of space-time points. To these quadruples [I added] sensory qualities, e.g., colors, are assigned first, and then numbers as values of physical state magnitudes. They are then made according to general rules of maximization.[39]

Edmund Runggaldier adds his interpretation of Carnap's approach to *Der Aufbau* [1928]

> It is commonly said that the 'logical construction of the world" outlined by Rudolf Carnap in this *Aufbau* is an attempt at a phenomenalistic system. One should, however, be sceptical of such an isolated characterization of the *Aufbau*. (page 61)
>
> In the case of the *Aufbau*, on the one hand, the characterization of the constructional system in it as "phenomenalistic" might be appropriate since Carnap does choose for it, to begin with, a "phenomenalistic" basis (§61-67). On the other hand, however, the elements of Carnap's basis differ strongly from the "sense data" in the tradition of British empiricism. The basic elements of the *Aufbau* are not - in Goodman's terminology - "qualia", i.e."non-concrete qualitative elements" or repeatable "presented characters". Besides this special difficulty the characterization "phenomenalistic" does not do justice at all to one of the most puzzling aspects of the *Aufbau*. This aspect consists in Carnap's attempt at the elimination of even the last non-logical concepts or objects from his constructional system (§153ff.) (p. 61)
>
> Carnap accepts and shares Poincaré's standpoint that not 'the given' itself, if taken as a complex of qualitative 'contents', but only the relations between these 'contents' of sensations have an objective value. Carnap, however, points out that Poincaré did not go far enough. From relations one has to go to structures.... Rudolf Carnap himself has called this process 'formalization' and 'dematerialization'. Thus all the objects of a mature science or of real knowledge are not 'content' but 'form'.[40]
>
> In fact, at the beginning of his phenomenalistic system Carnap places not phenomenal elements or classes of such elements but phenomenal relations . . . . The basic relations take precedence over the basic elements, but the latter cannot be fully made or constructed out of the former.[41]
>
> The basic relations take precedence over the basic elements, but the latter cannot fully be constructed out of the former. (p. 71)
>
> On the one hand, it is true that the elementary experiences are preferred as basic elements because they are in Carnap's view, the closest approximation to what is given, namely, a single and unbroken stram of experience; but on the other hand, Carnap's formal-logical" interests are

hard to overestimate. They are the ultimate reason for his choice of his narrow basis, (p. 81)

Within the actual constructional system, only the basic elements are genuine "objects" with a non-fictive status. All of the other "objects" are "quasi objects" or classes of primitive objects . . . . [Only] the elementary experiences are indivisible and incapable of analysis . . . (p. 90)

What can be proved intersubjectively and communicated from one person to another is only relations or structures, but not the elements as such between which they hold. "Sense impression contents" are intersubjectively not accessible. (p. 125)

Carnap's actual constructional steps do not reach beyond sense impressions, or, in more detail, beyond the visual places and colours of the visual field. (p. 132)

But on looking back from his perspective of almost forty years later Carnap makes a criticism of *Der Aufbau*: in favor of what he would call "physicalism", even though the criticism that relations take precedence over sensations also allegedly applies to objects. In short, by trivializing the differences between physical objects and sensory elements by subordinating them both to relations, he is also attempting to detach his methodology of science from both epistemological and ontological questions, something only his later attempt to reduce epistemological and ontological differences to words or "different languages" would ultimately seem satisfactory to him. Carnap writes:

The positivist thesis concerning the reducibility of thing-concepts to auto-psychological concepts remains valid, but the assertion that the former can be defined in terms of the latter must now be given up and hence also the assertion that all statements about things can be translated into statements about sense data.[42]

2. THE METAPHYSICAL PROBLEM OF REALITY (1928)[1]

by Rudolf Carnap

A

We now wish to deal with a problem of reality which is quite different from the one that we have discussed up to now. We have determined what constructional (empirically ascertainable) conditions must be fulfilled in order for an object to be called real in the customary usage of the empirical sciences. In addition to this "constructional" or "empirical" problem of reality, the question may arise whether or not we must ascribe "reality" in a special sense to these empirically real objects. For this special sense, there are various possible formulations; most commonly, it is characterized as *independence from the cognizing consciousness*. Thus, we are obliged to differentiate two different meanings of the word "reality". Wherever it is necessary, we shall indicate them by calling the one "empirical reality" and the other "metaphysical reality". Later on we shall give a justification for this second expression.

EXAMPLES. The difference between the two meanings becomes clear through the following two questions: "Was the Trojan War a real event or merely an invention?" and "Are those objects which are not feigned or simulated, for example, the perceived physical bodies, real, or are they merely contents of consciousness?" The first question is treated by historical science; it is to be resolved with empirical and constructional methods, and hence there is no divergence of opinion among the adherents of the various philosophical schools. The second question is customarily transacted within the field of philosophy; it is answered in different ways by different schools; we shall see later that it is extra-constructional and hence extrascientific; it is metaphysical.

REFERENCES. We commonly use the words "actual" and "real" as synonymous. Külpe [Realis.] distinguishes the postulated, inferred (i.e. constructed) objects from the processes of consciousness. He calls the former "real", and the latter "actual"; but this seems a little two far removed from customary usage.

B

The second concept of reality (in the sense of independence from the cognizing subject) indicates the point where the schools of realism, idealism, and phenomenalism part company. These schools are distinguished from one another by the fact that they ascribe reality in the second sense to object domains of varying extent (within the field of the empirically real). *Realism* holds that the constructed physical and heteropsychological objects are real. Subjective *idealism* holds that the heteropsychological, but not the physical objects are real. The more radical form of solipsism denies even the reality of heteropssychological objects. (Objective idealism ascribes reality to a superindividual, absolute subject [Hegel], which is not constructed within our system; hence, we shall not consider this school in the present contact.) Phenomenalism [according to Carnaps unusual definition] agrees with realism in maintaining that real entities exist outside of the domain of the autopsychological; on the other hand, it agrees with idealism in denying this reality to anything physical; according to phenomenalism, reality must then be ascribed to unrecognizable "things-in-themselves", whose appearances are the physical objects.

[In Carnap's view] the concept of reality (in the sense of independence from the cognizing consciousness) does not belong within (rational) science, but within the realm of metaphysics. This is now to be demonstrated. For this purpose, we investigate whether this concept can be constructed, i.e. whether it can be expressed through any objects of the most important types which we have already considered here, namely, the autopsychological, the physical, the heteropsychological, and the cultural. At first impression, it might appear as though this were possible. An object which I have recognized, that is, an object which has been constructed on the basis of my experience, will have to be called "independent of consciousness" if its constitution does not depend upon my will, i.e. if an act of volition which aims at a change of the object does not result in such a change. But this does not agree with the concept of reality as it is meant by realism and idealism (the former ascribing it to, and the latter denying it of, physical bodies). For, according to the definition which we have just attempted, a physical body which I hold in my hand should not be called real, since (even in the opinion of realists) it changes if I carry out an approximate act of will; this would then become contradictory to the realist position. On the other hand, this definition seems to require that any physical thing which lies outside of our technological reach, for example, a crater in the moon, should be acknowledged as real, since (even in the opinion of idealism) it does not change if I carry out an appropriate act of will; this then would be contradictory to the position of idealism.

One could try in various ways to give a definition of reality (in the sense of independence from my consciousness) in such a way that the concept becomes constructable. However, one can show in each such case that the concept which is so defined does not agree with the concept as it is meant by realism as well as by idealism. it must be noted that this holds, not only of a constructional system which has the system form represented in our outline, but for any experiential constructional system, even for a system which does not proceed from an autopsychological basis, but from the experiences of all subjects or from the physical. *The (second) concept of reality cannot be constructed in an experiential system; this characterizes it as a nonrational, metaphysical concept.*

REFERENCES. It seems that we agree with Bertrand Russell (Scientif.] 120ff. in the indicated conception that the concept of non-empirical reality cannot be constructed. However, this does not seem to be consistent with the fact that in Russell questions of the following kind are frequently posed, which (independently of how they are answered) imply a realistic persuasion: whether physical things exist when they are not observed, whether other persons exist; whether classes exist; etc. ([Scientif.], [Mind] 308, [External W.} 126, [Sense-Data] 157 and elsewhere). Cf. also Weyl [Handb.] 89.

C

The indicated conception of the concept of reality is related to that of positivism, which goes back to Ernst Mach. Cf., for example, Ostwald [Naturphil.] 101ff.; the concept of reality as it is defined there corresponds to the constructional concept of reality. The same holds for the concept of reality as it is defined by Bavink (Ergbn.] 26, 187); thus Bavink is right when he describes it as neutral relative to the realism problem.

The definition of the concept of a *thing-in-itself* goes back to the concept of reality (in the sense of independence from the cognizing subject). Thus, in our conception, this concept, too, must be placed *within metaphysics*, for metaphysics is the extrascientific domain of theoretical form (§182).

REFERENCES. If things-in-themselves are defined as real objects which are not given (as is done by Schlick [Erkenntnis.] 179), then they must indeed be counted among the cognizable objects and thus must be placed within the domain of (rational) science and not within metaphysics, for then they coincide with the constructed real objects. However, it seems to us that this definition is not very practical, since it deviates altogether too much from customary usage (cf. Külpe [Realis.] II, 213). The same holds also for the characterization of the constructed real objects as transcendent ([Erkenntnisl.] 180). The essential limit of transcendence, according to customary usage, lies between what is recognizable (in our language constructable) objects and the nonrecognizable (not constructable) objects. If one wishes to emphasize here, through a special expression, the limit between the given objects and those objects which are constructed but not given, then the term "transgression" {"transgredient" or *transgressive* objects) may serve for this purpose; this term has been introduced by Ziehen [Erkth.] 275; Ziehen justifiably makes a sharp distinction between this concept and the concept of transcendence.

D

The following concerns objects which are empirically real (i.e., objects) which are (in constructional language) placed in the total system of the object types in question; cf. (in realistic language) objects which have been "recognized" or "determined" as "real").

Relative to these empirically real objects of the various object types, construction theory and *realism* agree on the following points: 1. Such objects can be clearly distinguished from the unreal objects of the same object type (dreams, hallucinations, inventions, etc.). Only to the extent that they can be clearly distinguished, are they used in the formation of the system of knowledge. 2. They can be intersubjectivized, i.e., in principle, they can be placed in constructional systems which belong to other persons and can be confirmed or corrected through the reports of other persons; they are included in the system of knowledge only to the extent to which they are capable of being intersubjectivized. 3. They are independent from being cognized in the sense that they exist also at times when they are not represented in my experiences or in the experiences of another. 4. They are independent of me, in the sense that a wish to change them does not result in a change of their characteristics unless a physical causal chain connects an appropriate motion of my body with the object in question. 5. They are governed by their own regularities which makes it occasionally possible to make predictions: if I put my body in an appropriate position, then an experience of a certain predictable kind takes place, whether I want it or not. However, there is agreement, not only in the points just mentioned, but in all points in which assertions are made at all by both theories. *Construction theory and realism do not contradict one another in any point.*

Construction theory and subjective idealism agree with one another in the claim that statements about objects of cognition can, in principle, all be transformed into statements about structural properties of the given (with retention of the logical value, cf. §50). Construction theory agrees with *solipsism* in that the given consists of my experiences. Construction agrees with *transcendental idealism* in the conception that all objects of cognition are constructed (in idealistic language "are created in thought"); in fact, the constructed objects are objects of conceptual knowledge only *qua* logical forms which are generated in a certain way. Ultimately, this holds also for the basic elements of the constructional system. Even though these basic elements are initially introduced as unanalysable units, eventually, in the progess of construction, different properties are attributed to them, and they are analyzed into (quasi) constituents (§116). It is only through this type of procedure, that is, only as constructed objects, that they become objects of cognition in the proper sense of the word, that is, in particular, objects of psychology. Here again we have the situation that there is agreement between idealism in its different varieties and construction theory in all points where assertions are made by *both* theories. *Construction theory and idealism* (objective, subjective, and solpsistic idealism) *do not contradict each other at any point.*

The same holds for *phenomenalism*. For, aside from construction theory, and construction theory neither affirms nor denies the existence of things-in-themselves. Here again, we find agreement in all points where *both* theories make assertions. *Construction theory and phenomenalism do not contradict one another at any point.*

It is not particularly surprising that none of these doctrines - realism, idealism (in its different varieties), and phenomenalism - show in themselves any contradiction to construction theory, yet contradict one another. For, the three schools do, after all, agree with one another and with construction theory on the following points: ultimately, all knowledge goes back to my experiences, which are related to one another, connected and synthesized; thus, there is a logical progress which leads, first, to the various entities of my consciousness, then to the physical objects, furthermore, with the aid of the latter, to the phenomena of consciousness of other subjects, i.e. to the heteropsychological, that is, to the cultural objects. *But this is the theory of knowledge in its entirety.* Whatever else construction theory states about the necessary or useful forms and methods of construction belongs to the logical, but not to the epistemological, aspect of its task. The theory of knowledge does not reach beyond what has just been indicated. How cognition can proceed from one object to another, how in what sequence, and in which form the levels of a system of cognition can be formulated, - all this is contained in the indicated material. The theory of knowledge cannot ask any further questions.

But where do the contradictory aspects of realism, idealism, and phenomenalism belong, if not to the theory of knowledge? The assertions of these doctrines which stand in contradiction to one another are all related to the second concept of reality, and this concept, as we have already seen, belongs to metaphysics: from this it follows: *the so-called epistemological schools of realism, idealism, and phenomenalism agree within the field of epistemology. Construction theory represents the neutral foundation which they have in common. They diverge only in the field of metaphysics, that is to say (if they are meant to be epistemological schools of thought), only because of a transgression of their proper boundaries.*

It is occasionally said that there is a (usually tacit) realism at the bottom of the practical procedures of the empirical sciences, especially of physics. However, we must clearly distinguish between a certain kind of language usage and the assertion of a thesis. The realistic orientation of the physicist shows itself primarily in the use of realistic language: This is practical and justifiable. On the other hand, realism, as an explicit thesis, goes beyond this and is not permissible; it must be corrected so as to become "objectivism": the regular connections (which in natural laws are formulated as implication statements) are objective and are independent of the will of the individual; on the other hand, the ascription of the property "real" to any substance (be it matter, energy, electromagnetic field, or whatever) cannot be derived from any experience and hence would be metaphysical.

REFERENCES. The above-indicated standpoint is closely related to what Gätschenberger, [Symbola], says about the reconciliation between idealists and spiritualists on the one hand, and materialists on the other: "Spiritualism is a translation of spiritualism"; "All philosophers are thus correct, but they express themselves with varying degrees of ineptness, and they cannot help this, since they use the available language and thus consequently speak in a hundred sub-languages, instead of inventing one pasigraphy." This mental language is the goal of construction theory.

[Carnap's study of realism] contains detailed exposition of the difference between the empirical and the metaphysical concept of reality and more exact reasons why the realism debate should be banished from science and placed within metaphysics.

3. INCLUSIVE AND FAIR CLASSIFICATION

by the editor

Reasonable people who opposed George Berkeley's idea-lism, his attempt to identify the physical world with sensations which he called "ideas", held exactly the same poaition which Carnap alleges is "metaphysical" and "incompatible with science", namely, that there is a physical world and it exists independently of what can be made conscious to us, what has traditionally been called epistemological *not metaphysical* realism. By associating metaphysics with both common sense and the position of most great scientists from Galileo to the present Carnap not only misleads philosophers but falsifies the history of science and philosophy. People who deliberately mislead are dishonest and equating opponents of Berkeley with "metaphysical realism" or "metaphysics" is not fair.

One of the glories of the Scientific Revolution was Galileo's insistence[1] that the primary purpose of science was not merely to "save the appearances" but to understand the real external physical world, a world beyond everything sensory and conscious. The apparent motion of the planets and stars around the earth were not their real motions. Nor were the sun and the moon the same size as seemed to be the case from a strictly sensory or empirical perspective; nor did real physical objects become smaller as we moved farther away from them or the planets go backwards for several weeks in the summer as strict empiricists seemed forced to hold. One had to rise above strict empiricism and direct realism in order to understand and accept a representational type of indirect realism about the external physical world.

All of this became common sense in most western European countries, except Germany and later Austria, where the contrary ideas of Immanuel Kant increasingly dominated philosophy. Rudolf Carnap seems to have absorbed the basic idea-lism of Immanuel Kant, presumably from his university teachers. Carnap like many other Germans tried to reverse the basic approach of Galileo and the Scientific Revolution, by treating the physical world of Galileo, Descartes, Boyle, Locke, Newton,[2] Darwin, and many other great scientists as "metaphysical" as if incompatible with science. Only in Germany or with people similarly ignorant of the history of science could Carnap's trick succeed

The problem for fair and inclusive classification has been to distinguish between direct and indirect epistemology instead of rejecting everything consciously indirect as "metaphysical" or "incompatible with science". One has to learn that both direct and the indirect positions with respect to consciousness are or can be epistemological, as if it were better to use inclusive rather than exclusive classification, since the latter approach by limiting reasonable alternatives seems to reduce our potential understanding of the positions in question and epistemology in general. Also, it did not seem to be fair to the excluded points of view..

Nor does direct epistemology include only what is being sensed or conscious now. It traditionally means what in principle could be consciously experienced. Similarly, indirect epistemology does not mean what is not conscious now but what in principle cannot be made conscious to us. Consciousness has traditionally been the touchstone which distinguishes between basic epistemological positions. Were one to deny that consciousness exists at all, then one is a presentist or holds a direct epistemology, much as if one thought that consciousness were all-inclusive. Nor could using a physical object language in later years change Carnap's underlying phenomenalist assumptions. Meaning is not idetermined by linguistic expression, but by the intent of the speaker or writer. This is obvious to most people. Nor does objectifying meaning make it genuinely objective. Intended meaning is private and subjective, and this appears to be an objective fact about that privacy and subjectivity. Reality and fact are not the same in this case.

Carnap wrote in the English translation of his 1928 book which he personally approved in 1961):

> For this special sense, there are various possible formulations [of what he called "metaphysical realism"]; most commonly, it is characterized as *independence from the cognizing consciousness*.[3]

Epistemology is *not* about unconditional knowing. All knowing is conditional and hence probabilistic at best, but is about the limits of possible consciousness which distinguishes epistemological presentism from epistemological representism. Representists often affirm and presentists deny that a trans-conscious physical world can be referred to and understood by science. Carnap seems to recognize here both positions and set of beliefs. He seems to have understood at least one of the three basic factors we have mentioned in the introduction to this book which could contribute significantly to mature understanding of epistemology. But by equating indirect epistemology with "metaphysics" he exposed his own unfairness and mental limitations.

To be sure, one of his problems was a tendency to equate epistemology with "theory of knowledge" instead of with theory of understanding. For in the real world human beings are indeed finite and fallible and no knowledge seems to be truly infallible, whether about the "given" or the allegedly "absolutely certain". Genuine knowledge would have to be *unconditionally* certain and that is too pretentious and unrealizable a standard, hence the need to change the definition of "epistemology" to transform it into theory of probably correct information and understanding, where even misunderstanding is a kind of data or understanding.

Carnap's understanding of epistemology seems to have been narrow and historically unsophisticated even before his attempt to reduce all epistemological questions to linguistic ones. By restricting epistemology to presentist positions he naturally thought that ultimately these differences were merely linguistic, but had he taken into account that epistemology as a field of study includes both presentist and representist positions, both what is direct and indirect with regard to consciousness, then he could not have logically fallen into linguistic reductionism. Far from all epistemological positions merely being linguistically different from each other it is epistemological difference which forces us to use basic philosophical and scientific expressions such as "matter", "mind", "atom", "fact", and "phenomena" differently from people with a different epistemology. Nor is it possible to be clear without epistemological assumptions. All clear reference is either to what can be made conscious or to what cannot be made conscious.

One of the many uses of that prejudicial term "metaphysics'"is theory of reality or ontology. Another common meaning is the study of what is allegedly trans-sensory or trans-conscious, a meaning close to that Carnap seemed to prefer during the 1920's. But indirect epistemology is not the same as indirect ontology. Ontology as normally understood is primarily concerned with whether what is real is ultimately reducible to one type of thing or to two or more types. Epistemology is concerned with what we can understand and how we understand.[4] By dismissing all indirect epistemology as "metaphysical", Carnap revealed his failure to properly distinguish between epistemology and ontology in general as well as how epistemology influences what we mean by ontological terms such as "matter" and "mind", which people holding different positions in epistemology almost always use in different ways.

From the perspective of most historians: first, the same idea can normally be understood in terms of different languages such that meaning is primary and not language. Second, artificial languages are not real languages. And third, epistemological and ontological differences cannot be reduced to linguistic differences. Carnap's later philosophy with its alleged primacy of language fails to properly distinguish meaning from language and mere modes of expression from a genuine linguistic system.

In spite of the work of numerous representist philosophers and scientists in many lands during the 1920's as well as the classic work by E.A. Burtt , *The Metaphysical Foundations of Modern Science,* which first appeared in 1924, Carnap refused to seriously consider a position as accessible to science which characterized reality as what existed independently of consciousness or as an epistemological position.[5] He insisted on limiting the field of epistemology to presentist alternatives while scorning representism as metaphysical and inconsistent with science, and this in spite of the fact that most scientists starting with the Scientific Revolution had very largely been representists in their work. As for his claim that metaphysics was incompatible with science while treating metaphysics as if it included representism, he was in effect arguing that the representist theories of Galileo, Descartes, and Newton were incompatible with science. No reasonable person can accept Carnap's gross distortion of such facts of history. Nor should one be surprised that Carnap's elaborate annotated bibliography at the end of *Aufbau* (English version) includes *nothing* on the history of science, not even Mach's book on the history of mechanics where Galileo is treated as an "empiricist"[6] and Newton as a "lapsed empiricist". In short, young Rudolf Carnap seemed to *reverse* the facts, as if the Scientific Revolution with its very heavy reliance on the representative theory of perception had never taken place.

To be sure, pejorative and misleading definitions of "metaphysics" had long been widespread in Germany, and Mach had popularized his own as well, but even in Germany and Austria there were enough scientists like Max Planck and Albert Einstein who were warning against the prejudicial identification of fruitful indirect epistemological positions with something as abusive and pejorative as using the term "metaphysics" to label indirect positions which were needed especially in later years in order to accommodate the reality of atoms and molecules in a reasonable way.

Even many of the scientists and historians who may share presentist inclinations with Carnap, would surely support inclusive and fair classification which acknowledges the position of *most* other scientists. Regardless of which epistemological position will or ought to prevail in the future, surely presentism should not "win" just by employing restrictive classification which denies relevant status to its strongest opponent or by claiming that all such differences are "merely linguistic". Carnap was aware that honoring the atomic theory was not really the same as believing in the reality of atoms, but by encouraging this misleading conflation and other intellectual tricks he misled generations of followers. His oversights were not all unintentional. Many were deliberate.

It is almost always unacceptable for a mere commentator to impugn the motives of famous philosophers or to suggest that they were not fully fair or honest. But Carnap has to be an exception. He has misled too many people for too long. This is obvious to serious research historians but perhaps may still shock a few young scientists and philosophers. Is rhetoric legitimate? Not if it misleads people, and Carnap consciously misled many people, largely by means of clever re-definitions and deceptive classification as well as by falsely alleging that accepting the atomic theory is the same as believing in the reality of atoms or makes one an atomist. Carnap was a phenomenalist, a type of idea-list who could not accept the reality of atoms *as most people normally understand them.*

With all due respect, Rudolf Carnap cheated and should have been thoroughly ashamed of himself. Indeed, as the childhood rebuke goes: "Cheaters never win"! But alas, philosophy may sometimes be different. But it still requires gullible philosophers, people susceptible to slanted and unfair re-definitions and other types of rhetoric to preserve his reputation, which from a historian's point of view has long been in tatters, and deservedly so. Definitions and classification should possess bedrock neutrality and not be used for purposs of persuasion. But as important as they are, they are no substitute for evidence. To imply that the underlying epistemological assumptions of many of the greatest scientists were "metaphysical" as if incompatiblre with science if witting is an attempt to persuade by employing an invidious distinction, that is, without presenting any legitimate evidence. It is normally a form of cheating, and if not, it shows such an appalling ignorance of history as to disqualify the philosopher from serious consideration. Why have we given him consideration? Because we would try to be fair regardless of how severe our criticisms; admittedly he by no means cheated all the time; but even when honest, his ignorance of the history of science and philosophy of science was so great that it should serve as an object lesson for the future.

How could any person who claims to be educated not know that Galileo and most leaders of the Scientific Revolution rejected "save the appearances" science in favor of "understand reality science"? That is, they generally rejected presentism and direct epistemology in favor of indirect realism and representational epistemology? A shift which constituted the heart of the Scientific Revolution? Many Germans of that time may have overlooked this change because of the Thirty Year's War, but over two hundred years later, Carnap should have known enough history of science and philosophy to have discovered what had actually happened. He clearly did not.

Perhaps Carnap's most unusual reversal was to revive a meaning of the term "phenomenalism" which was almost completely opposite to the normal one, even though many Austrians including Höfler and the University of Vienna Philosophical Society had been using the proper, non-misleading one for years. Epistemological phenomenalism in our current sense has had a long history, but Carnap's curious re-definition is almost as contrary and inconsistent with the normal one as Kant's definitions of realism and idealism were with the normal 18th and 19th century ones used outside of Germany.. Nevertheless, concerning "Phänomelismus" even in Germany what is now the normal use was familiar and gradually became primary especially after H. Kleinpeter's book *Der Phänomenalismus* (1913) helped popularize Ernst Mach's philosophy. In what has long been the normal epistemological sense, phenomenalism means that we can only "know" or "reliably understand" sensory appearances or "what has been empirically given". Indeed, Rudolf Carnap himself in his later writings sometimes approached the normal definition as if aware of how misleading and unsatisfactory the old definition had been, though he seems to have never summoned the courage or honesty to admit that he himself had long been a phenomenalist, first, in largely Machian terms as gilded with aspects of conventionalism, and later, when his epistemologivcsal position can best be described as *phenomenalism expressed in a physical object language*. Here is Carnap's misleading definition of "phenomenalism" in the English translation of *Aufbau*:

> *Phenomenalism* agrees with realism in maintaining that real entities exist outside of the domain of the autopsychological; on the other hand, it agrees with idealism in denying this reality to the physical; according to phenomenalism what is real or exists must be ascribed to "things-in-themselves" which are unrecognizable, whose appearances [to us] are the physical objects.[7]

Things-in themselves? One might naturally wonder, what do those have to do with phenomenalism? Mach would surely have been surprised to learn that he believed in things-in-themselves. But the contrarian definition which Carnap used, while not as entirely as idiosyncratic as it may seem at first, surely would have been misleading to many people by 1928. But perhaps it is difficult for a historian of science to enter into Rudolf Carnap's inner sanctum. He seems to have been identifying phenomenalism with Kant's position rather than with that of Hume or Mach, and that is both confusing and abnormal. Let us list how Carnap deviated from normal philosophical definitions.

BASIC CLASSIFICATION

(Ferrier introduced the term "epistemology" in the middle of the 19th century. The German term "Erkenntnistheorie" was introduced at about the same time. Latin countries prefer Hamilton's term "gnosiology" while using "epistemologie" to mean systematic history of science or theory of science.)

Rudolf Carnap	Hamilton[8] (& editor)
(as author of his own brand of Classification)	(Hamilton used "gnosiology",[9] not Ferrier's "epistemology")
1. *Realism* holds constructed physical and heteropsychological objects real. 1. Distinguishable from dreams. 2. Can be intersubectivized. 3. They can exist when not noticed. 4. Governed by own regularities.[11] ("Empirical Realism")	*Direct Realism* holds that we can directly notice external physical objects, but that they continue to exist when not noticed. (Hamilton's "Presentative Realism".)[10] "New Realism" circa 1910.
2. *Metaphysical Realism* alleges that there is a trans-conscious mental or physical world.[12] (Carnap normally dismisses this position.)	*Indirect Realism* holds that external physical objects are real, even though we can *never* notice them, but must infer them. (also called "Epistemological Dualism"[13])
3. *Epistemological Representism* is the position that conscious symbols or concepts can represent conscious constructions. (This is more of a methodological than an epistemological position.)	*Epistemological Representism* holds that mentally edited images partly represent the size, shape, and number of external physical objects which are unconscious. (Hamilton's "Representative Realism"[14]; see also Critical Realism"[15] cir. 1920.)
4. *Causal Realism* accepts the causal law that like causes will lead to like results. See Kant. (In Machian moods Carnap rejects this claim.)	*Causal Realism* holds that real physical objects contain mass energy, and normally weight, which they use to exert or resist pressure. (This notion is inked with "*vera causa*")
5. *Subjective Idealism* holds that the heteropsychological, but not the physical, objects are real.[16]	*Epistemological Idealism* holds that only what is conscious or sensory can be "known" or

	reliably understood. This differs from direct or presentative realism by excluding *potentially* conscious or sensory "matter".
6. *Solipsism* denies even the reality of heteropsychological objects.[17]	*Solipsism* holds that only I or my conscious experience exists.
7. *Epistemology* is theory of knowledge and includes only what can be made sensory or conscious. (Carnap rejects indirect, representist, and dualist epistemology as "metaphysical" and as not part of epistemology).[18]	*Epistemology* includes both presentist and representist positions, i.e. it includes possible understanding of both conscious and trans-conscious phenomena. (Hamilton's "Gnosiology")
8. *Metaphysics* includes everything outside of "empirical realism" and science i.e. outside of what can be made sensory or conscious. It is incompatible with science.[19]	*Metaphysics* is another word for ontology or theory of reality. It generally studies our deepest assumptions about what is real, including the basic assumptions of science.
9. *Reality* means empirical reality or metaphysical reality. Science studies the former and the latter is rejected by science.[20]	*Reality* is commonly understood either as what makes a causal difference, i.e. possesses mass or energy; as what can be made sensory or conscious; or as what is formal or religious
10. *Science* has as its first aim the formation of a constructional system, not in a temporal but in a logical sense.[21]	*Science* has as its first aim to understand the real world, mental and physical, past and present, and in both idealized and non-idealized ways.

Rudolf Carnap alleges that what he calls "construction theory" which is apparently another name for "the logical structure of the world" does not in any way contradict realism or "idealism (objective, subjective, or solipsistic) at any point".[22] In short, he has attempted to re-define them to be compatible with his own epistemology, namely, epistemological presentism. But since he does not recognize that what he calls "construction theory" rests on or is itself an epistemology, he may in some way be approaching his later position that all epistemological and ontological positions are "merely linguistic" or "external to science", though it is not obvious how he can get from here to there. We have heard that Wittgenstein's linguistic philosophy intervened at this point, but it is hard to accept that presentism as an epistemological *belief* or *assumption* disappeared either for Rudolf Carnap or other members of the Vienna Circle and Logical Positivism. For historians, the positivist habit of condemning or treating as "merely linguistic" precisely those views which are probably the ones most responsible for the "metaphysical" or even "meaningless" seems as historically misleading as it is irrational.

4. CARNAP'S BASIC EPISTEMOLOGY

by the editor

In arguing that his rejection of indirect epistemology as "metaphysical" implies that his own gnosiology or epistemology is direct or presentist, does not clarify what *type* of presentism he held. By "realism" (when not talking about "metaphysical realism") he clearly meant a kind of direct realism, that is, one which fully excluded what we have called indirect, representist, and causal realism. But closer delineation is needed. There is a difference between conscious and sensory directness, even though all types of sensory presentism are or can be made conscious. That is, we may sometimes be conscious of what is *not* sensory, but which could still fit under the classification of "conscious experience". Similarly, we all experience in our lifetimes much that never was conscious, such that experience and conscious experience are no more identical terms in meaning than sensory and conscious directness or phenomenalism and empiricism, even though there is often overlap. (Empiricism is properly a methodological term meaning interest in or reliance on sensory evidence and is compatible with numerous different types of epistemology and ontology. For example, Bacon and Locke were "empiricists", but this does not mean that they were phenomenalists like Berkeley, Hume, Mach, or Carnap.)

Rudolf Carnap emphasized both logical construction and contact with what was sensory or empirical and tried many different ways to reconcile them, even alleging that what he called "construction" somehow included both of them. He also alleged that "construction theory" was compatible with a totality of epistemological theories and yet was not self-contradictory. He could argue this way of course because he denied that epistemology included indirect and representist positions[23] But even if epistemology were fully bifurcated to exclude everything consciously indirect, "construction theory" would still change meaning with almost every difference in epistemology as to remind most serious epistemologists of "round squares" or "the three body problem". Rudolf Carnap's so-called "Construction Theory" thus becomes an all-purpose, epistemologically unfalsifiable, white elephant.

For example, phenomenalism (in the normal sense) and direct realism are both presentist positions in the field of epistemology, but the differences between sensations and "sensed objects" are by no means merely linguistic, as Carnap himself would later acknowledge when denying that statements about objects could be reduced to statements about sense perceptions.

As for his attack on epistemology, there were times when in the process of rejecting everything trans-conscious as "metaphysical" he appeared to be presupposing that consciousness was all-inclusive, a position which even his own classification which allowed for direct realism seemed to categorize as "idealistic, subjectivistic, and solipsistic". On the other hand, if by "realism" he meant direct realism and considered that compatible with "constructed objects", then this position goes well beyond de facto consciousness, since it recognizes that "constructions", whatever they are, *may* exist or be real when not conscious, even if in principle they can always be made conscious. In short, his talk about "construction theory" being compatible with any and all epistemology, even in his restricted sense, is clearly false.

But if *Der Aufbau* of 1928 was Rudolf Carnap's last or virtually his last open in his restricted sense, is clearly false.

But if *Der Aufbau* of 1928 was Rudolf Carnap's last or virtually his last open discussion of epistemology before he went into his linguistic cocoon, doubly flawed as that discussion was because of his tendency to underestimate the real differences between phenomenalism (in the normal sense) and direct realism and because he saw the contrast as one between science and metaphysics or positivism and metaphysics and not between presentist and representist epistemology, his discussion of epistemology, regardless of how confused, is still a godsend to historians who want to try to classify his basic beliefs in a reasonable manner which is properly connected with both past and current trends and movements.

Of course he would reject our conclusion that like A. J. Ayer, a well-known English Logical Positivist from later years, he oscillated between phenomenalism and direct realism in epistemology and rather unlike Ayer between matter-presentism and form-presentism in what in the introduction to this book we called "onto-epistemology". But that would probably be because the normal definitions of these labels would be partly uncomfortable for him and he might be inclined to underestimate the qualifications we have added. No one always follows his best understanding even in disciplines such as epistemology or ontology. Almost everyone is inclined to use that particular operative understanding which seems to best solve the problem at hand, even if that means provisionally shifting our philosophical assumptions. Carnap himself being often inclined toward relativism and pragmatism would surely have agreed with this part of our reasoning only to initially oppose our conviction that almost everyone has a preferred or best epistemological and ontological understanding, even if he doesn't use it all the time or even most of the time. But no opposition can change the fact that reference is unclear unless the epistemological identity of what is referred to is clear, i.e. whether it is within or beyond conscious limits.

One's "best" outlook can often be discovered both in the process of human reference and in the identification of physical objects, where no practical problem seems to force our hand. Representists commonly assume that we can refer to whatever we want, and that with only a few exceptions, wanting to refer to something *is* referring to it, such that by wanting and assuming we can actually refer beyond all conscious limits. Carnap himself uses this type of reference on occasion; especially when he attacks metaphysics, and it is a common operative understanding for almost everyone, but in this author's late interview with Carnap (1969) he still fiercely rejected it as constituting his best understanding, not necessarily because he didn't have preferences, but because he didn't want to believe that he employed a "subjective" theory of reference. But many historians naturally suppose that everything real is literally subjective except trans-conscious energy, matter, and space, and if anything is idealized or otherwise unreal it can neither be objective nor subjective. On the other hand, part of being fair and reasonable is to attempt to be as *relatively* objective as possible, even if absolute objectivity in our judgment is quite impossible. Similarly, Carnap still refused to admit that what we mean by indirect or representist epistemology (and Hamilton by "representationism") was part of epistemology.

(On the aesthetic side, Carnap looked very different in 1969 than in his earlier photographs. He had a pointed beard and was very much thinner, and even looked a bit aristocratic. Did our disagreements during the interview kill him? Or harm his frail health? Probably not. We were told the next year at UCLA that he died of peritonitis.)

Nor should we regard Carnap's restricted understanding of epistemology as unique. Seemingly in ignorance of Galileo Galilei's rejection of "save the appearances science" and his preference for "understand reality science" during the Scientific Revolution, many German thinkers both before and after Kant, opposed what we have called indirect epistemology about the external world and condemned it as being "metaphysical". Many German thinkers would also interpret the "modern physics" of Planck and Einstein as confirming their presentist views (even if Planck and Einstein themselves rejected their approach, and the reality of atoms was inconsistent with strict presentist epistemology).

In England and America on the other hand, the field of epistemology with its earlier tradition of gnosiology was generally understood to include both presentism and representism (See the Appendix, especially Montague and Lovejoy) and numerous books in the history of science have commonly treated most physical science as having been generally representist starting with the Scientific Revolution and continuing up to the present.[24] Indeed, from the perspective of enlightened hindsight, Rudolf Carnap, much German philosophy, and Logical Positivism would appear to have been out of step with the larger picture of scientific progress, though fashionable in some theoretical circles and occasionally contributing something of importance.[25]

Yet, Carnap must have known that many of his scientific colleagues thought matter was indirect, but he still called that position "metaphysical" as if "unscientific". That was neither fair nor honest. His various systems of basic classification in philosophy *were harmful* since they did mislead many people, especially students. Expression and belief are different. Physical object language is compatible with phenomenalism in belief, as Carnap proved. Nor is "physicalism" the same when held by presentists and representists. It means direct realism in the first case and indirect realism in the second. Carnap and most Logical Positivists ignored these distinctions. *They* were naive, but *he* was worse. He deliberately misled. He cheated. Nor if comments from his colleagues are correct did he fully avoid other character defects. [Friedrich Herneck while a student and follower in Prague learned about his sexual misdeeds. Charles Hartshorne writes in his Schilpp volume (p. 35) about Carnap's "inadequate heed to the concerns of his wife" while they were in Chicago. She committed suicide later in Los Angeles, his last teaching post.]

Carnap surely tried to contribute something valuable to methodology of science, but he identified science too fully with form and construction when most science remained observational, experimental, historical, and applied..As a child of German presentist philosophy in general and Kant, Frege, and Mach in particular, his horizon like his types of classification were simply too narrow and restrictive to make a valuable contribution. Even scientists have epistemological and ontological assumptions that cannot other than influence their work, and most of them are indirect or representational in charadcter..

In *The Logical Structure of the World* (1969), the English translation of *Der Aufbau* (1928), his first major work in philosophy of science, he cites 130 books, which he mentions in his bibliographical index, but *none* of them were on the history of science. What is perhaps his best-known book in English *Meaning and Necessity* (1947 & 1956) lists 73 books, but *none* on history of science. As for his own writings in the Schilpp anthology (1963 & 1991) it lists 214 entries,[26] but again, *none* on the history of science, though perhaps to use a "metaphysical" expression: "There is more than meets the eye".

A defender of Carnap once defiantly argued that he was "a towering figure" and he hoped that no more criticisms would be made of his weakness in history of science. No one, not even a historian, likes to see a tower fall, even a potential Ozymandias, or would hurt anyone's feelings, but perhaps the time has finally come for presentists to face facts. But if that is still too hard, then perhaps they can at least be persuaded by fairer-minded presentist predecessors like Sir William Hamilton to help reform their philosophical classification to make it more inclusive and less prejudicial.

NOTES

1 See Klaus Göbel, "Die Wuppertaler Familie Carnap", *Romerike Berge. Zeitschrift Für Heimatpflege*, **20** (1970-71) 9-14. See also "Johann Sebulon Carnap 1826-1898, Ibid. **21** (1971) 112-121.

2 The educator was Friedrich Wilhelm Dörpfeld (1824-1893) whose complete works were published in 1913, and the prominent archaeologist who discovered much about Ancient Greece, particularly Athens, was Wilhelm Dörpfeld (1853-1940).

3 Rudolf Carnap, "Carnap's Intellectual Autobiography" in *The Philosophy of Rudolf Carnap*, edited by Paul Arthur Schilpp, Open Court: La Salle, 1991 (1963), p. 3.

4 Ibid., pp. 3-7.

5 Ibid., pp. 4-6 & 11-13.

6 Ibid., p. 5.

7 See 19th century physics textbooks starting with Adolphe Ganot, *Cours de physique purement expérimentale, `a l'usage des personnes étrangères aux connaisances mathématiques* . . . [published by the author himself], Paris, 1859. Ganot's physics textbooks were translated into the languages of many countries including England, Australia, The Netherlands, The United States, India, Japan, and Russia and seem to have been the most popular and widely used introductory textbooks in physics until the end of the century and in some coutries well beyond.

8 Ernst Mach, *Popular Scientific Lectures*, Open Court: La Salle, 1943 (1895), pp. 241-242.

9 Carnap 1991, op. cit., p. 9.

10 Carnap 1956, op. cit., p. 43

11 Ludwig Boltzmann, *Theoretical Physics and Philosophical Problems*, edited by B. McGuinness, Reidel: Dordrecht, 1974, p. 275.

12 Like many mathematicians and physicists, Carnap seems to have rejected the notion of efficient causes as powers or forces. This is understandable given the imprecise number of variables as relevant causal factors which has to be taken into account in order to set up constant or necessary relations between them, but given the indispensable role of efficient causes in non-idealized science and in practical life, such an attitude seems to be very short-sighted. It is too informative a notion to ignore or treat as "unscientific".

13 Epistemological phenomenalism is the position that we can only know sensations. Ontological phenomenalism is the view that only sensations are real. Mach supported both positions. See his book *Analysis of Sensations*, Dover: New York, 1959, p. 12. But on page 363 in the same book Mach also accepts the reality of relations, a concession which Carnap as we shall see will eagerly accept and develop further.

14 For a discussion of "economy of thought" see J. Blackmore, *Ernst Mach*, University of California Press: Berkeley, 1972, pp. 169-174.

15 The allegation was that phenomenalism was neutral between realism and idealism, but epistemologically, phenomenalism has been considered part of idealism ever since Bishop Berkeley identified the physical world with sensations which he called "ideas".

16 Rudolf Carnap, *Der logische Aufbau der Welt*, Weltkreis: Berlin-Schlachtensee, 1928.

17 Rudolf Carnap, *Der Raum. Ein Beitrag zur Wissenschaftslehre*. Inaugural Dissertation zur Erlangung der Doktorwürde . . . Universität Jena, 1921. 87 pages. [This work was published the following year in *Kant-Studien* as Ergänzungshefte # 56.]

18 Rudolf Carnap, "Drei Dimensionalität des Raumes und Kausalit†ät: Eine Untersuchung über den logischen Zusamenhang zweier Fiktionen" in *Annalen der Philosophie und philosophischen Kritik* (Leipzig), **4** (1924) Heft 3, pp. [105]-130.

19 Edmund Runggaldier, *Carnap's Early Conventionalism - An Inquiry into the Historical Background of the Vienna Circle*, Rodopi: Amsterdam, 1984.

20 Ibid., p. 18.

21 Ibid. p. 19.

22 Ibid. p. 18.

23 Hans Reiner Sepp, *Edmund Husserl und die phänomenologische Bewegung*, Verlag Karl Alber: Freiburg/München, 1988, p. 425.

24 Hugo Dingler, *Die Grundgedanken der Machschen Philosophie*, Leipzig, 1924. See also Gereon Wolters, "Früher Konventionalismus. Der Carnap-Dingler Briefwechsel", in Peter Janich (ed.), *Methodische Philosophie. Beiträge zum Begründungsproblem der exakten Wissenschaften in Auseinandersetzung mit Hugo Dingler*, Mannheim, 1984, pp. 60-76.

25 Rudolf Carnap, *Der logische Aufbau der Welt* (1928), 2nd edition, Felix Meiner Verlag: Hamburg, 1961.

26 *Kantstudien*, **28** (1923) 90-107.

27 Runggaldier 1984, op. cit., p. 30.

28 Henri Poincaré, *Science and Method*, New York, 1958, pp. 15-24.

29 Philipp Frank, "Das Kausalgesetz und Erfahrung", *Annalen der Naturphilosophie*, Volume **6**, 1907. See also his book: *Das Kausalgesetz und seine Grenzen*, Julius Springer Verlag, Wien, 1932. Many years ago, Gerald Holton once mentioned to this author that Philipp Frank's *Nachlass* was located in a Boston bank, but neither this author nor anyone elese has apparently seen fit to follow up this lead.

30 See Hans Hahn, "Superfluous Entities or Occam's Razor [1930]", in *Hans Hahn - Empiricism, Logic, and Mathematics*, edited by Brian McGuinness, Kluwer: Dordrecht, 1980. While Hahn who died in 1934 seems at least from his blunt philosophical writings to have been severe and humorless in his personality, his wife Lili (or Lily), who was still alive in 1968 when interviewed by this writer, had also studied mathematics and subsequently would become well-known in Vienna for her humor and ability to remember anecdotes about Viennese celebrities, especially her professor in physics, Ludwig Boltzmann. Several of her stories can be found in Blackmore's 1995 books on Boltzmann. For more recent material by and about Boltzmann see: John Blackmore (ed.) "Ludwig Boltzmann - Troubled Genius as Philosopher ", *Synthese*, **119** (1999) nos. 1-2, pp. 1-232.

31 Runggalder 1984, op. cit., p. 38. By "causality" ("*Kausalität*") Carnap and Ruggaldier mean "law-determined regularity" not cause and effect activity as in normal English. The usual word for "cause" in German is "*Ursache*".

32 Hans Vaihinger, *Die Philosophie des Als Ob*, Berlin, 1911. (The first two parts of the book were apparently written from 1876 to 1878. Vaihinger alleges that illness made it difficult to finish the book earlier. Mach corresponded with Vaihinger. He also seems to have given quite diverse opinions about "As If philosophy".)

33 Franco Volpi and Julian Nida-Rümelin (eds.), *Lexikon der Philosophischen Werke*, Alfred Kröner Verlag: Stuttgart, 1988, pp. 549-550. (Carnap lists Vaihinger's book [8th edition, 1922] in his bibliography to *Der logische Aufbau der Welt* in 1928, but it is not clear if or how much he may have been influenced by it.)

34 W.L. Reese in *Dictionary of Philosophy and Religion*, Humanities Press: New Jersey, 1980, pp. 602-603 maintains that Vaihinger and Schmidt founded the journal *Annalen der Philosophie* in 1919 in order to help spread Vaihinger's As If philosophy.

35 Edmund Runggaldier, "The Secondary World as Fiction, and the Failure of its reduction to the primary World", *Carnap's Early Conventionalism*, Rodopi: Amsterdam, 1984, pp. 38-43.

36 Rudolf Carnap, *The Logical Structure of the World* [1928] *& Pseudo Problems in Philosophy* [1928], translated by Rolf A. George, University of California Press: Berkeley and Los Angeles, 1969, p. vii.

37 Rudolf Carnap, "The Development of my Thinking", in *The Philosophy of Rudolf Carnap*, edited by Paul Arthur Schilpp, Open Court: La Salle, 1991, p. 16.

38 Ibid., p. 18.

39 Ibid., p. 19.

40 Runggaldier 1984, op. cit., p. 66.

41 Ibid., p. 71.

42 Carnap 1969, op. cit., p. viii.

2

1 Rudolf Carnap, *The Logical Structure of the World & Pseudoproblems in Philosophy*, Translated by Rolf A. George, University of California Press: Berkeley & Los Angeles, 1969 (1967), pp. 281-287.

3 & 4

1 Edwin Arthur Burtt, "The Subjectivity of Secondary Qualities", in *The Metaphysical Foundations of Modern Science*, A Doubleday Anchor Book: Garden City, 1954 (1924), pp. 83-90. This selection also includes a translation of Galileo's own words from *The Assayer* (1624) as found in *Opere IV*, pp. 335-336.

2 "The Metaphysics of Newton" , ibid., pp. 207-302.

3 Rudolf Carnap, *The Logical Structure of the World...*, University of California Press: Berkeley and Los Angeles, 1969, p. 281.

4 Arthur Lovejoy was perhaps the deepest American epistemologist. Unfortunately, his major work, *The Rrevolt Against Dualism* (1930), is not easy to read.

5 Like Rudolf Carnap, Burtt opposed representism, and also like Carnap he used the term "metaphysics" to describe it, but first, the English term was not as opprobrious as in German at that time, and second, having studied the history of science he was well aware that most of the main contributors to the Scientific Revolution were representists about the external physical world as were many or most later physicists, something which Rudolf Carnap seems not to have realized, or at least did not admit. A broad education is important for a philosopher and Carnap seems to have been lightly read in works outside of formal and systematic disciplines. His understanding of history of science seems to have been especially weak,

6 Ernst Mach, *The Science of Mechanics*, Open Court, La Salle, 1960, pp. 151-190. Mach also seems to criticize Galileo for his interest in the atomic theory as if that were merely a relic of "ancient and mediæval influences". p. 187.

7 Carnap 1969, op. cit., p. 282.

8 Not to be confused with the mathematician of the same name and era. The philosopher Sir William Hamilton (1788-1856) distinguished between knowledge about what is presented and reasonable probability or belief about what is represented. Hamilton's value here is evidence that even presentists are capable of employing reasonably inclusive and fair terminology and definitions. (James Ferrier is normally thought to have introduced the term "Epistemology" into English during the 1850's and in time it replaced Hamilton's earlier expression "gnosiology".)

9 The term was apparently introduced by Hamilton in 1836-37 in a lecture which was later published in *Lectures on Metaphysics and Logic* edited by H.L. Mansel and John Veitch, volume I, Edinburgh and Boston, 1859, pp. vii & 122.

10 The word "presentative" has a long history but seems to have first made its debut as a philosophical term in 1842 with the sense of immediate or intuitive and was included in *The Works of Thomas Reid*, two volumes, which Hamilton edited, and which first came out in Edinburgh in 1846, p. 804. It was used later by Herbert Spencer.

11 Carnap 1969, op. cit., pp. 284-285.

12 Ibid., p. 282-284.

13 Ledger Wood, "Dualistic or Representative Realism" in *A History of Philosophical Systems*, edited by V. Ferm, The Philosophical Library, New York, 1950, pp. 525-527.

14 bid. See also J.M. Baldwin (ed.), *Dictionary of Philosophy and Psychology*, Volume **II**, Macmillan: New York, 1925, p. 464: "Hamilton (*Lectures on Metaphysics*) classifies theories of know-

ledge as immediate and representative."

15 Ibid (Wood).

16 Carnap 1969, op. cit., p. 282.

17 Ibid.

18 Ibid., p. 286.

19 Ibid., pp. 282-284.

20 Ibid., p. 281.

21 Ibid., p. 288.

22 Ibid., pp. 282-284.

23 Ibid., pp. 284-287.

24 Perhaps the most widely read book on history of science during the last half of the 20th century has been James F. Mason's work *A History of the Sciences*, Collier Books: New York. (The editor's copy went through nine printings between 1962 and 1975 alone.) See pp. 500-502. There have also been many translations into other languages.

25 Ibid. This is certainly Mason's opinion.

26 Paul Arthur Schilpp (ed.), *The Philosophy of Rudolf Carnap*, Open Court, La Salle, 1991, pp. 1018-1056.

REFERENCES

Baldwin, J.M. (ed.) *Dictionary of Philosophy and Psychology*. Vol. II. New York: Macmillan. 1925.

Blackmore, John. *Ernst Mach*. Berkeley: University of California Press, 1972.

Blackmore, John. *Ludwig Boltzmann - His Later Life and Philosophy 1900-1906*. two volumes. Dordrecht: Kluwer, 1995.

Blackmore, John. (ed.) "Ludwig Boltzmann - Troubled Genius as Philosopher". *Synthese*, 119 numbers 1 & 2, pp. 1-232.

Boltzmann, Ludwig. *Theoretical Physics and Philosophical Problems*. edited by B. McGuinness. Dordrecht: Reidel, 1974.

Burtt, Edwin Arthur. "The Subjectivity of Secondary qualities". in *The Metaphysical Foundations of Modern Science*. Garden City: Doubleday, 1954. (earlier editions in 1924, 1930, & 1952)

Carnap, Rudolf. "Der Raum. Ein Beitrag zur Wissenschaftslehre". *Kant-Studien - Erganzungshefte #56*, 1921. (Jena doctoral dissertation, 1920. 87 pages.)

Carnap, Rudolf. "Drei Dimensionalität des Raumes und Kausalität: Eine Untersuchung über den logischen Zusammenhang zweier Fiktionen". *Annalen der Philosophie und philosophischen Kritik* (Leipzig). 4 (1924) Heft 3, pp. 105-130.

Carnap, Rudolf. *Der Logische Aufbau der Welt*. Berlin: Schlachtensee, 1928. 2nd edition, Hamburg: Felix Meiner Verlag, 1961.

Carnap, Rudolf. *Meaning and Necessity*. Chicago: The University of Chicago Press, 1956 (1947).

Carnap, Rudolf. "Carnap's Intellectual Autobiography" in *The Philosophy of Rudolf Carnap*. edited by Paul Arthur Schilpp. La Salle: Open Court, 1991 (1963), p. 3 - 84.

Carnap, Rudolf. "The Development of my Thinking". in *The Philosophy of Rudolf Carnap*, edited by P.A. Schilpp. La Salle: Open Court, 1991, pp. 3 - 39.

Carnap, Rudolf. *The Logical Structure iof the World & Pseudo Problems in Philosophy*. translated by Rolf A. George. Berkeley & Los Angeles: University of California Press, 1969. (original German edition, 1928)

Dingler, Hugo. *Die Grundgedanken der Machschen Philosophie*. Leipzig: 1924.

Frank, Philipp. "Das Kausalgesetz und Erfahrung". *Annalen der Naturphilosophie*. volume 6, 1907.

Frank, Philipp. *Das Kausalgesetz und seine Grenzen*. Wien: Julius Springer Verlag, 1932.

Ganot, Adolphe. *Cours de physique purement expérimentale.* Paris: 1859.
Göbel, Klaus. "Die Wuppertaler Familie Carnap". *Romerike Berge. Zeitschrift für Heimatpflege,* 20 (1970-71) 9-14.
Göbel, Klaus. "Johann Sebulon Carnap 1826-1898". *Romerike Berge. Zeitschrift für Heimarpflege,* 21 (1971) 112-121.
Hahn, Hans. "Superfluous Entities or Occam's Razor (1930)". in *Hans Hahn - Empiricism, Logic, and Mathematics.* edited by Brian McGuinness. Dordrecht: Kluwer, 1980.
Hamilton, Sir William. *Lectures on Metaphysics and Logic.* edited by H.L. Mansel & John Veitch. volume I, Edinburgh & Boston: 1859.
Lovejoy, Arthur. *The Revolt Against Dualism.* New Brunswick & London: Transaction Publishers, 1996. (Ist edition 1930. Second Edition 1960)
Mach, Ernst. *Popular Scientific Lectures.* La Salle: Open Court, 1943 (1895).
Mach, Ernst. *Analysis of Sensations.* New York: Dover, 1959 (1st German edition, 1886).
Mason, James F. *A History of the Sciences.* New York: Collier, 1975. (9 printings from 1962 to 1975)
Poincaré, Henri. *Science and Method.* New York: Dover, 1958 (1905). 1st French edition, 1903.
Reese. W.L. *Dictionary of Philosophy and Religion.* New Jersey: Humanities Press, 1980.
Rungaldier, Edmund. *Carnap's Early Conventionalism.* Amsterdam: Rodopi, 1984.
Sepp, Hans Reiner. *Edmund Husserl und die phänomenologische Bewegung.* Freiburg/München: Verlag Karl Alber, 1988.
Vaihinger, Hans. *Die Philosophie des Als Ob.* Berlin: 1911. (8th edition, 1922)
Volpi, Franco & Julian Nida-Rümelin (eds.). *Lexikon der Philosophischen Werke.* Stuttgart: Alfred Kröner, 1988.
Wolters, Gereon. "Früher Konventionalismus. Der Carnap-Dingler Briefwechsel". in Peter Janich (ed.) *Methodische Philosophie. Beiträge zum Begründungsproblem der Exakten Wissenschaften in Auseinandersetzung mit Hugo Digler.* Mannheim: 1984, pp. 60-76.
Wood, Ledger. "Dualistic or Representative Realism". in *A History of Philosophical Systems.* edited by V. Ferm. New York: The Philosophical Library, 1950, pp. 525-527.

Chapter Nine

MUSIL BETWEEN MACH AND STUMPF

by Michio Imai

Section Guide
1. Introduction
2. Dissertation
3. Closer Analysis
4. Musil on Mach
5. Ethics & Mysticism

1. INTRODUCTION

Mach appears to have exercised a measurable influence on the Austrian Robert Musil (1880-1942). This seems most conspicuous in Musil's doctoral dissertation on Mach which was titled in its final published form: *Beitrag zur Beurteilung der Lehren Machs*, Wilmersdorf, 1908. But it is possible that the influence was deeper in parts of the later work which made Musil famous: *Der Mann ohne Eigenschaften*, 3 volumes, 1930, 1933, & 1943. This book is an ironic and satirical commentary on the very last years of the Habsburg Monarchy as presented in a clear, unsentimental, and analytic manner, which in a stylistic way could almost be called "scientific".

Robert Musil was born in Klagenfurt south of Vienna. His father was a professor of engineering. Musil himself went through training for several professions including the military, engineering, and philosophy,[1] but the clear success of his short novel *Die Verwirrungen des Zöglings Törless* in 1906 encouraged him after finishing his dissertation in 1908 to reject an invitation by Alexius Meinong to come to Graz and make philosophy his life's work. He married in 1911 and held several minor posts such as editor, military officer during the war, librarian, and briefly even *Ministerialrat* but they did not allow him to concentrate completely on literature,[2] so in 1922 he "tossed caution to the winds" and became a full-time writer, but as a perfectionist who continually revised his material he could not publish much and had to be financially helped by admirers of his work in order to survive.

Robert Musil would later deeply regret throwing away three professions for something as unstable as literature, which was made more precarious by the political turmoil which started with the Great War in 1914. Indeed, like himself, the "hero" of his famous book, namely, the man without qualities, would also be unable to choose a profession. Furthermore, in Machian terms a man with no qualities at all would not exist, even as a mere grouping of sensations, a pessimistic or masochistic application of Mach's theory of the ego. But most commentators identify Ulrich's alleged lack of qualities in his great novel with an exaggerated conception of possibility such that he had difficulty making decisions or fully accepting that what seemed evident was real.[3]

But in this paper, let us concentrate on Musil's dissertation on Mach, not because it is more important than his majestic novel, but because, first, we are on somewhat firmer

J. Blackmore, R. Itagaki, and S. Tanaka (eds.). Ernst Mach's Vienna 1895-1939. 187-209

ground concerning Musil's opinions about Mach's ideas, and second, while Mach had significant influence on many literary and musical figures in Vienna, including Musil, this book is focused on Mach's influence on philosophers with a deep interest in science, which for a while seems to have been Musil's primary interest as well. Furthermore, in order to help understand Musil's thought within the context of his times, almost all of the sources which we will consult here have been taken from primary and secondary works written before Ernst Mach's death in 1916.* This approach has advantages and disadvantages. So, dear reader, let us return to yesterday, and attempt to comprehend his ideas from a historical perspective which he himself might have found familiar. On the other hand, however, some of the larger issues dealt with in the last pages of this paper will require at least some discussion of his masterpiece *The Man without Qualities* and will unavoidably mention some aspects from a more recent perspective.

Starting in year 1903, Musil would live in Berlin where he would study philosophy, finish his novel on young Törless (1906), and his main advisor in philosophy would be Carl Stumpf (1848-1936), a former student of Franz Brentano, gifted musician, professor of philosophy, and who had contributed much to psychology. Stumpf had also been a colleague of Mach in Prague for a short time and would remain a fierce opponent of Mach's particular type of phenomenalism.

Given Carl Stumpf as an advisor it is most surprising that he would have allowed Musil to write his dissertation on Mach, but the young man was Austrian, Mach's philosophy was popular at the time, even if Mach denied that he was a philosopher at all, and Stumpf could always control the result by asking Musil to revise his manuscript, which he did. By soft-pedaling Ernst Mach's phenomenalism and concentrating on his philosophy & methodology of science, to the extent such was possible, Robert Musil seems to have both followed a methodological interest of his own and touched lightly on the epistemological and ontological points Stumpf was most sensitive about. But this also means that the dissertation may not reflect Musil's *full* views on Mach, especially since we do not know what Musil changed in order to make the equally unknown corrections which Stumpf requested.

But before we can describe the contents of the dissertation in detail, we should make clear that in those days most German doctoral dissertations, at least in philosophy as judged from several published results in libraries and archives, were generally rather modest and hasty affairs with little of the research and wide reading so often required today. Over the years there have been at least twenty-seven dissertations written on Mach, and in German alone.[4] Why were there so many written on Mach's philosophy? Possibly because his writings were easier to read than most philosophy books, and many of his thoughts had become popular, especially outside of professional philosophy.[5] Another point to keep in mind is that positivists were often considered to be atheists, and few if any German or Austrian universities would knowingly offer a university chair to an atheist or anyone suspected of being a positivist, hence writing a dissertation severely criticizing such an influential positivist like Ernst Mach could well have "warmed the cockles" of many professional philosophers, especially at German universities. This general exclusion of positivists and Jews from full professorial chairs would change briefly after the First World War when both Germany and Austria "enjoyed" socialist governments, though each case is different and requires research. In Vienna, Ernst Mach's old friends Wilhelm Jerusalem and Heinrich Gomperz would belatedly become full professors and the mathematician Hans Hahn would come to Vienna and in turn would help invite Moritiz Schlick to occupy Ernst Mach's old position in philosophy, but by the middle twenties nationalism would generally take precedence and "Machists" and Jews would once again become largely unwelcome in top academic posts, especially in Vienna.

Hahn and Schlick reacted by forming their own special private group which evolved into both the Vienna Circle and by offering adult education lectures which were given under the auspices of the new *Ernst Mach Verein*.

Most dissertation writers on Ernst Mach would adopt what they thought to be a modified Kantian point of view like their neo-Kantian professors and would then describe and criticize E. Mach's perspective accordingly. Musil would also appear to do the same thing, at least concerning law-based causal explanation. In the course of our own paper, we will compare Robert Musil's Mach dissertation with some of the other philosophy dissertations on Mach which were published before the First World War, that is, well before Robert Musil would become famous as a novelist. This procedure should help us place Musil's apparent understanding more clearly within the context of his youth. This also means that much as we appreciate Hans Henning's biography of Ernst Mach published during the War in 1915, that is, a year before Mach's death, we cannot fully agree with his conclusion about "the innumerable Kantian dissertations on Mach's epistemology" which "are all worthless".[6] Actually, Hans Henning selects five works to criticize. In addition to Musil, Henning discusses dissertation/books by Buzello, Hell, Hönigswald, and Reinhold.[7] Some of them, of course, do have value for particular purposes, and for our purpose that includes Robert Musil. Nor is it likely that a philosopher as judicious as Alexius Meinong would have asked Musil to come to Graz as an assistant or potential *Privatdozent*, if he had shared Henning's blanket condemnation of him.

In addition to Musil's academic background, we must of course search for the inner motives which led him to write on Ernst Mach, and ask how he himself judged his own dissertation. But unfortunately, there are only a few fragments in his diaries which seem relevant plus a few letters. Therefore we will first examine the contents and character of the dissertation, and give an estimate of it within the framework of academic philosophy. Only after that will we proceed to attempt to get at the inner philosophical aspects.

2. CONTENTS OF THE DISSERTATION

Precisely because German philosophy dissertations of this period tended to lack significant research into unpublished documents and included little reading in secondary sources, they will often reflect the occasional university courses which the student has taken and the ideas of a few world-famous philosophers. But there is also at times some measure of originality even if it may seem a bit immature. All of these characteristics can be found in Musil's dissertation. But Musil was already a published author by 1908 and was older than most other *Doktoranten*, hence, his dissertation should have been able to rise above Hans Henning's judgment.

Musil's dissertation consists of only five chapters: The contents are briefly: Introduction, Mach's epistemology & economy of thought, his opposition to "mechanistic physics", Mach's polemic against the concept of causality, and in the final chapter reflections on "functional connections", the denial of natural necessity, elements, and what are seen as additional contradictions in Mach's thought.[8]

The *Introduction* describes Musil's main purpose which is one of testing whether Mach's philosophy is genuinely the result of natural science, since: "Today it is the word of the natural scientist which carries weight wherever epistemological or metaphysical questions come under the scrutiny of exact science. The times are past when a picture of the world could spring full-blown from some philosopher's forehead."[9] Another point to note in the Introduction is Musil's manner of characterizing Mach's philosophy. Musil enumerates Ernst Mach's principal theories like the theory of description, functionalism, and the theory of economy. Musil thinks that none of Mach's theories is particularly original in itself but may be in terms of how they are woven together into a coherent tapestry. For this reason his purpose is to understand Mach's thought deeply and systematically, and to judge thereafter whether it is the legitimate result of current natural science.

In *Chapter Two* Musil takes up *Denkökonomie*, according to which the task of science lies primarily in the economy of thought. Science aims at comprehending sensual experience by the least amount of valid thought. That is to say, it simplifies sensations in both their physical relations and psychological or physiological relations. Human life requires an economy of thought, and so does science. In fact, that is how science originated. Science is also subject to the self-preserving biological principle which is contained in the principle of evolution. Thus, Musil calls Mach's view "evolutionary-historical", "cognitive psychological", and "thought-economical" in its way of cognizing how science should approach sensations.[10]

Musil then considers whether Mach's philosophy is epistemologically sound. His conclusion is that *Denkökonomie* is basically "sceptical". This is so because it does not lay the foundation of science as the search for truth and because it cannot even convince itself that such an ambition is needed. Musil thinks that epistemology must give a foundation to the search for truth. Therefore his judgment is that while Mach's theories provide a worthy view of characterizing some science, it does not decisively or satisfactorily lay the foundation of science.

In *Chapter Three*, Robert Musil examines Mach's attitude to traditional Newtonian mechanics or what many philosophical opponents have insisted on calling "mechanistic science". This concern was a main theme of Mach's well-known book *Mechanik in ihrer Entwicklung* (1883). Musil claims that in Mach's opinion modern physics had largely started during the Scientific Revolution with an attempt to model processes behind the appearances, but that under the rising influence of epistemologically-minded philosophers and scientists has shifted increasingly to mathematical models of the appearances themselves. According to Mach, the criterion which decides the superiority of a resulting model is its economical value.

But at this point Musil takes up Mach's theory of description. There is a difference between a description by means of a model or picture which is indirect, hypothetical, and idealized and clear, straightforward, conceptual description of a sensory fact which is direct and employs no model. But in spite of Mach's own theory of economy, it is evident from his own writings that he obviously much prefers direct description of sensations. Still, by not opposing an idealized, model-based but "economical" approach which he himself seems reluctant to follow, Mach has the merit of presenting a criterion, namely, *Denkökonomie*, which is useful for much science, and indeed, hypotheses in mechanical physics are often proposed and estimated according to this principle, especially when understood in logical rather than psychological terms, and are increasingly viewed as superior to the "ontic" approach used in Newtonian science which was still concerned with what existed beyond the appearances. But for Musil, the principle of economy, whether understood in logical or psychological terms, is insufficient since it cannot be an epistemological principle, as Mach's own preference for direct, non-idealized description reveals.

Another topic discussed in Chapter Three is the nature of key physical concepts such as mass, space, time, motion, and temperature. It is Mach's standpoint that these concepts contain more than we are able to empirically demonstrate and that therefore we should eliminate any surplus beyond what we can directly experience. If fully done, this would eliminate any remaining "metaphysical" aspects in the concepts of contemporary physics. But Musil criticizes Mach's antimetaphysics at this point, for it overlooks the fact that since scientific concepts *do* in fact go beyond experience, eliminating the non-empirical excess may risk destroying physics in its current form since it *needs* that excess, and this is not least because of its simplifying, idealizing, mathematical responsibility, its need to economize. In short, Mach's phenomenalism is not fully compatible with Mach's theory of *Denkökonomie*.

In *Chapter Four*, Musil discusses the concept of causality. In physics it is normally related to deterministic laws. However, such laws oversimplify natural phenomena, which cannot be onesidedly related from cause to effect. Ernst Mach proposes instead the mathematical concept of function, by which one may be able to comprehend the various interpretations of nature in its true empirical manifestation.

The equations in physics are typical examples of functional presentation. Musil thinks that this functional view shows an indifference to metaphysics, or better put, there is solid ground to assume that in addition to functional relations that there are also real causal relations, especially in the sense of deterministic laws, which is the type of causality which most physicists seem to prefer, at least when such are available. In short, on this point Ernst Mach is clearly not in agreement with the attitude of most physical scientists, though frankly, it is not always easy to plausibly distinguish between the necessary or deterministic parts of an equation and the merely functional or truly non-necessary aspects. Nor is it always evident which equations are laws and which are not, not to mention which rest on deterministic assumptions and which not.

Musil also discusses the concept of substance which Mach firmly rejects. Ernst Mach seemed to hope that functional description would gradually help eliminate the notion of substance, but if there is no causality without some notion of substance and most scientists insist on causality, then Mach's approach will seem, and to some extent already does seem, inadequate to them..

Chapter Five is the concluding chapter of Musil's book and it occupies a third of it. And it is here that Robert Musil examines Mach's opposition to natural necessity, that is, his denial that necessity exists in what Mach means by "nature", that is, the world of sensations as related in a "physical" manner. Mach accepts logical necessity but not natural necessity., But as Musil points out his belief that logic like everything else originated from sensations and can at least in theory be reduced to them is opposed by many mathematicians who want to think of logic and math as possessing an ideal or idealized character, that is, as if Mach's approach - to use Husserl's language - were "psychologistic" and unsatisfactory.

Furthermore, when someone denies natural necessity, either in a Kantian sense of a priori existence which organizes sensations or in a more idealized or trans-mental sense, then it should be asked wherein lawful knowledge lies. According to Mach, scientific laws as deterministic or necessary laws, are more the product of mental abstraction from sensory regularities, that is, from mere empirical laws, than something really factual in nature. They are idealized, and idealization is not a fully sensory or empirical concept. The alleged necessity of deterministic laws are thus imposed on nature only in a fictive manner. It is at this point that Musil finally analyses Mach's theory of elements, by which Mach allegedly means colors, sounds, and other sensory complexes, which during the 1880's and possibly the 1890's even included spaces and times, which republication of *Analyse der Empfindungen* kept "alive" well after Mach had changed to his 1895? distinction between metric, physical, and physiological spaces and times.

But when physical scientists use symbols in their equations which refer to these elements, the latter are rather conceptual and distinct from sensory elements. In short, there is an idealized aspect even within Ernst Mach's conception of elements, and for some theorists and mathematicians there is often little else. The sensory aspect then virtually disappears into abstract qualities and even more abstract relations.

Musil now comes to his concluding remarks which focus on how Mach's denial of natural necessity allegedly leads to contradictions in his philosophy of science. In fact, the last chapter of Robert Musil's dissertation is titled "The Final Contradictions", a negative phrasing which presumably pleased his doctoral advisor, Carl Stumpf, who was almost certainly considerably more opposed to Mach's philosophy than Musil actually was.

Musil's key point is his insistence that Mach like other physicists presupposes the uniformity of nature, a presupposition which lends to mere Humean conjunctions among appearances, that is, to empirical laws about sensations, a necessary or deterministic aspect. If Ernst Mach indeed presupposed the uniformity of nature, then it would seem his criticism of deterministic laws as mere abstractions or simplified versions of sensations, whether true or false in terms of Mach's overt understanding, risked self-contradiction, and indeed, could mean that he held or presupposed a fictitious difference between empirical and deterministic laws.

Musil's own view was that necessity coming from abstraction and idealization has its own foundation in fact, though whether he means strict sensation as opposed to some other type of fact, needs further study. His conclusion is that Mach has neither reached a solution of this problem nor even suggested a way to find it. One might add that Musil's inclination toward natural necessity finds a social and political analogy in his book *The Man without Qualities*, as if given the alleged follies of the Habsburg Monarchy that its dissolution was "inevitable", even if events never reach that stage in his book, but he of course had the advantage of hindsight while virtually no history books or novels about the Empire written *before* World War I even hint at the possibility of such a fate. Musil ends his thesis with the following:

> [My conclusion] is true of course only of the final metaphysical and epistemological results as these have been considered here. As has been widely recognized, Mach's writings are full of sparkling comments and the most fruitful suggestions. But considerations of these do not fall within the scope of our task.

3. CLOSER ANALYSIS

As previously mentioned, Hans Henning in his book, *Ernst Mach als Philosoph, Physiker, und Psycholog*, Leipzig, 1915, included Musil's thesis within the Kantian tradition (or better Neo-Kantian) of epistemological criticism of Mach. At first glance, Robert Musil seems to be where Henning puts him. He did defend natural necessity and it was in terms of lawfulness, but the *types* of necessity and lawfulness were not at all Kantian. There was nothing about a priori categories of the mind, and there was a strong defence of what might be called the positivistic conception of science as something beyond the theological and metaphysical "stages".[11] Nevertheless, if one looks for easy pigeonholes, the Neo-Kantian one is probably as good as any other, since the conclusion if not the reasoning and assumptions largely go along with that perspective. In order to clarify this a little bit more, let us compare it with Richard Hönigwald's work, *Zur Kritik der Machschen Philosophie - eine erkenntnistheoretische Studie*, Berlin, 1903.

This work, one of the very first on Mach, is remarkable, not only as a pioneering effort from a young thinker who has been fortunate enough not only to receive some attention after his death but even to have his *Nachlass* published (ten volumes, 1957-1977). Richard Hönigswald (1875-1947) in his first book, focuses on epistemology almost to the exclusion of methodology of science, almost an exact opposite of Musil's approach.[12] He then tries to show that Mach's theories as an epistemology break down. He points out that the principle of economy only holds as a regulative principle, and that Ernst Mach's standpoint results in *Heraclitian pantha rei,* that is, an uncontrollable rush of mere appearances, such that it is incapable of constituting anything like a steady epistemological foundation.[13] But while Robert Musil also makes this criticism, in spite of that he still makes a determined effort to follow Ernst Mach's reasoning as if that remained worthy of close attention independently of Hönigswald's and his own severe conclusion. Part of this attempt at rigor and respect for philosophical exactness comes from the fact that unlike Richard Hönigswald, Musil seems to be more interested in Ernst Mach's methodology of science than in his epistemology.

But while the differences between the two critics is not so great on most issues, it clearly comes out in their treatment of Mach's theory of the subject, self, or ego. For Hönigswald the epistemologist such is *the* central issue in Mach's philosophy, while for Robert Musil the lover of science it is hardly mentioned at all. For Höningswald, who is sometimes held to be a Kantian in the narrow sense, epistemology presupposes a subject, while Mach does his best to exclude presuppositions from epistemology as something not known but merely inferred (though contrary to his own phenomenalist epistemology he does allow inference about the sensations of other people, even if he tries to reduce all human "subjects", "selves", or "egos" including his own to mere internally related collections of sensations). Musil's relative silence on the problem of solipsism could be interpreted either as possible agreement with Mach or the feeling that the issue is not relevant for methodology of science. In short, there is a strong suggestion that Musil did not follow the Kantian or Neo-Kantian line on this issue.

On the other hand, when Musil turned some years later to the writing of his famous book it seems possible that he had become satirical about Mach's rejection of the ego, self, or subject. That is, he may have come to regret some of the minor independence of judgment in philosophy which he dared to show and which Hans Henning had brusquely denied when he made his blanket denunciation of all of the doctoral dissertations written on Ernst Mach prior to his own book (1915). Nevertheless, with Musil's concern with methodology of science does not mean that he had reached a novel position about it at that time or ever. Also, from a strict Neo-Kantian standpoint (if there is one) Musil's thesis was one-sided and unsatisfactory, though still compatible in a number of respects with a Neo-Kantian critique of Mach's epistemological position.

With the help of documents included at the end of his dissertation let us attempt to evaluate Robert Musil's thesis.[14] It is known that it was originally presented to Carl Stumpf under the title *"Studien zur erkenntnistheoretischen Grundlage der Physik mit Bezug auf die Anschauung E. Machs"* on June 27, 1907, but it was re-written according to Stumpf's suggestions. But since the work after revisions was re-submitted many months later, in fact early in the next year, one may reasonably suppose that perhaps the alterations were considerable, perhaps even drastic. But we do not know, and speculation may not help us much, but the possibility still exists. But given Stumpf's negative and even polemical attitude toward Mach's perspective, it is hard to believe that the revisions made the dissertation "less anti-Mach". Indeed, the opposite seems more likely.

The work was examined and judged by a professorial committee of the University of Berlin on February 27 and accepted on March 14, 1908. The main judge was Carl Stumpf, but another philosopher was also present, Alois Riehl (who in turn is also alleged to have influenced Hönigswald). Perhaps because Musil's *Nebenfächer* were mathematics and physics, two other men were included, the mathematician Amandus Schwarz and also the physicist Heinrich Rubens.[15] The grades given to the dissertation and examination were about average, and in all fairness hardly deserved anything higher.[16] He had wanted to use his critical mind in order to produce a quality critique, but he could not carry it through. Given his subsequent reluctance to accept a career in philosophy he may well have not been satisfied with the thesis. But one must still know more about why Alexius Meinong wanted him in Graz.

Carl Stumpf's brief comments reflect opposition to Mach's philosophy more strongly than praise of Musil's work, though the latter is also implied with respect to the second submission.

> [The dissertation] was initially sent back to the candidate for re-writing . . . . [Though] criticism is everywhere immanent, the author starts first with a positive attitude [toward Mach's philosophy] so as to show how one is unhesitatingly driven from one position to another such that [Mach] finally falls into inconsistency. [But] at various points he will still have to clarify and elucidate many matters, and make the form moderate.[17]

This appears to be a rather brief and appropriate comment, and the last sentence suggests the reason for the average grade. One may still wonder, however, whether Stumpf was fully satisfied with the revised dissertation, or whether he suspected that Musil was still trying to insert his own possibly more positive opinion of Mach's point of view. But to judge this we really need to take much more into consideration, especially Stumpf's own attitude toward Mach. This can be partly understood by a long quotation from Ernst Mach's article "Sinnliche Elemente und naturwissenschaftliche Begriffe" from the year 1910 which generously includes within it a long quotation from Carl Stumpf.[18] Mach writes:

> To open the International Congress on Psychology in Munich on August 4, 1896, Prof. Dr. Carl Stumpf gave a speech in which he criticized my theory of psychological knowledge. I had indeed inscribed my name as a participant in the Congress, but on discovering the strongly hypnotic-telepathic program I chose not to attend. Thus, his speech did not come to my attention for some time, especially since I was busy with other things and soon after was afflicted by a serious illness. To be sure, I did give a brief answer to the objections of Stumpf and others in a series of editions of my *Analyse der Empfindungen*, objections which I do not regard as either personal or wilful, but only as typical. But since Stumpf's speech has recently appeared in a third edition I would like to present his main passage along with my own. Stumpf's contribution to this debate can only profit from being placed in relief against my own [in this article]. He writes:
>
> > One could almost envy the follower of [Mach's] theory who believes that he has found a shortcut to the heights of his epistemological and psychological standpoint. But neither of the propositions on which his theory rests can be supported by the facts. The object and goal of natural science to construct lawful relations can never be found at all in sensory appearances. The regular succession and coexistence which scientists require in scientific laws cannot be found in the confines of an individual's

> own consciousness. It only exists within the processes which we would ordain are beyond sensory appearances and completely independent of consciousness, that is, if we would speak about lawfulness [in nature] at all. We do not experience [*erkennen*] what is real at all and only know [*erkennen*] its relations in the totally abstract form of equations which we attempt to picture in a kind of visualized space [*Raumanschauung*] but which only serves as a dispensable symbol. These lawful relations and how they are constructed constitute the 'physical world' of science, while the sensory appearances from which the physical world of [alleged] common consciousness is built merely serves as starting point for the exploration of the purely mathematical, I would say even algebraic, world. But it is hard to speak one's mind fully about such an expert on the history of science as Ernst Mach to the effect that he has failed to understand the true direction of physical investigation, in fact turned it on its head. But even the greatest personal and scientific respect cannot change convictions.
>
> And second, about the mental or psychological world which seems to be so fully dissolvable into sensory appearances which we experience such as thinking, feeling, and willing, at least up to now the history of doing psychology provides us with no guarantee [of finding certainty]. On the contrary, All of the attempts since the days of Condillac to really carry through a reliable analysis of them have failed. This of course does not unconditionally prove the impossibility of such a project for all future time, but one should still be willing to confess that less dogmatic confidence appears to be justified in maintaining that sensations can be reliably analyzed as if such were the same as a logical axiom which needs no proof and should even be given priority to it.
>
> So if I see correctly, this sensualistic monism [of Mach] has dissolved itself [into nothing]. The real course of science has certainly contradicted his claims about the physical world, and about the mental, not confirmed him in the least.[19]

Mach replied in the same article:

> A prominent figure on the philosophical side [Carl Stumpf] has now objected that lawful relations "never exist at all" among sensory phenomena and that lawfulness as rightly understood by natural scientists is something completely different. But unprejudiced readers will have already recognized [*erkennen*] the presence of lawfulness in appearances themselves in the examples which I have already given . . . .[20]
>
> Therefore, in any case, the physicist [like Ernst Mach] has nothing to look for "beyond sensory phenomena". But whether philosophers [such as the psychologist Stumpf] will always find it necessary to determine the existence of reality independent of consciousness which does not exist in itself at all but can only be known in the completely abstract form of equations is a question which philosophers must decide among themselves. Perhaps they will one day ask if such dubious relations make any sense at all. Perhaps they will even raise the question if this approach is necessary and where it is most likely to lead? Hopefully, physicists of the 20th century will not be forced to pay attention to this kind of intrusion. Nor is it impossible that the great anti-metaphysician from Königsberg [Kant] in his efforts to purify his discipline may not himself have left a little metaphysical mold [*Schimmelflocke*] of his

own behind, which since then has spread in all directions.[21]

The main point of controversy between Mach and Stumpf is directly related to the problem of natural necessity which Musil discusses in his dissertation. One might think he must have known about Stumpf's critique of Mach [1896}, but his dissertation does not exactly fit Stumpf's point of view, which leads one to think that Musil chose his dissertation topic and wrote the first version in ignorance of how severely Stumpf opposed Mach. In other words, Musil may originally and perhaps "really" have been somewhat more sympathetic to Mach than the revised version - the only one which we seem to have - indicates.

In addition, Robert Musil's way of handling the issue of natural necessity suggests that he was aware that for all of the fiery polemics between the two men that Mach and Stumpf were not really that far apart. Both men were "presentist" in an epistemological sense by being deeply concerned with what could be directly noticed or experienced as if they were denying the existence of a physical world which existed independently of or behind the appearances, though Mach tried to hang representism around Stumf's neck while also implying that Carl Stumpf was too much influenced by Kantian "metaphysics", but in fact Stumpf's position at this time, while it has been alleged to have been largely "realist" in a sense close to that of the early Franz Brentano who had been Carl Stumpf's teacher, appears at least in this particular passage to have been closer to that of Plato. Like the mighty Greek thinker, Carl Stumpf seems to have treated sensations or what Mach called "elements" as partly untrustworthy appearances which were subject to "right opinion" but not "knowledge", while virtually identifying what is real with Platonic Forms in general and with mathematical equations in particular.

Musil when defending natural necessity did not adopt either Immanuel Kant's identification of natural laws as something given or determined by a priori categories of the mind, or Platonic Forms, or Stumpf's own emphasis on abstract equations, but with a basic assumption which all law-focused scientists seem to take for granted, but which has little or no true status in the field of epistemology, namely, what is called "the uniformity of nature". The main point is that Musil's manner of supporting natural necessity while compatible with Carl Stumpf's position is actually quite different, and while seeming to be opposed to Mach's more epistemological point of view, could easily be handled by Mach as a useful heuristic tool to fit into his theory of *Denkökonomie* as if applied to science, but also as something not having any real epistemological, ontological, or causal significance.

But we must not be too hard on Musil. The above remarks by Mach had appeared in 1910 two years after the young man, who was not so young any more, had published his dissertation. Musil was simply not in a position to write all that he really thought and have it accepted by Carl Stumpf. This naturally depresses a writer, and Musil in spite of his deep concern with science was already a successful author in literature. Why obey philosophers when he could say what he wanted in literature?

Nevertheless, Musil's later regret at abandoning all of the fields in which he had been originally trained, but especially philosophy, is clear in a letter to Josef Nadler from December 1, 1924:

> Before I took my doctor's degree, I had published *Verwirrungen des Zöglings Törless*. The success of this book unfortunately contributed to the effect that I estranged myself from my teacher Stumpf in Berlin and refused indignantly to come to Graz as an assistant and to qualify myself as a university lecturer there.[22]

It might seem unusual that the subsequently famous novelist could so quickly regret that he was bound to such an unpredictable field as literature, but such things happen, especially if the writer is naturally gloomy. But he was fortunate that especially-created Musil-societies first in Germany and then in Austria would try to keep him afloat financially, though toward the end of his life things became so difficult that "afloat" is the wrong word. He was sinking. The Austrian society disappeared with Hitler's conquest of Austria in 1938 and Musil was forced to retire to Switzerland, but with no German language publishing house as an outlet, and very little remaining German-language readership. Nor could he even raise his spirits sufficiently to finish his book. Nor could he reasonably hope at the time of his death (1942) that his works would ever be properly appreciated in the future.

The letter above suggests that there may have been a break with Carl Stumpf. Was it intellectual or emotional? And if so what was it? It is also conceivable that Robert Musil was unhappy not just because the dissertation didn't entirely represent his own views, but that perhaps his own views were not themselves "clear and distinct". But if so, he largely only had himself to blame. He had not carried out enough research either to clarify his own views or to write a first class dissertation. He "imagined that it is more important to think out one's own position and to ask for advice only for examining and completing it",[23] a common opinion among youth, but which tends to underestimate what one can learn from extensive reading and which blithely ignores how many other philosophers have already written on what one is interested in. In short, there is too little appeal to *other* works.

4. MUSIL ON MACH

The finished dissertation was neither formally nor materially very close to perfect. Its character becomes more apparent when we compare it with the Mach dissertation by Reinhold Schultz of 1907 (which was published under the name of Ferdinand Reinhold in Leipzig in 1908).[24] Let us paraphrase the first two lines and then explicate parts of *Die Machsche Erkenntnistheorie. Darstellung und Kritik*, 1907:

> Ernst Mach's philosophical views are under lively discussion today, but they lack a systematic connection, nor does Mach aim at it. Nevertheless, there are people who now appear to be trying to construct a comprehensive world view from Mach's basic thought.

After considering several different versions of Mach's thought, the young author then tries to give a definitive description of where Mach really stands before turning to a critique. But he cautions the reader that since Ernst Mach lacks a system, and since even his doctrine of elements is not rigorous or free from ambiguity, that while one can indeed refute individual aspects (as Schultz then claims to do), that those critics who think that Mach's *overall* position is clear enough to be subject to exact and complete refutation are badly mistaken.[25] Indeed, It was not without reason that Mach denied he was a philosopher. In short, Schultz showed a degree of sophistication missing in Musil's account and one can agree with most of his comments pretty well.

Though the Preface by itself appears to contain the author's principal conclusions, one should also note that the main text which follows the line expressed in the Preface is divided into two parts: 1) "Representation" and 2) "Criticism of Mach's Epistemology", and that there is a lot of rather detailed elaboration of earlier points in the text. There is, to be sure, a suspicion that this repetition plus elaboration technique may appear monotonous for some readers?[26] But has a doctoral dissertation ever been rejected for that reason? Especially by a professor who may even be more monotonous?

It is also interesting that the thesis is on a contemporary figure who has only recently "burst upon the academic scene". Such dissertation topics may have been a paradigmatic habit of the time or simply may have been the kind of thing which seems to attract so many young scholars "who were born yesterday". The notion that what is new is probably significantly better than what is old is a natural belief to have when one is young, especially if one is as interested in science as Musil apparently was, but in philosophy it is often Plato or Kant who disabuses one of this common illusion, hence choosing "new" philosophers on whom to write one's dissertation and then becoming disappointed is surely a widespread and natural experience.

But equally significant disappointment has probably been felt by most of the known twenty-seven authors of German dissertations on Mach, and not just by Hönigswald or Schultz/Reinhold. They were indeed published as was required by academic rules of the time, but while we cannot say how many people have subsequently read them, it is clear that almost no Mach scholars, including Heller, Blackmore, Wolters, Hoffmann, or even other contributors to this very volume have ever commented on any of them, except on Musil's, who later became famous. Was Hans Henning correct that dissertations on Mach's philosophy were all "worthless"? Probably not, but we will never know about most of them until more scholars deign to examine them.

Perhaps most useful in connection with Robert Musil's dissertation was his gradual realization (which was not always or fully mentioned *in* the dissertation), first, that both Mach and Musil had studied mathematics and physics but turned to philosophy and increasingly to psychology, second, that psychological considerations were actually predominent in Mach, at least in *Populär-wissenschaftliche Vorlesungen*, as they were or would be in Musil, third, that with respect to Mach his deep interest in psychology was probably a factor in his rejection of natural necessity, and fourth, that Mach like Musil increasingly adopted and practiced a literary writing style.[27]

There are only a few passages in Musil's published diaries and letters which refer to his dissertation, and therefore, it is difficult either to describe its genesis in detail or to present a clear picture of his own response to the work. But let us present the little data we have. Musil wrote in his diary on May 26, 1902:

> Mach's *Populär-wissenschaftliche Vorlesungen* fell into my hands today at the right time. The contents seem reasonable and yet have high consequence. To be sure, I have not tried to examine it critically, and I must remind myself again to be cautious.[28]

What is remarkable about this reference, however, is that it is the *only* book by Mach which Musil mentions in his diaries, letters, and publications subsequent to his dissertation.[29] But it must have had some influence, since we discover that two year's later Musil himself is cheerfully writing popular scientific articles.[31] Also, since many of Mach's articles in that very readable book are on psychology it is interesting to find out that Musil also wrote on applied psychology early in the 1920's. In other words, while Mach's ideas in his popular and psychological writings *probably* influenced Musil, it is *virtually certain* that Mach's writing style influenced the younger man.

Musil seems to have selected Ernst Mach's philosophy of science for the theme of his dissertation because of his interest in science as well as his sympathy with several of the topics in Ernst Mach's book on popular lectures which have scientific, technical, philosophical, and especially psychological aspects. It seems that Robert Musil was attracted to philosophy by a desire to develop a 20th philosophy which went beyond the traditional scientism and positivism of Auguste Comte and the early 19th century with its opposition to psychology, and he found the germ in Ernst Mach's thought. What Musil intended was to was to put together, so to speak, a key to the 20th century. Ernst Mach provided some of the strands, but Robert Musil apparently made it his task at least in the first version of his dissertation to discuss it acutely but fairly, though one cannot be certain, but definitely not in the harsh, neo-Kantian manner of Hönigswald or the subtle but almost equally negative style of Schultz/Reinhold in the two previous dissertations on Mach. Indeed, Musil was already beginning to express reservations about the academic style in philosophy as the proper approach for expressing his ideas. (Diary entry for May 13, 1905).[32]

Thus, even well before Carl Stumpf criticized the initial version of the dissertation in the Summer of 1907, Musil had revealed signs of becoming disheartened, presumably because of some philosophical problems which he had difficulty solving, resolving, or dissolving. More explicitly:

> Diary: February 17, 1907:
>
> I am very depressed. My work will not go ahead. My day ends so quickly and without solid results. I am afraid I shall lose control of myself. Besides the eternal struggle with my work, there is simply too much of it. I am surfeited.[33]
>
> Diary: March 11, 1907:
>
> I am no more than a mummy with a mere tic for work.[34]

By the middle of the year Musil had finished the first version and by the beginning of the next the revised dissertation so that he was finally able to obtain his PHD at last. But from what his scarce letters reveal,[35] he seems to have had no feeling of satisfaction, but merely one of freedom or release. The dissertation was done. It was vaguely compatible with neo-Kantianism via his own approach to "natural necessity", but on most issues only by not discussing them. Most of his own views only surfaced in general remarks or in formally unimportant passages. His attraction towards Mach's ideas was objectified into neutral description coupled with the apparent development of criticism based on rejecting Mach's reduction of scientific laws to mere simplified and idealized constant conjunctions among appearances.

But if the dissertation was little more than a hurdle-jumping exercise to please Carl Stumpf, it did open the opportunity to obtain a starting position with Alexius Meining in the Styrian capital of Graz.[36] He also seems to have had a chance at the Technical University in Munich. But he rejected the first and was apparently rejected by the second.[37]

Years later he would reflect in his diary about the possibility of having had an appropriate academic career if he had accepted the Graz invitation.[38] But it seems more probable that he could not have devoted himself fully to traditional philosophy. Like Meinong himself, he was somewhat drawn to experimental psychology, but unlike his mostly philosophical mentor he might well have engaged in it more completely. This also hints that perhaps his philosophical immaturity as revealed in his dissertation might have turned out to have been a positive rather than a negative factor. For what was most in demand in psychology in those days was an empirical or experimental attitude rather than a systematic or transcendental approach. But we are guilty of speculating, which becomes especially questionable when we keep in mind how little true satisfaction he had obtained from experimental psychology when he had been at the University of Berlin under Stumpf, who was a philosopher with considerable interest in and who had done some important work in the field of experimental psychology. In fact, Musil has reminisced that had he gone to Graz he would have left himself out and still turned to literature. But it is hard to be certain.

Indeed, at some point Musil must have realized that for all of his interest in reason, science, and an "objective" style of writing that his own personality was basically gloomy and that his own life had been determined more by emotional factors such as Stumpf's dislike of Mach's philosophy, his own emotional decision to follow a literary career rather than one in professional philosophy, and by feelings "caused" by war, inflation, and depression. While his Mach dissertation shows little interest in ethics, this apparently was a main concern of his, and given the temper of the times and Franz Brentano's followers in the field of phenomenology he apparently chose to base ethics more on feeling rather than on reason, means-end logic, high ideals, or even the Ten Commandments.

5. ETHICS AND MYSTICISM

While Musil's interest in science had originally helped him become interested in Mach and vice versa., it is still a bit speculative how he became interested in ethics along with mysticism. A feeling for justice and a touch of despair may have been factors.

> It was important that I always keep my attention on and engagement with ethics, but I had difficulty finding a gateway which suited me. In other words, I had not studied enough.[39]

At first glance, the timing of the above observation might suggest that his concern with ethics could be related to his opposition to some of the political developments in conflict-ridden Austria and Nazi Germany, but there is an earlier remark from the less tumultuous 1920's which indicates that ethics had been a major concern even then. He writes in a letter to Josef Nadler dated on December 1, 1924 that he had studied philosophy, mainly logic, epistemology, and psychology "though my inclination turned to ethics".[40]

But since it is difficult to grasp how studying these fields helped him in his desire to understand morality better, we are in something of a dilemma. Apparently, he thought that preparation in logic, math, physics, and psychology was not utterly futile but could at least help make his ethical thinking more rigorous. Indeed, he would write in a late letter:

> I would like . . . then to add that I cannot think of fruitful philosophical activity going on in the long run without a penetrating study of mathematics and psychology, and this is the case regardless of how one wants to distinguish the essence of philosophy from other types of . . . knowledge.[41]

Nevertheless, perhaps it was lack of *obvious* connection of these fields with ethics that was a factor in henceforth expressing ethical ideas by means of literature. But this too is only speculation.

Robert Musil, however, was not the only Austrian who emphasized the importance of ethics while in fact training himself in other domains. Ludwig Wittgenstein did the same thing. For example, when queried about his book *Tractatus logico-philosophicus* he produced the well-known comment to Ludwig von Ficker that "The meaning of the book is an ethical one."[42] Such a comment may seem only partly applicable at best but perhaps Musil and Wittgenstein were referring to ethics in a rather symbolic or tenuous sense. Whereas for Wittgenstein ethics was inexpressible, for Musil it was only too difficult to express. Another possible explanation is that perhaps in German or "Austro-German" the word "*Ethik*" had come to mean at least for some intellectuals something other than what "ethics" means in English.

The puzzle may become even more interesting when we discover that another book published on Mach during the period when Musil was writing his dissertation also connects Mach if not explicitly with ethics at least with value [*Wert*]. The volume was by Bernhard Hell and was titled: *Ernst Machs Philosophie. Eine erkenntniskritische Studie über Wirklichkeit und Wert*, Stuttgart, 1907. But it is not certain what if any impact it had on Musil.

But if we return to the first quotation up above but include the very next sentence, then we have what may be an even stronger clue. At least by the late 1930's if not earlier Musil seems to have developed an interest in the ideas of the recently deceased (1928) German thinker Max Scheler, who was attempting to develop a "phenomenological" theory of ethics. Therefore, let us repeat the quotation but this time add the next sentence:

> It was important that I always keep my focus on and engagement with ethics, but I had difficulty finding a gateway which suited me. In other words, I had not studied enough. *But it was Scheler who passed through the gateway.*[43]

Unfortunately, Max Scheler (1874-1928) is one of a group of thinkers who were well-known in the German-speaking Europe of his time, but who for some reason seem not to have drawn much attention in English-speaking philosophy, hence some clarification is in order. Scheler reached maturity at a time in Jena and then in Munich when the ideas of Franz Brentano (1838-1917) and his student Edmund Husserl (1859-1938) were very influential. From Brentano Scheler took the idea that ethics was primarily a matter of feeling and from Husserl the method called *Phänomenologie*. Combining the two, Max Scheler believed that many ethical values could be intuited from a conscious show or "immediacy of essences" [*Wesenschau*]. He placed these values within five categories and then attempted to rank them. These might be translated as *sensory values, useful values, life values, spiritual values*, and *holy values*.[44] His most important book on ethics carried the title: *Der Formalismus in der Ethik und die materiale Wertethik*, Halle, 1913-1916.

The final word in the title, namely, *Wertethik* seems to be the key to our problem. Ethics has traditionally been understood since the Greeks as largely a matter of intent and action. Actions are "good" or "bad" in Christian terms depending on how compatible they are with The Ten Commandments, The Sermon on the Mount, and in large measure the long lists mentioned by Dante Alighieri in his tripartite work on Heaven, Purgatory, and Hell called *The Divine Comedy*. But by far the most conspicuous values are the sins and they are viewed as negative, and often expressed as "Thou shalt nots", as if one could be moral or ethical simply by avoiding doing evil. Also, because most of these sins are also punished by civil authorities as crimes against the state, the matter of court trials and judges becomes very important. And it is normally thought to be crucial for judges and juries to be fair and objective as if they should never let themselves be influenced by mere feeling or prejudice. This intent- and action-based concept of ethics reached its extreme form in Immanuel Kant's doctrine of duty for duty's sake and in his so-called categorical imperative, as if no action were truly moral if it were motivated by mere feeling or self-interest or were not suitable for everyone.

Max Scheler, had been strongly religious himself, long a Catholic and later supporting a kind of pantheism, though it has been alleged that shortly before his death he became an atheist, perhaps to help justify putting human feeling first, though he seems to have largely rejected traditional "objective" ethics in favor of his radical theory of *Wertethik*, as if ethical *values* could be determined by our intuitive feelings. Scheler often used half-ethical, half-emotional expressions like *Mitgefühl, Miteinanderfühlen,* and *Nachfühlen* to help make his point.[45] This drastic shift from traditional ethical judgment to emotional preference, while not total, meant that Scheler and possibly Musil and Wittgenstein *also* were talking about ethics in a very different sense than is normally understood in the English-speaking world. At least for these few gentlemen, German and Austrian *Ethik* now seemed to mean something akin to *Wertethik*. It also seems that Scheler's feeling-first philosophy could be regarded, like existentialism of a later day, as a defence of or justification for expressionaist art, sculpture, music, and literature, as if self-expression were more important than the rational beauty of its results, a perspective which Robert Musil may have secretly shared underneath his stylistic objectivism.

It is now possible to understand how Robert Musil by studying logic, epistemology, and psychology *may* have thought that he was better preparing himself for understanding what would soon be called *Wertethik.* By studying "the flow of appearances" and looking for the right emotional features in them in a manner close to that which Max Scheler would spell out only a few years later as if one could really "intuit ethical values". Furthermore, for Max Scheler they were absolute and as essences could be regarded as universals. While it is conceivable that Musil could have read Max Scheler's doctoral dissertation and *Habilitationschrift* and even his withdrawn book on logic published in 1903,[46] it seems probable that he could only have noticed the partial similarities in their value-ethical outlook with the publication of Max Scheler's major work which came out from 1913 to 1916.[47] Nevertheless, anyone familiar with Edmund Husserl's *Logische Untersuchungen* (1900-1901) might have been capable of drawing the same value inferences which Max Scheler had done and which Musil apparently sympathized with. One says "apparently" because in Scheler's case his tendencies toward dogmatism and exaggeration were so conspicuous and the changes in his philosophical positions so many that it may be difficult to find anyone who ever accepted his full philosophy,[48] but perhaps his emphasis on feeling in *Wertethik* kept loyal followers longer.

But there are differences of opinion. Some observers want to bring Musil closer to Vienna Circle positions of the 1920's and to Logical Positivism by emphasizing his attraction to Mach, science, and phenomenalism. And they could also mention that many Logical Positivists have also tried to identify ethics and value theory with feeling and relativism, but they were optimists about science and apart from Moritz Schlick relatively indifferent to ethics and value theory which they tended to dismiss as unscientific and irrational, but Musil was a pessimist and the emotional world *did* matter to him, even if his writing style was cold and analytical.

Indeed Musil probably came to the conclusion that most human beings can understand and appreciate *Wertethik* better through novels and literature than through any abstract, academic, or technical philosophy or anything as variable and provocatively written as most of the very numerous works of Scheler. This means that if Robert Musil was primarily interested in ethics as as he said he was, then *Wertethik* could help explain both his attraction to Mach and his switch away from philosophy to literature.

In short, Musil's interest in Mach the psychologist who studied feeling was indeed relevant for ethics, even if the final version of the dissertation by stressing methodology and philosophy of science partly reflected his own and Mach's study of math and physics while respecting or overrespecting Professor Carl Stumpf's own psychology-oriented sensitivities. Nor - since Stumpf had been a student of Brentano like Husserl - is it impossible that it was Stumpf who initially placed Brentano's tendency to think of ethics in terms of feeling in Musil's mind rather than Husserl's student, Max Scheler? But regardless of the initial source, Robert Musil did later sympathize with aspects of Scheler's *Wertethik.*

But is using "rational" writing to favor something as emotional as *Wertethik* rational or irrational? Furthermore, William Johnston in his book *The Austrian Mind*, Berkeley, 1972, p. 392 has written: "Like Broch, Ebner, and Rank, Musil blamed rationality for thwarting self-fulfillment." Does *Wertethik* as the conception of intuiting value from "the flow of feelings as conscious universals" seem to lend itself in spite of Max Scheler's claims to the contrary to value relativism? since different people have different feelings and naturally rank values based on feeling differently? Relative value and relative truth are different, but claims to absolute truth about value relativism is hardly a problem-free position. But we also have to keep in mind that relativistic truth for many Germans is contrasted with Hegel's "Absolute" and is not understood as lack of correspondence with reality.[49] We don't know why Musil's interests drifted away from reason and science. Nor do the emotional events which influenced his career seem to justify the view that using reason or "overemphasizing" it was the culprit. Indeed, the opposite seems true.

Still, there is a strong suspicion, though admittedly it is not more than that, that the emotional proclivities which Musil attributes to some of his protagonists like Walter and Clarisse and especially to his antagonists such as the murderer Moosbrugger in his novel by suggesting that there is no objective right and wrong, may have had some connection with Scheler and *Wertethik* and may be at the heart of the negativism in the book about the active or rational life itself. But on the other hand, it has been alleged that Musil first refers to *The Man without Qualities* as a possible novel as early as 1905,[50] which is after he first read Mach and probably before he was acquainted with Scheler's *Wertethik.*, but as earlier suggested this of course does not mean that he could not have learned it from Stumpf or reading Husserl or that he would have been as satirical about Austria-Hungary in 1905 as he became after the collapse of The Dual Monarchy in 1918, especially about what he came to regard as the immense amount of false optimism in pre-war times. Indeed, as an officer with a strong military background he had apparently fought during the First War for many of the very same "illusions" which he subsequently chose to mock. But pessimism was the fashion in many circles even prior to the War. Musil and the Catholic writer Hermann Broch were friends,[51] and Broch was surely no optimist about "this world" either.

But still, Scheler and *Wertethik* could also have had some connection with the well-known "mysticism" which many commentators have claimed to find in Musil's literary masterpiece and which he himself was not only aware of but defended, and which was designated by him as "bright as day mysticism".[52] This view with its pantheistic implications which are not fully dissimilar to some of Scheler's final visions can be found in the following passage:

> These Christian, Jewish, Indian, and Chinese texts and testimonies span

> more than a millenium. Still, one can see in all the same . . . a united structure of inner evolution. They are only different in their external relations, the unique contexts in which all theology and heavenly wisdom have found identity under a single protective roof.[53]

It has been claimed that Musil was not an extensive reader and relied on some of the materials in *Ekstatische Konfessionen* as edited by Martin Buber in describing mysticism in his novel,[54] but the critic Bernd-Rüdiger Hüppauf has paid more attention to the "coincidence" between Musil's own inner ideas and mysticism than to the influence Buber's book had on Musil.[55] He also alleges that that there is something in the novel which one can relate to German mysticism of the 13th and 14th centuries, perhaps even to its best known representative, Meister Eckhart, but if so Robert Musil was not fully conscious of it in his correspondence.[56]

But if most forms of mysticism seek peace of mind or ineffable kind of quiet "inner salvation", then it might seem that *Wertethik* by accentuating emotion and allowing its strongest forms to determine some of the strongest values would be grossly incompatible with mysticism. But if we think in terms of the mystic's common appeal to several stages of understanding, then the contradiction vanishes. The notion that feeling can intuit or determine value would be an initial stage such as Schopenhauer suffered from which results in misery, and which troubles some of the characters in Musil's novel. A second stage might well be the realization that feeling, especially if it includes desire, can risk disappointment or other pain and as part of an active or materialist life needs to be overcome. The third stage would be a much more meditative and less painful path leading to peace, Heaven, Paradise or Nirvana.

Bernd-Rüdiger Hüppauf has added that mysticism naturally prefers immanence to transendence, since immanence allows a union of God with oneself. But the immanence in Musil has lost the moment of union with God, and is, instead, another innerworldly kind of union, as if there were a desire to combine the mystical with enlightenment.[57] In *Der Mann ohne Eigenschaften* "mysticism has lost its religious context and its spirituality can therefore move toward the intellectual intention of enlightenment. Mysticism and enlightenment, which seem so opposed to each other, have lost their original antagonism and are united into a synthesis."

But can the mysticism which Musil was concerned with actually link Musil with Mach? We can find in spite of Mach's immanentism almost no remark favorable to mysticism in his own writings. It is very clear that the experimental physicist, historian of science, and methodologist of science *wanted* to have nothing to do with mysticism, but like it or not, his theory of elements, his phenomenalism when viewed as "maya" fit the foundation of early Theravada Buddhism like a glove, and this was recognized rather early by Buddhists on Ceylon and later by Mach himself as well as several biographers.[58] Furthermore, his theory of *Denkökonomie* or economy of thought has parallels in Zen Buddhism, the Buddhist rationale being that if there is no consciousness, then there is no suffering, and thinking less can reduce consciousness. But if Musil's mysticism was not of a Buddhist or religious sort, then he may have tried to transform Ernst Mach's theory of elements, economy, and description into a type of mysticism more fitting Ernst Mach's scientific intent than what Mach himself had done.

It is interesting that Logical Positivists almost to a man while honoring Ernst Mach the phenomenalist and methodologist of science have *refused* to acknowledge how close his epistemology was to that of Buddhism and much mysticism, but had they checked the later life of many phenomenalists and positivists they would have discovered that Mach was not alone. Perhaps this irrational refusal is connected with the tendency of most

Logical Positivists and analytic thinkers to persuade themselves and others that their position is not just *a* philosophy but *the* philosophy of science, and then pointing out the similarities in epistemology with Buddhism, mysticism, and several religions runs counter to their self-image and their unending propaganda campaigns. They want to think of themselves as rationalists (while of course meaning "pure" reason rather than practical reason), as if they were well beyond all unreason and mysticism, but of course they have deceived themselves, just like they refuse to face the similarities in epistemology with idealists such as George Berkeley which both Lenin and Sir Karl Popper have pointed out.

How about Mach's theory of the elements? Both subject and object are unified into them. Subject or ego is approved by Ernst Mach but only as a collection of sensations and not anything substantial. This is both a Buddhist view and the opinion of many modern scientists and philosophers. Also many mystics agree with many physicists that personality may be nothing substantial, but merely a series of fleeting impressions, no more real than the ego or self. Ernst Mach's preference for description over explanation is also paralleled in Zen Buddhism as well as in law-focused science, provided that one interprets laws as Mach does as merely descriptive in character. Thus, Ernst Mach can be seen as rejecting causality,[59] which is also rejected by mysticism. But since the famed wheel of Karma is a causal theory, where the amount of Karma or sin which we acquire in one life affects what kind of creature we become in the next life, it might seem that Ernst Mach's position is incompatible on this point with Buddhism and Hinduism, which are surely types of mysticism, but actually, both religions contain a series of levels of understanding with the higher levels transcending the notion of causes. Can science do without causes? Mach and Musil seem to have leaned toward that goal, even if outside of science "human survival of the fittest" and economy of thought seem to presuppose or require a conception of causes, and in fact Musil often refers to causes, especially in his correspondence,[60] though it is a fact that the characters in his book often seem to believe in false causes or misunderstand causes, which provides much of the irony and satire in his long work.

But Musil may not have been alone in inferring from a reluctant Mach any kind of scientific mysticism as a positive philosophical position. Let us remember Fritz Mauthner (1849-1923), the Austrian philosopher of language who in his own words had also been under Ernst Mach's influence "since he heard Mach lecture in the Casino in Prague". Fritz Mauthner advocated a kind of godless mysticism, which also had similarities with Buddhism.[61]

But most mysticism in its rejection of an active solution of our problems tries to transcend almost all forms of ethics, but *Wertethik* would probably be more congenial as an initial stage with its emphasis on human feeling and *Anschaulichkeit*, though at the higher levels of mystical understanding it too would be transcended. But what stands out in the case of Robert Musil and perhaps also Mauthner was the insistence that science and mysticism were compatible,[62] and that like with Buddhism, one did not need to believe in a God or Divinity in order to be religious or mystical, and of course such belief was not necessary in order to do science either.

NOTES

[1] E. Wilkins & E. Kaiser, "Foreword", in Robert Musil, *The Man without Qualities*, Volume I, Capricorn Books: New York, 1965, p. iii.

[2] Ibid., p. v.

[3] Robert Musil, *The Man without Qualities*, ibid., chapters 2-4, pp. 6-14.

* We have of course read several recent works on the relation between Mach and Musil even if they are not specifically cited in this paper. They include G.H. von Wright, "Musil and Mach", in *On Mach's Theories*, translated by Kevin Mulligan, Philosophia Verlag: Munich, 1982, pp. 7-14; Paul-Laurent Assoun, "Musil, lecteur d'Ernst Mach", in *Pour une évaluation des doctrines de Mach*, Presses universitaire de France: Paris, 1985, pp. 5-48 and a postscript pp. 177-194; and Silvia Bonacchi, "Robert Musils Studienjahre in Berlin 1903-1908" in *Musil-Forum*, Beilage I, 1992, pp. 1-58.

[4] The figure of twenty-seven comes from one of the editors who in 1967 saw 27 listed in a repository for unwanted publications in Chicago.

[5] Joachim Thiele in his article "Ernst Mach-Bibliographie", *Centaurus*, **8** (1963) 230-237 lists the following dissertations on Ernst Mach: Richard Hönigswald (1903), under a pseudonym Reinhold Schultz (1907), Robert Musil (1908), Mordechai-Ber Rudajew (1913), Rudolf Thiele (1914), Karl Gerhards (1914), Eugen Stock (1920), Adolf Jensen (1923), Julius Becker (1927), and Franz Kallfelz (1929).

[6] Hans Henning, *Ernst Mach als Philosoph, Physiker, und Psycholog*, Leipzig, 1915, p. vi. (Henning was largely a clarifier and defender of Mach who in turn corresponded with Henning from 1912 until 1915. There are nine letters from Henning to Mach which were in the Ernst-Mach-Institut and which are now in the Deutsches Museum in Munich.)

[7] Three dissertations on Mach were published as books as follows: First: Bernhard Hell, *Ernst Machs Philosophie. eine erkenntniskritische Studie über Wirklichkeit und Wert*, Stuttgart, 1907. Second: Ferdinand Reinhold, *Machs Erkenntnistheorie*, Leipzig, 1908. And third: Herbert Buzello, "Kritische Untersuchung von Ernst Mach's Erkenntnistheorie", *Kantstudien*, Ergänzungsheft, Nr. 23, 1911.

Hönigswald's book, *Zur Kritik der Machschen Philosophie*, Berlin, 1903, was not a published version of his dissertation but seems to have preceded it, i.e. *Über die Lehre Hume's von der Realität der Aussendinge. Eine erkenntnistheoretische Untersuchung*. (1904). Recent works on Hönigswald include E,W, Orth & B. Alexandrowicz (Hrsg.), *Studien zur Philosophie Richard Hönigswald* (1996) and R. Grassl, *Der Junge Richard Hönigswald* (1998). See also W. Schmied-Kowarzik, *Erkennen-Monas-Sprache* (1997).

[8] Robert Musil, *Beitrag zur Beurteilung der Lehren Machs*, Wilmersdorf, 1908. The page numbers for the chapters are as follows: **1**) 5-12, **2**) 12-32, **3**) 32-56, **4**) 56-79, and **5**) 79-124. For readers of English see Robert Musil, *On Mach's Theories*, translated by Kevin Mulligan, Philosophia Verlag: Munich, 1982: **1**) 15-19, **2**) 20-30, **3**) 31-43, **4**) 44-56, and **5**) 57-80. Henceforth, for Musil's dissertation page numbers in German will be introduced by an "S," and English pages by a "p." or "pp."

[9] Ibid., S. 5, p. 15.

[10] Ibid., S. 21, p. 24.

[11] Auguste Comte's "Three stages of progress" put the Positive Stage on top.

[12] R. Hönigswald, *Zur Kritik der Machschen Philosophie. Eine erkenntnistheoretische Studie*, Berlin, 1903, Chapter One.

[13] Ibid., pp. 13 & 17-18.

[14] Musil, *Beitrag*, op. cit., S. 135-140.

[15] Ibid.

[16] The marks given to the dissertation and the examination were *laudabile* and *cum laude* respectively, but they tend to mean different things in different countries. We are told that in the

United States where such accolades are rare that *cum laude* is a rather high mark, but it seems that in the Germany of Musil's day that *cum laude* was average, *magna cum laude* above average, and *summa cum laude* was superior. (See *Beitrag*, ibid., S. 138f and a letter from Musil to Johannes von Allesch dated March 13, 1908 in which he writes about the examination, a letter which can be found in Robert Musil, *Briefe 1901-1942*, edited by Adolf Frisé, Rowohlt: Hamburg, 1981, pp. 51f.)

17 *Beitrag*, ibid., S. 138.

18 This article by Mach first appeared in *Pflüger's Archiv für die gesamte Physiologie des Menschen und der Tiere*, **136** (1910) 263-274 as part of a Festschrift für Ewald Hering. It was published with another essay in 1919 with Friedrich Adler as the anonymous editor, and afterwards collected in Mach's *Populär-wissenschaftliche Vorlesungen*, 5. Auflage, J.A. Barth: Leipzig, 1923 with Ludwig Mach as the anonymous editor.

19 Ibid., Enst Mach, "Sinnliche Elemente und naturwissenschaftliche Begriffe", *Populär-wissenschaftliche Vorlesungen*, Böhlau: Wien-Köln-Graz, 1987, pp. 577-578.

20 Ibid., pp. 576-577.

21 Ibid., pp. 587-588.

22 Robert Musil, *Briefe 1901-1942*, hrsg. Adolf Frisé, Rowohlt: Hamburg, 1981, p. 367.

23 Robert Musil, *Tagebücher, Aphorismen, Essays, und Reden*, hrsg. von Adolf Frisé, Hamburg, (1955) 1976, p. 919.

24 Hans Henning also discusses the Schultz/Reinhold dissertation on Mach, but it is still not clear why Schultz published his dissertation using a pseudonym. As far as can be determined, Schultz was born in 1883 and entered the University of Munich to study history in 1903. He had very broad interests including art and literature which he pursued further in Leipzig from 1904. He then turned to philosophy and studied especially under Max Heinze. His judges were Heinze and Wilhelm Wundt, and the degree was given on March 30, 1907.

25 Reinhold Schultz, *Die Machsche Erkenntnistheorie. Darstellung und Kritik*, Berlin, 1907, p. 5f.

26 Later, Franz Kallfelz adopted the same method. See his thesis: *Das Ökonomieprinzip bei Ernst Mach. Darstellung und Kritik. Das Prinzip der Maximalleistung des Denkens*, Munich, 1929.

27 Ernst Mach's literary style of writing is most conspicuous in his psychological and anthropological writings and least present in his physical articles. His books on the history of physics combine aspects of both, but with the scientific style predominant. Like Mach, Musil could also use both styles, but in the dissertation on Mach he stuck to a dry academic manner.

28 *Tagebücher*, op. cit., p. 20.

29 Furthermore, Musil would have read at that time the 1896 or 1897 editions which still had only 15 chapters. Starting with the 1903 edition the chapters begin to increase until by the 1923 edition there are 33 chapters. One might also add that the American edition which actually first appeared a year before the first German edition, that is, in 1895 never contained more than the original 15 chapters even in the later editions of 1910, 1943, and 1986.

31 Most of these can presumably be found in volumes two and three of *Robert Musil - Gesammelte Werke in Einzelausgaben*, hrsg. von Adolf Frisé, Rowolt: Hamburg, 1955 and 1957. A bibliography of these is listed in *Tagebücher*, op. cit., p. 941-944.

32 *Tagebücher*, Ibid., pp. 149f.

33 Ibid., p. 208.

34 Ibid., p. 209

35 Robert Musil to Johannes von Allesch, March 13, 1908 and to Franz Blei, April 12, 1908. See *Briefe*, op. cit., pp. 51ff.

36 See Robert Musil to Alexius Meining, December 30, 1908 and January 18, 1909 in *Briefe*, Ibid., pp. 61ff.

37 Ibid.

38 See Musil's diary at the end of the 1930's, *Tagebücher*, op. cit., pp. 918f.

39 Ibid., and Wilfried Berghahn, *Robert Musil in Selbstzeugnissen und Bilddokumenten*, Rowohlt: Reinbek bei Hamburg, 1967, p. 57.

40 *Briefe*, op. cit., p. 367.

41 Robert Musil to Karl Baedeker, August 16, 1935. Ibid., p. 652.

42 B.F. McGuinness & G.H. von Wright (editors.), *Ludwig Wittgenstein - Briefwechsel*, Suhrkamp Verlag: Frankfurt am Main, 1980, p. 96.

43 See footnotes 38 and 39. We have added the italics.

44 Wolfgang Stegmüller, "Angewandte Phänomenologie: Max Scheler" in his volume: *Hauptströmungen der Gegenwarts-Philosophie*, Band I, Kröner: Stuttgart, 1978, pp. 96-134.

45 Ibid., pp. 106-108.

46 Wilhelm Mader, *Max Scheler in Selbstzeugnissen und Bilddokumenten*, Rowohlt: Reinbek bei Hamburg, 1980, pp. 140-141.

47 Max Scheler, *Der Formalismus in der Ethik und die materiale Wertethik*, Halle.

48 Even the quotations from obituaries after Max Scheler's death, at least those in Mader's book, op. cit., pp. 143-145, do not reflect agreement with Scheler's ideas. Perhaps the remarks which come closest were from Martin Heidegger (p. 144).

49 See Rom Harré and Michael Krausz, *Varieties of Relativism*, Blackwell, 1996.

50 Berghahn, op. cit., p. 159.

51 Ibid., p. 164.

52 R. Musil, *Der Mann ohne Eigenschaften*, in *Gesammelte Werke in Einzelaufgaben*, Bd. 1, hrsg. von Adolf Frisé, Rowohlt: Reinbek bei Hamburg, (1952) 1987, p. 1089.

53 Ibid., p. 766.

54 *Ekstatische Konfessionen*, gesammelt von Martin Buber, (1909), 5th Auflage, 1984.

55 Berned-Rüdiger Hüppauf, *Von sozialer Utopie zur Mystik. Zur Robert Musils 'Der Mann ohne Eigenschaften'*, 1971, p. 13.

56 Ibid., p. 14.

57 Ibid., p. 15.

58 John T. Blackmore in his biography, *Ernst Mach - His Work, Life, and Influence*, University of California Press: Berkeley & Los Angeles, 1972, in Chapter 18, "Mach and Buddhism", mentions some of the authors who discuss Mach's attraction to Buddhism including: Theodor Beer, Robert Bouvier, Paul Carus, W. Fred, Heinrich Gomperz, Carl Haas, Richard Hönigswald, Hermann Keyserling, Anton Lampa, and Kurt Schmidt.

59 Ernst Mach, *Erkenntnis und Irrtum*, 5. Auflage, J.A. Barth: Leipzig, 1926, p. 248.

60 Berghahn, op. cit., p. 57.

61 Fritz Mauthner, *Der Atheismus und seine Geschichte im Abendlande*, 4 Bde., Stuttgart, 1920 bis 1923.

62 William M. Johnston, *The Austrian Mind*, University of California Press: Berkeley, 1972, pp. 196-199.

REFERENCES

Assoun, Paul-Laurent. "Musil, lecteur d'Ernst Mach". in *Pour une évaluation des doctrines de Mach*. Paris: Presses universitaire de france, 1985, pp. 5-48 & 177-194.

Berghahn, Wilfried. *Robert Musil in Selbstzeugnissen und Bilddokumenten*. Reinbeck bei Hamburg: Rowohlt, 1967.

Blackmore, John. "Mach and Buddhism". in *Ernst Mach - His Work, Life, and Influence*. Berkeley: University of California Press, 1972, pp. 286-299.

Bonacchi, Silvia. "Robert Musils Studienjahre in Berlin 1903-1908". in *Musil-Forum*, Beilage 1, 1992, pp. 1-58.
Buber, Martin (ed.) *Ekstatische Konfessionen*. 5. Auflage. 1984 (1909).
Buzello, Herbert. "Kritische Untersuchung von Ernst Machs Erkenntnistheorie". *Kantstudien. Ergänzungsheft*, Nr. 23, 1911.
Harré, Rom & Michael Krausz. *Varieties of Relativism*. Oxford: Blackwell, 1996.
Hell, Bernhard. *Ernst Machs Philosophie. eine erkenntniskritische Studie über Wirklichkeit und Wert*. Stuttgart, 1907.
Henning, Hans. *Ernst Mach als Philosoph, Physikar, und Psycholog*. Leipzig: 1915.
Hönigswald, Richard. *Zur Kritik der Machschen Philosophie*. Berlin: 1903.
Hönigswald, Richard. *Über die Lehre Humes von der Relaität der Aussendinge. Eine erkenntnis-theoretisxhe Untersuchung*. doctoral disseration, 1904.
Hüppauf, Berned-Rüdiger. *Von Sozialer Utopie zur Mystik. Zur Robert Musils 'Der Mann ohne Eigenschaften'*, 1971.
Johnston, William M. *The Austrian Mind*. Berkeley: University of California Press, 1972.
Kallfelz, Franz. *Das Ökonomieprinzip bei Ernst Mach. Darstellung und Kritik. Das Prinzip der Maximalleistung des Denkens*. Munich: 1929.
Mach, Ernst. "Sinnliche Elemente und Naturwissenschaftliche Begriffe". *Pflüger's Archiv für die gesamte Physiologie der Menschen und Tiere*. 136 (1910) 263-274. Also in *Populär wissen-schaftliche Vorlesungen*. Wien-Köln-Graz: Böhlau, 1987, pp. 570-588.
Mach, Ernst. *Erkenntnis und Irrtum*. 5. Auflage. Leipzig: J.A. Barth, 1926 (1905).
McGuinness, B.F. & G.H. von Wright. (editors). *Ludwig Wittgenstein - Briefwechsel*. Frankfurt am Main: Suhrkamp Verlag, 1980.
Mader, Wilhelm. *Max Scheler in Selbstzeugnissen und Bilddokumenten*. Reinbeck bei Hamburg: Rowohlt, 1980.
Mauthner, Fritz. *Der Atheismus und seine Geschichte im Abendlande*. 4 Bde. Stuttgart: 1920-1923.
Musil, Robert. *Beitrag zur Beurteilung der Lehren Machs*. Wilmersdorf, 1908.
Musil, Robert. *Robert Musil - Gesammelte Werk in Einzelausgaben*. hrsg. von A. Frisé. Hamburg: Rowolt. 1957-1959.
Musil, Robert. *The Man without Qualities*. Volume I. New York: Capricorn Books, 1965.
Musil, Robert. *Tagebücher, Aphorismen, Essays, und Reden*. hrsg. von A. Frisé. Hamburg: 1976 (1955).
Musil, Robert. *Briefe 1901-1942*. edited by A. Frisé. Hamburg: Rowohlt, 1981.
Musil, Robert. *On Mach's Theories*. translated by Kevin Mulligan. Munich: Philosophia Verlag. 1982.
Musil, Robert. *Der Mann ohne Eigenschaften*. in Gesammelte Werke in Einzelaufgaben. Bd. I. hrsg. von Adolf Frisé. Reinbeck bei Hamburg: Rowohlt, 1987.
Reinhold, Ferdinand. *Machs Erkenntnistheorie*. Leipzig, 1908.
Scheler, Max. *Der Formalismus in der Ethik und die materiale Wertethik*. Halle: 1954 (1913).
Schultz, Reinhold. *Die Machsche Erkenntnistheorie. Darstellung und Kritik*. Berlin: 1907.
Stegmüller, Wolfgang. "Angewandte Phänomenologie: Max Scheler". in *Haupt'Strömungen der Gegenwarts-Philosophie*. Band I. Stuttgart: Kröner, pp. 96-134.
Thiele, Joachim. "Ernst Mach-Biographie". *Centaurus*, 8 (1963) 230-237.
Wilkins, E & E. Kaiser. "Foreword". In *Robert Musil. The Man without Qualities*. Volume I. New York: Capricorn Books, 1965, p. iii.
Wright, G.H. von. "Musil und Mach". in R. Musil. *On Mach's Theories*. trans. by Kevin Mulligan. Munich: Philosophia Verlag, 1982, pp. 7-14.

Chapter Ten[1]

HUSSERL VS. JERUSALEM

by themselves

Section Guide
1. Husserl Criticizes Mach's ideas
2. Husserl Attacks Jerusalem
3. Jerusalem Responds to Husserl
4. Mach-Husserl Correspondence
5. Mach-Jerusalem Correspondence

1. HUSSERLCRITICIZES MACH'S IDEAS[2] (1900)

§ 55 The meaninglessness of an economy of thought for pure logic and epistemology, and its relation to psychology[3]

To the extent that these thoughts coincide with those of R. Avenarius and E. Mach, we have no differences; I gladly agree with them. I am genuinely convinced, in particular, that we owe a vast amount of logical illumination to the historical-methodological labors of Ernst Mach, and that this is the case where one cannot fully agree with his conclusions. Mach, unfortunately, does not appear to me to have tackled what are the most fruitful problems of the economy of thought, problems I tried to formulate above in a somewhat brief but quite sufficient fashion. His failure to do this is, in any case, partly due to the epistemological misinterpretations that he thought of as necessary foundations for his investigations. Mach's writings have, however, had an immense influence with respect to these very misinterpretations. This is the side of his thought that he shares with Richard Avenarius, and it is in respect to this which I must here oppose him.

Mach's doctrine of an economy of thought, like Avenarius's doctrine of least action, relates, as we saw, to certain biological facts; ultimately we are dealing with a branch of the theory of evolution. This fact makes plain why such researches throw light on practical epistemology, on the methodology of scientific research, but not at all on pure epistemology, and especially not on the ideal laws of pure logic. *Per contra,* the writings of the school of Mach and Avenarius seem to aim at an epistemology rooted in an economy of thought. Against such a conception, or such a utilization of the economy of thought, we must train the whole arsenal of objections which we above opposed to psychologism and relativism. Attempts to found epistemology on an economy of thought ultimately reduce to attempts to found it on human psychology; we do not need to repeat here nor emphasize the application of our arguments against this kind of thing . . . .

To underpin psychology with teleological principles as its 'basic laws', principles devised to explain our varied mental functions, does not promise much for the progress of the science. No doubt it can be instructive to point out the teleological significance of our

J. Blackmore, R. Itagaki, and S. Tanaka (eds.). Ernst Mach's Vienna 1895-1930. 211-235.

mental functions, and even of our more important mental structures, to show in detail, therefore, how and why the actually formed combinations of our mental elements have the utilitarian relation to survival that we expect *a priori*. But to treat the descriptively given facts as 'necessary consequences' of such principles, so as to seem to be giving a real explanation of these facts, and that particularly in contexts whose main aim is to lay bare the last foundation of psychology, all this can only breed confusion.

A psychological or epistemological law concerned with an *endeavor* to achieve as much as *possible* in this or that respect, is a chimera. In the pure sphere of [conscious] fact there is no maximum possibility, in the sphere of law no endeavor. What happens in each case, as psychological matter of fact, is quite definite; there is so much of it and no more.

The factual element in the principle of economy reduces to the fact that there are indeed such things as presentations, judgments, and other thought-experiences, which are further connected with feelings that, in the form of pleasure, promote certain directions of thought-formation, in the form of pain, discourage them. One can then assert, with some rough generality, the existence of a progressive formation of presentations and judgments which lead first to the formation of single experiences out of originally senseless elements, and then to the combination of such experiences into a *single more or less orderly, unity of experience.*

Psychological laws determine the emergence from our first roughly agreeing mental collocations of the presentation of the *single* world common to everyone, and of an empirically blind belief in its existence. One should, however, note that this world is not the same for everyone, but only so 'on the whole'; it is the same only to an extent which can afford a sufficient practical guarantee for our common presentations and actions. It is not the same for the ordinary man and the scientific research-worker; for the former it is a system merely approximate in its regularity, and shot through with countless accidents, whereas for the latter it is nature ruled throughout by absolutely strict law . . . .

We are in other words, *not* interested in the origins and changes of our world-presentation, but in the objective right which the world-presentation of science claims as against any other world-presentation, which leads to call *its* world the objectively true one. Psychology looks for understandable *explanations* of the *formation* of world-presentations. World-science (the sum total of the different sciences of the [consciously] real.) wishes to know by lucid *description* what obtains in [conscious] reality, what makes up the true, the actual [conscious] world. Epistemology,[4] however, wishes to grasp the understandable [only] from an objectively ideal standpoint in which the *possibility* of lucid knowledge of [what is consciously] real consists, thus the possibility of science and of knowledge in general. We *seem* in the case of the principle of economy to be dealing with a psychological or even an epistemological principle: but this appearance mainly stems from confusion between the factually given and the logically ideal,[5] the former being tacitly substituted for .the latter . . . .

Before all economizing of thought, we must already know our ideal, we must know what science ideally aims at, what law-governed connections, what basic laws and derived laws, etc., ideally are and do, before we can even discuss and assess the thought-economical function of knowing them. Certainly we have some vague notions of these ideas before we explore them scientifically, and thus we can talk of economizing thought even before a science of pure logic is in itself prior to all such thought-economics, but it remains absurd to base the former on the latter. . . .

The errors of this trend toward thought-economics are due in the end to the fact that those who engage in it, like all psychologistic thinkers, have an interest in knowledge which stops short at the empirical side of science. They fail in a certain manner to see the forest for the trees. They concern themselves with science as a biological phenomenon, and do not see that they are touching upon the epistemological[6] problem of science as ideally unified objective truth. The epistemology of the past, which still reckoned the ideal among its problems, seems to them in error. It can only be made a worthy object of scientific concern for them in one fashion: by showing it to have had, *relatively* speaking, the function of economizing thought at a more profound level of philosophical thinking. But the more that it becomes a philosophical fashion to judge the main variations of epistemology in this manner, the more must research protest against it, and the more necessary it becomes to discuss the basic points of dispute in as many-sided a manner as possible. We must, in particular, go as deep as we are able into the basically different thought-trends in the spheres of the ideal and the [consciously] real, thereby opening the way for a lucid clarification which is the presupposition for a final foundation of philosophy.[7] The present work would serve as a small contribution towards this end

2. HUSSERL ATTACKS JERUSALEM[8] (1903)

The intentions of the author go far beyond a psychology of judgment. Such a psychology is regarded by him as the foundation and presupposition for the whole of theoretical philosophy . . . . He seems to believe himself to have opened the way to important insights in general grammar, logic, and the theory of knowledge. Among such insights he counts the proof that we cannot maintain the evidence of inner perception, that sense perception is the ultimate and most certain source of knowledge, as well as the proof that the categories of substance and causality are not, 'as Kant supposed', innate and original concepts of the understanding, present prior to all experience, but rather are thought forms that are jointly conditioned physically and have been acquired through experience. And so forth . . .

Whoever reads the announcement of such goals, and at the same time considers the instrumentalities by which they are supposed to be attained, will not exactly enter upon the closer reading of the book with high hopes. However widespread the tendency to assign to psychology a fundamental importance for philosophy as a whole, very few will expect much from a metaphysics that has presumed to find the key for the solution to the ultimate questions about God and the world in a genetical-biological theory of human judgment.

Yet greater offense will be felt by all those who have gotten clear upon the relationship between the psychology of knowledge and the critique of knowledge - and consequently have grasped the basic distinction of essence that separates the *explanation* of thought processes (objective and theoretical analysis) from the *illumination* of the units of human thought (analysis of signification and validity).[9] The former locates us in the context of objective things and events: thus speaking of individuals as psycho-physical individuals, of their life circumstances, of their social unites and the psychical formations which develop therein. It delves into the elementary psychical states of these individuals, investigating their causal connections and their natural laws. Critical illumination by contrast to all of this, does not theorize. It asks no questions about mental or corporeal *things*, formulates no natural laws for their properties, and states and explains nothing upon the basis of natural laws. It does not explain, but clarifies. It inquires from whence in general, the idea of theory, as a unity of truths, is to be derived - along with all of the ideas and Ideal laws constituting it. It asks what is really meant by "validity" and "being", and what can make their meanings intelligible.

That we have here distinctions and difficulties and problems whose clarification and resolution is the work of millennia, is something of which the author [of the book we are now reviewing] cannot possibly have even the most remote of conceptions - as has already been made clear from his "promises" and from all of his further expositions.

But if, now, we consider the basic ideas in his theory of the judgment itself, leaving aside their epistemological and metaphysical applications, we then find that they have about them a naïveté hardly comprehensible. The author advances a theory of the judgment without any serious attempt at a description of judgment - within which there obviously would also belong a purely descriptive analysis of all those lived experiences from which judgments (of the essentially different descriptive types) are built up. At bottom his theory consists of an idle fancy hounded to death over a few hundred pages. The author is guided by the conviction that every judgment is a sentence [*Satz*], expressed or unexpressed and that if you think away the linguistic expression of the act of judgment, then nothing is left over[10] which one could call a judgment . . . .

We have chosen to distinguish between natural science which is abstract and nomological, oriented in terms of elemental laws, and natural science which is concrete and ontological, directed towards explanation of concrete natural formations. The latter has been grounded in the former. Its theoretical explanations are carried through by application of elementary laws arrived at by the abstract sciences. Would it not be an absurd undertaking to reverse this order and ground the abstract sciences upon the concrete ones? Precisely so, just as it would be absurd to ground genetic psychology, as an elemental and nomological science upon child psychology, cultural psychology, or upon anthropology and concrete biology in general. Naturally, this does not rule out the reception, from these concrete spheres, of motives that stimulate inquiry; nor does it rule out indirect testimony for what can, however, be realized in a definitive and decisive manner only through direct and analytic investigation into the elements. Now as to the present book, this perversion of principle [subordinating abstract science to what is concrete] constitutes its pervading character. The entire thinking of the author is moving within anthropological considerations, within those of concrete linguistics or similar matters . . . .

The to some extent, very favorable reception which Jerusalem's book has enjoyed from other quarters may make amends to the author for this harsh judgment. But it must be strongly emphasized, and expressed in the surest of terms, that there also are professionals who cannot allow this book to stand as *the* representative for contemporary efforts toward a theory of judgment.[11]

[Other harsh expressions in Husserl's critique:]

"The ever recurring clichés"

"The idle fancy by means of which the author thinks he has resolved 'the problem of judgment'"

"But enough of citations! The author's theory is amazing enough"

"In his presumed insights we can, unfortunately, only discover errors"

"Now, there are books of a psychologistic tendency which in spite of extreme errors, have rendered service - even an undying service - Unfortunately, the present book does not fall within this class"

"Nothing essentially new or worthy of notice relevant to this book is offered by the author's essay"

"In the author's presentations I have found nothing in the way of new ideas, and just as little independent work in his treatment of older views"

"Also, his critiques offer nothing worthy of notice"

"In what respect the author believes himself to have broadened the theory of deductive inference I am unable to say, in spite of an attentive reading of the book."

3. JERUSALEM RESPONDS TO HUSSERL[12] (1905)

What is Phenomenology? Husserl describes his version as follows: "Pure Phenomenology represents a neutral field of investigation in which different sciences have their roots. It would function as a preparation for psychology as an empirical science." Therefore, he considers Phenomenology to be a kind of critical introduction [*Vorarbeit*] to psychology. But the second part of his book *Logische Untersuchungen* shows clearly that it is not an introduction or prior stage but is only a segment of psychology itself. He really does make a descriptive investigation into our psychic phenomena, and every psychology must start from that. Analysis of what has been found by such introspective observation is an essential task of psychology as a science and it is by no means merely a preparation for it. In short, what Edmund Husserl has called Phenomenology is nothing other than mere descriptive psychology minus the genetic or historical approach. When interpreted in this way, it is clear that Husserl has actually done something important, and that it is here that his greatest talent manifests itself. Thus, we must regret that he has not stuck to his psychological investigations, for if he could learn genetic and biological ways of thinking he could advance the field considerably.

But if he insists on trying to separate phenomenology from psychology by force and refuses even to want to discover psychological foundations for logic, then he will fall into as much self-deception as has already occurred with his metaphysical views. He fights against "psychologism" but remains its practitioner. When he examines phenomena, he does so as a psychologist and indeed a very respectable one. But when he draws or infers from his psychological analysis conclusions, he reveals himself to be as John Stuart Mill says "abandoned by all of the Gods".

By his deliberate rejection of everything genetic and historical, he blocks his own path to the discovery of truth, and instead of understanding human experiences [*Erlebnisse*] in their own historical evolution he reads into these experiences "Ideas" and "Ideal Laws" which he dogmatically declares were already there before he began his investigations. Thus, what Husserl claims to be the foundation of "Pure Logic" is not free as he says from metaphysical and psychological presuppositions. He has offered us no more than Anselm's "ontological proof [for the existence of God]" as a model for a dogma which he had held before he had considered any proof. The supposed absolute validity of its truth is not drawn from the facts but from earlier "established" concepts of a purely dialectical nature. The basic problems of epistemology are not even touched upon.

Husserl is so deeply consumed by his own brand of dogmatism, that he is in no way even accessible to criticism.

Husserl's attempt to reach a pure logic by reviving Scholastic methods can therefore be seen as a complete failure, because he has built his world on dogmatic [unexamined] presuppositions. But let us proceed, since we can learn something from Husserl by doing so, namely, that without examining psychological and ontological factors more than he does one cannot even approach a satisfactory epistemology . . . .

It is interesting that Husserl in his *Logische Untersuchungen* defends adopting the methods of the Scholastics, He himself writes:

> But the objection that the question is about restoring the Scholastic-Aristotelian logic whose slight value in the history of judgment has been commented upon, shouldn't bother us in the least. Perhaps it should be emphasized that the questionable discipline is by no means of such a narrow circumference and so poor in profound problems as has been brought against it. Perhaps the old logic is nothing more than an incomplete and troubled realization of the Idea of pure logic. but it still deserves respect as a first beginning and skillful effort.*

Husserl shares with the majority of mathematicians the persuasion that the basic concepts of mathematics, especially numbers, are given *a priori* and neither stem from experience nor need to be controlled by experience . . . . Everything which Husserl brings against psychologism is included within the following formal syllogism:

> 1) The theory which explains the truth of logical laws and which is known to be a priori and self-evident is correct.
>
> 2) Psychologistic theories do not explain the truth of logical laws.
>
> 3) Therefore, psychologistic theories are not correct.

Everything in the first volume of his work is basically contained in this syllogism. His complete polemic against psychologism can also be found in it. Nor is his major premise anywhere grounded, but is merely dogmatically asserted. Thus, where Husserl believes that he has found impenetrable truth, where the solution lies, is precisely where the real problem begins. It is his dogmatically asserted major premise which we dispute. Where he stops, we start.

An especially instructive example of the dogmatic character of Edmund Husserl's argumentation can be found in his criticism of Ernst Mach's principle of the economy of thought. Husserl shows us considerable understanding for its practical significance. Indeed, he even supplies us from his mathematical knowledge some new perspectives for the application of this principle. But then he makes his astonishing criticism, which in fact is the same one he made against my theory of judgment. Namely, as it were, we must already have a conception of the perfect Form of economy of thought *before* we apply it as a principle, therefore we cannot use the principle to establish or justify the Form, and therefore, logic as the science of Forms cannot either in part or whole be based on psychology. But Husserl admits "We must thank Ernst Mach's historical-methodological publications for a wealth of logical information [*Belehrung*]." But if economy of thought can inform us about "logical information" by explaining it, why cannot economy of thought as well explain logical laws? Or as Husserl prefers to put it "enlighten" us about those laws? [Don't logical laws contain logical information?]His answer is apparently that if the laws were to contain such information it would deprive the laws of their *a priori* character. We supporters of psychologism believe that not just logical information but also logical laws are the product of scientific thinking, because it agrees with our conception of mental life. [i.e. because logical laws *consist* of information] . . . .

In fact, I have the impression from Husserl's argumentation in his books against economy of thought that while one can learn a lot about logic from it, that logic itself cannot emerge from it, because it is not mentioned in any approved Bible of Logic. Indeed, Husserl treats my position with the same inconsequence as he does that of Mach, except that his abuse is much greater and more personal against me. Husserl insists that unlike Mach nothing can be learned about logic or anything else from my writings . . . . He accuses us both of "anthropomorphism" as if we can never break loose from it. But he overlooks what his accusation implies. If we can never escape it, then how can Husserl do it? Where did he get the capacity which he denies to us? Who sets up the powers and laws which are necessary? . . . . Mach's mistake and mine apparently come from our "relativism" and "scepticism",[13] that is, our refusal to believe in Husserl's dogma.

The dogmatic character of Husserl's philosophy can also be discovered elsewhere. He insists that logical laws are never natural laws and says: "No natural law is *a priori* recognizable" [But is any law a priori recognizable?] . . . . He goes even further, however, and claims that he is in a position to prove that logic possesses "truths-in-themselves"[14] . . . . Now, I also consider Bernhard Bolzano [from whom these ideas come]to be a significant logician, but "sentences in themselves" and "truths in themselves" in the manner in which Husserl applies them include great unclarities and even contradictions . . . . But Husserl couples "in-itself" talk with the concepts of "Idealities" and *Evidenz*. Both concepts have a very different status and meaning with him than is usual with other philosophers, and it is therefore necessary for us to pin down exactly his conception and way of speaking . . . .

Evidenz is nothing other than the "experience [*Erlebnis*] of truth". Truth is naturally only experienced as something Ideal in a real act. In other words, according to Husserl: "Truth is an Idea, whose individual character is being currently experienced in evident judgment" . . . . Thus his "truth" is only an "Idea", that is, a mere species of concept which combines what is common to many contents of experience [*Erlebnisinhalte*] and which we may, indeed we must, give a name, because otherwise we could not keep it firmly in mind. To this Idea, to this species of concept, we therefore give the name "truth". [i.e. truth is not correspondence with reality, but an internal enlightenment gained from intuition.] . . . . Against the kind of *Idealismus* to which Husserl would persuade us, I defend the healthy realism of science . . . .

We normally designate everything "ideal" which moves within the realm of thinking or of thoughts and never leaves that realm. And we then classify everything which exists independently as "real", whether we happen to be thinking about it or not. Accordingly, we designate everything which seems to be consciously "immanent" as "Idealism" in epistemology or within theory of knowledge. And on the other side, we identify any position which holds that our sensory organs enable us to perceive the world as it exists independently of our thinking [as realism]. In short, Idealism sticks completely with what is immanent to consciousness or the senses while Realism allows reference and understanding to transcend consciousness or the senses.

But Husserl's contrast between "real" and "ideal" is completely different. For him, factually present psychic experience with all of the neighboring conditions which accompany it is "real". Contrasted with that, the general laws of thought which manifest themselves within experience [*Erlebnis*] he considers to be "ideal". Therefore, for him the

"*Reale*" is at the same time the Subjective, while the "*Ideale*" is the general and the "Objective". Thus, he contrasts psychological and "real connection" with the "objective or ideal". The [Scholastic notion of] species is regarded as "Ideal" when it is contrasted with the individual case which is "real". Under "Idea" [as contrasted with "Ideal" and "real"] he means the generic concept of the species. But generic concepts are not abstractions for him which can help complete our thinking in order to overcome the manifoldedness of individual things, no, they are for him an *a priori* given. [Like Platonic Forms] they stand in contrast to empirical reality, and the relations which prevail between "Ideas" are in his opinion knowable *a priori* and do not need to be confirmed by experience at all.

His "Idealism" is therefore not an *epistemological*[15] view as some critics of Husserl seem to believe. Rather, his "Idealism" is totally characterized by the conception of the [ontological] "Realists" of the Middle Ages, whose slogan ran *Universalia ante rem.* Only from this point of view does Husserl's utterances about "real" and "ideal" and "natural law" and "Ideal law" become understandable. Therefore, for him what is "ideal" [even if it is not even real] is objective . . . .[16]

In Section 7 of the second volume of *Logische Untersuchungen* Edmund Husserl discusses what he calls "the principle of the presuppositionlessness of epistemological investigations". In the very first paragraph one is astonished to read the following sentence from this decided opponent of Psychologism: "If my point of view is to be more than a mere opinion about the meaning [*Sinn*] of knowledge [*Erkenntnis*], but is to become a strict requirement for judicious knowing [*Wissen*], then it must be based purely upon perceptual and conceptual experience [*Erlebnis*]." Doesn't this mean when translated from Scholastic thinking into German that epistemology must be based on a psychological foundation? Yes, and don't the following words from Husserl teach that the realm of the psychical is not really abandoned at all [when reference passes beyond it]?

> The act of thinking is occasionally directed toward transcendent and even non-existent and impossible objects, and this does no harm. For this worldly [*gegenständliche*] direction toward the imagination and opinion of phenomenalogically not-realized objects is naturally a descriptive characteristic relating to human experience [*Erlebnis*], and thus it must clarify and establish the meaning [*Sinn*] of such opinion on the basis of that experience [*Erlebnis*]; indeed, in no other way would such be possible.**

Thus, Husserl believes that only a penetrating analysis of the act of knowing itself done without stepping beyond the realm of experience can succeed in developing a sound epistemological foundation. This also means [implies] that he must be counted as an epistemological idealist now, who regards the existence of an extramental world as unprovable and as superfluous for the grounding of theory of knowledge. But at the same time, if we can take him at his word here, we can classify him to be a believer in psychologism for he also considers thinking about transcendent objects to be a feature concerning human experience [Erlebnis].

Hence, at least to this point, Edmund Husserl could it seems be counted among the epistemological idealists in the usual sense. But scarcely have we constructed this picture on the basis of the simple signification of his words, than we detect at once from the following utterances how falsely we have understood him. He goes on to say:

> The question whether it is justified to assume that 'psychic' and 'physical' realities are different from the Ego has been fully excluded

> from epistemology. What the essence of these realities is and what laws they come under, whether the atoms and the molecules of the physicists belong to them, and likewise the question of the existence and nature of the 'external world' are metaphysical questions.[17]
>
> On the contrary, the theory of knowledge can offer us general enlightenment on ideal essences or comprehends the meaning of thinking as knowing *[erkennenden]*, indeed the general question whether and how far knowing or reasonable supposition about objects is possible which if not given in the experience of our thinking are also not known in the most pregnant sense.[18]
>
> But epistemology does not consider the particular question of whether on the basis of factually given events we can really acquire knowledge of them or even if it is possible to realize this knowledge at all.***

Edmund Husserl therefore disputes whether the question of the existence of the external world and whether we may assume that psychic and physical realities are different from the Ego or Self are epistemological [*erkennistheoretisch*], or which appears to be the same for him, whether they are critical epistemological [*erkenntniskritische*] questions. His very reluctance to adopt or accept this proves that Husserl does not understand either the deepest character of epistemological problems or even the true province of critical epistemology.

To examine the possibility and the frontiers of human knowledge means above all asking whether we have decided for immanence or transcendence [whether the physical world is within or beyond sensory or conscious limits], since it must be asked as a preliminary question before we can accept any ontological or metaphysical position. This most important of all of the many epistemological questions must be asked, since on the answer even the very possibility of metaphysics [or ontology] depends, or even whether it has content or meaning.

If the answer to the epistemological question is to be given like most phenomenalists want that we *cannot* ever escape from immanence [in our understanding], then, since we may not do what we cannot do, we have no justification for carrying out any metaphysical investigations. Only if epistemological investigation has taught us . . . that transcendence is critically justified are we then justified in asking whether the world is monistic or dualistic or whether we should interpret it materialistically or spiritualistically.[19]

But when Husserl says that the question about the existence and nature of the external world is a metaphysical question, then he confuses the tasks of epistemology and metaphysics in a way which is not legitimate. [In short, it is not an honorable procedure.]

Questions about the existence of the external world belong to theory of knowledge or epistemology, while those which are about what is the nature of the external world belong on the one hand to science and on the other to metaphysics or ontology. Because of Husserl's confusion about where to distinguish epistemology from ontology, it is no wonder that he falls into contradiction in the very next sentence. For him epistemology includes the general question whether and how far knowledge [*Wissen*] or reasonable supposition is possible when it is not given in the experience [*Erlebnis*] of the thinker. But what does that mean except that objects may exist outside of experience, which reflects an epistemological position more liberal than the one he emphasized in the quotation we gave? One which allows a measure of transcendence [which of course is also implied by his religious position]

How little Husserl is clear concerning the metaphysical character of "pure logic" is shown by the following consideration, which he puts at the conclusion or end of the second volume of *Logische Untersuchiungen* . . . . By "Ideas" Husserl understands as we have previously pointed out nothing other than generic concepts, but which do not result from abstraction but are independent essences which exist before individual things and are therefore viewed as *Universalia ante rem*. Whoever in this sense hypostasizes Ideas as [the essence] of sensibilia and the understanding is practicing metaphysics as certainly as did Plato, who was also a metaphysician. If Husserl does not notice this and simply denies that he is practicing metaphysics, then he has proved that he is still dominated by dogmatism, and has not yet adopted a critical or self-critical approach . . . .

But Herr Husserl is surely right about one thing. He doesn't need any metaphysical theories, because in the most naive way he simply decrees his far-reaching metaphysical claims in the most dogmatic way he can. His "Ideal laws", his "Idea of truth" not only presume to be valid for all human consciousness, and for every possible consciousness, but even outside of and independent from its realization in "absolute" consciousness and in so-called consciousness "in itself". Thus he goes beyond every possible experience [*Erfahrung*] and maintains something so decidedly transcendent that it is hard to conceive how any thinker of the 20th century, a thinker who feels so close to Kant, has not been able to become clear that what he has been doing is metaphysics, and a completely unjustified, arbitrary metaphysics it is too.

4. MACH-HUSSERL CORRESPONDENCE[20]

#1 Edmund Husserl to Ernst Mach[21] Halle, June 18, 1901

Highly Respected Herr Professor:

Please accept my heartiest thanks for sending me the new edition of your *Mechanik* but also for the detailed consideration with which you have valued my book *Logische Untersuchungen*. I cannot avoid the persuasion that the differences between us do not go as deep as may seem at first. Nothing lies farther from me than the opinion that my endeavors focused on "pure" logic have in any way "invalidated" your methodological approach in the scientific investigation of experience [*Erfahrung*]. In no manner do I intend to place in question the appropriateness [*Recht*] of using the biological and genetic-psychological way of considering scientific problems. What I am defending myself against is the subordination of the epistemological enlightenment that pure logic can give to science[22] to the standpoint of psychological genesis and biological adjustment [*Anpassung*].

I am fighting against the sceptical "psychologism" of our time, which, as has already taken place with John Stuart Mill, has dissolved the boundary between matters of fact and the relations of ideas. Sentences which belong "to the form of thinking", that is, to the ideal meaning [*Sinn*] of thinking as such (for example, to meaning of sentences generally, conclusions generally; the meaning of yes and no, if and then, of one, some, and all, of how much and a definite amount . . .) cannot be conceived as expressions of empirical generalizations.[23] They do not belong to the accidental accompaniment and development of human thinking, but to the essence[24] of thinking generally. They can therefore not be enlightened for us at all by genetic psychology (of presentation, judgment, or knowing and also not by economy of thought).

My chapter on economy of thought is directed primarily at the school of Avenarius and most especially against Hans Cornelius, against his analysis of logically fundamental ideas and principles, against his distinction of natural and logical theories and so forth. The objection was meant for *him* that any alleged enlightenment by economy of thought of the purely logical would level down the difference between blind and logical thinking. Because he believed that he would be able to call on you to support his main points, most respected Sir, and as he seemed to me at that time not to be in error about you, this caused me in my critique to bring your name into the matter.

Since I represent a scientific standpoint,[25] may I in all modesty take the liberty of saying openly. The total dedication with which you applied your interest in science as biological, epistemological, and economic phenomenon has certainly been extraordinarily fruitful for the illumination [*Aufhellung*] of science as a practical function of knowing [*erkenntnispractischen Function*] and therefore for scientific theory in the sense of methodology.

Like all the world, I too have found rich suggestions from your works. I have never been able to put one of your books in my hands without being instructed by the wonderful historico-methodological analyses with which you have provided for us as a gift and which through the originality and freshness of your representations has deeply edified us. I had become so absorbed all the time in what was new and fruitful that I scarcely noticed the collision with your way of thinking which seemed so one-sided and did not do justice to the ideal, pure-logical content of science. In fact, your view was so very one-sided that it might even appear as if the genetic and economical approach to epistemological enlightenment could be sufficient by itself for science.

But in any case, some young philosophers, above all Hans Cornelius,[26] have so understood your position as to position it in the foreground and sought to defend it, which signified to me nothing else than *a breach of the frontier* . . . But in *this*, by the daily growing effect of their writing, there lay a danger, which I have tried to meet with criticism. However, also my way of thinking is obviously *one-sided*, but as I would like to believe, it is a *justified* one-sidedness.

I perceive the incompleteness of my investigations so much so that I tend to attribute to them an excessive worth. (Serious occupation with great questions does not let me escape immodesty.) Still, I retain the persuasion to have attacked a problem area, where previously little had been done and have brought some clarity to its own self-justification and to its natural boundaries, but without having achieved sufficient success. Formalistic treatment of the purely logical is not lacking, nevertheless, it would be well to provide phenomenological enlightenment of the same. The goal of a strict description of the "origin" of logical ideas which is fully free of all metaphysics and of all presuppositions from particular sciences has for a long time been unachieved. In fact, the essence of this goal has not at all been sufficiently clarified with the result that its importance has not been acknowledged.

I have gained so little real enlightenment about these relations from the literature, that in order to escape from most intense "intellectual discomfort" I have first had to seek the way through several years of work to find a new beginning. But admittedly I myself have not been able to overcome the extraordinary difficulty of carrying out strict phenomenological analysis of thinking experience [*Erlebnis*] by using raw and groping efforts. The series of investigations presented in Part II of *Logische Untersuchungen* has shown entirely that the issue is not one of the mere restitution of gloomy old formal logic. On the other hand, I do go back in part to Leibniz and to others like Locke and Hume (if also combined with other views). In fact, I may now say that as long as the border or boundary between the epistemological and methodological ways of considering pure logic and practical logic is not disturbed, then I do not think that a conflict essentially exists between our respective ways of investigation.

In conclusion, I add my heartiest wish for the full restoration of your health. May it be permitted to you, most honored Herr Professor, to spread your very important theoretical investigations into current and ever newer branches of science and to help send us the abundance of your teaching by preserving the serious and pure joy of learning which you have offered in your previous works. With the expression of sincerest respect

Your Complete Servant

E. Husserl

#2 Ernst Mach to Edmund Husserl Vienna, June 23, 1901

Most Honored Colleague:

Through a regrettable oversight, your letter lay concealed under some incoming printed matter, so that only today was I able to learn the contents. Please excuse the lateness of my answer. But let me hasten to say that I have nothing against your comments to object to, and therefore I hope for full agreement between us.

Given the nature of things, I cannot hope for an essential improvement in my health, and therefore, I also cannot place any special hope upon my future work. With heartiest thanks and best wishes your success, my most respectful greeting

Your Servant

Dr. Ernst Mach

5. MACH-JERUSALEM CORRESPONDENCE

#3 Ernst Mach to Wilhelm Jerusalem München-Vaterstetten, June 8, 1913

Highly Honored Friend:

I have just received your letter of June 7, and am answering it at once. What you say about Foerster[27] has pleased me a lot. That he is already an unusually gifted teacher, can be considered as pure profit.

I became acquainted with Husserl through his *Logische Untersuchungen*. I cannot discover in it anything other than psychological investigations. Nor can I understand how it could be regarded as anything else. Concerning Driesch's *Die Logik als Aufgabe*, his consideration of abstract thinking and his conception of logic as theory of organization [*Ordnungslehre*] have also pleased me alot.[28] I am not unsympathetic towards these things.

I would be most happy to contribute something about your new book. Since it corresponds to my inclination, a request from the editor is not necessary at all. Eisler[29] will hopefully not become upset that a non-professional helps out. Dr. Anton Thomsen seems reasonably satisfied. I am very happy that our long correspondence [concerning philosophy] has not ended in offence or insult, which I was afraid would happen.

I am now learning for the third time in my life how to come and go through the room by using a ledge. I go down one side of the room and come back on the other. I would never have learned this with human guidance, because it is always too active. I hope to be able to write the review soon in a way which corresponds with your wishes. But I would gladly like to hear about the Japanese translation.[30] With hearty respectful greetings

Your Old Friend

Dr. Ernst Mach

#4 Ernst Mach to Wilhelm Jerusalem Haar bei München, July 17, 1913

Highly Respected Friend:

I accept with thanks and pleasure your proposal to write here and there about philosophical phenomena [*Erscheinungen*]. My wife[31] and my sister[32] thank you for your friendly recollection [*Erinnerung*]. My sister has even read your *Einleitung* [Introduction to Philosophy] with interest. I understand your objection to monism, even if I myself cannot agree with you. But it will not divide us.

Thank you very much for mediation concerning the *Volksbildungsverein* [Adult Education Society].

It is perhaps good that [Josef] Popper pauses for a while in his X-ray treatments, since he can then observe in an unbiased way how effective they are. Please greet Popper heartily for me and my people [*Leuten*]. From house to house with the most respectful greetings

Your Old Friend

Dr. Ernst Mach

#5 Ernst Mach to Wilhelm Jerusalem Haar bei München, Sept. 25, 1913

Honored Friend:

I am very pleased to hear from you again. I'm still alive. But I am not yet working very actively, though I have begun and almost finished a short little work *Mechanik und Kulturgeschichte*.[33] Unfortunately, I still cannot stand or walk independently. I am happy to hear of your book on Pragmatism. Please greet Anton Lampa[34] and his wife and thank him for me in my name for his promotion to the Royal Bohemian Society. [Anton] Marty[35] has obviously sought a reason to justify himself. I have heard nothing from Popper,[36] but Prof. W. Pauli[37] has recently told me about his health.

I also recently saw the advance proof from [Hugo] Dingler,[38] who is a *Privatdozent* in Munich. The book has been given the title: *Grundlagen der Naturphilosophie*, and for the moment the title does please me. Also, [Hans] Kleinpeter[39] may let loose some good books soon: a translation of [William] Jevon's *Logic* with a mathematical appendix and a volume on psychology which emphasizes theory of volition.

I enjoyed Rosegger's *Waldschulmeister*[40] a lot, even if I am by no means his unconditional admirer. I don't know Gottfried Keller's *Grüne Heinrich*. [Your proposed book] *Erkenntnis und Gesellschaft*[41] has raised my curiosity. Thank you for the report from Braumüller;[42] I will write to him. My heartiest greetings to you, yours, and all friends

Your Old and True

Dr. Ernst Mach

[P.S.] According to the Vienna newspapers *Neue Freie Presse* and the *Neue Wiener Tagblatt* there is an interesting sociological book called *Prestige* by Ludwig Leopold, a landowner from Seksard in Hungary. Perhaps you or one of your friends could review it. The volume was intended for me, but I am not sociologist enough to dare to attempt such a thing. I don't know the author, but I have already apologized.

[P.P.S.] Dr. Josef Jacob's book *Methodik des mathematische Unterrichts* is outstanding.[43] Of all my teachers, only *one* had as much understanding and *heart* for young people.

#6 Ernst Mach to Wilhelm Jerusalem Haar, August 3, 1914

Dear Friend:

I have the feeling concerning the Congress[44] that the best I can expect is a verbal account in order to learn about what went on. You will have already received several impressions by rumor one after the other which seem to contradict each other. For example, I have no idea at all what I should make of Husserl's *Wesenerschauung* [intuition of ideas as if such could form a non-factual science].[45] I think it is still psychology or perhaps psychological deep insight [*Tiefblick*]. It also appears to have been prepared with a heavy touch of [Joseph] Eisenmeier's metaphysics.[46] You see that I agree with your position toward psychology. From where else could one draw cognizance [*Kenntnisse*] or knowledge [*Wissen*], other than from facts, and indeed, from psychical or psychological facts?

Professor Dr. Pauli[47] was here for some days with his son.[48] He thinks that he is a great mathematical genius - But I believe that one has to wait out the development of puberty before one can judge about that. I also suffered from serious illusions about my second son [Heinrich] and his relations to chemistry. I was quite deceived by a misrepresented psychosis - I mean he committed suicide in Göttingen.[49]

Are you going to Salzburg this year? I would gladly speak with you. My daughter Carolina has already announced that she will come for a visit in August with her two children. My heartiest thanks for your letters and the articles in *Zukunft*.[50]

Your True Servant

Dr. Ernst Mach

#7 Ernst Mach to Wilhelm Jerusalem Haar bei München, March 28, 1915

Most Honored and Dear Friend:

Thank you very much for so quickly sending the Apocryphal chapters of the Bible, and also for the beautiful gift which accompanied it. I put it next to my edition of the Bible in seven languages, which I used for some words about comparative linguistics in preparing my Dortmund speech [of 1886 on educational reform]. But I am far from having finished reading the chapter.

Your report on Popper pleased me a lot, but it does not completely agree with a letter which my son received from him.

Your Loyal and Thankful

Dr. Ernst Mach

[P.S.] We had in my father's house[51] a magnificent edition of the Bible in folio with beautiful woodcuts and copperplate engravings in Latin with a German translation. Thus, the Bible became a memory of my youth.

#8 Ernst Mach to Wilhelm Jerusalem Haar b. München, April 5, 1915

Most Honored Dear Friend:

Today I am somewhat "abler", or, let us say, in a better mood. After a thoroughgoing cleaning, my typewriter also types with less effort, therefore, I am directing the Word once more toward you in order to give thanks for your beautiful present, which I have now fully enjoyed. The Apocrypha is a necessary enlargement of the Bible, which up to now have been missing.

The book Judith and Esther in fact contains magnificent short stories [*Novella*] which you informed me about at the time. How does it come to be that these Apocryphal segments are lacking in all editions of the Bible? With the heartiest greetings

Your True Servant

Dr. Ernst Mach

#9 Ernst Mach to Wilhelm Jerusalem (Haar b. München), July 1, 1915

Dear Honored Friend:

I am always happy to receive news, especially from you. Congresses will not be given during these times. Nowadays, I almost never read philosophy, which makes your reports and stimulation even the more welcome! I have partly read [Theodor] Ziehen[52] and am of your opinion about him. For the branding together [*Umstempelung*] of Plato and Goethe, if I may be allowed to give an opinion, strikes me as rather comical. Please write me again about [Ludo] Hartmann's essay.

I'm happy that Popper is doing well and hope the same for you.

Your True Servant

Dr. Ernst Mach

#10 Ernst Mach to Wilhelm Jerusalem (Haar b. München), July 20, 1915

Dear Honored Friend:

What you wrote last interests me a lot. Your judgment is also quite right about [Richard] Wahle.[53] I will come back to the matter in a few days about your comments. I am sending you a dissertation[54] about [Henri] Bergson, which I have only partly read and which I have not understood. Is there a pre-established harmony between the a priori and experience as one would assume if one had read Bergson? And yet the *a priori* shall again determine nothing about concepts of experience!

Your True Servant

Dr. Ernst Mach

#11 Ernst Mach to Wilhelm Jerusalem Haar b. München, August 16, 1915

Dear Honored Friend:

I come back once again to the possibility of a pre-established harmony between apriorism and the concept of experience, something that seems so hard to make thinkable.[55]

If we think of an arbitrary [*beliebigen*] case of a concept of experience, for example, the equivalence of heat or the conservation of energy, which have both gone through long historical ages and stages, halts and faults, such that the intellect can hardly paint any logical "structure", but where precisely here an [unhistorical] a priori concept can yield at once what is right, [then how can we reconcile the two approaches?] You spoke in your last letter about "the dissolution of a priorism". I am curious about what you would like to say about it.

Your True Servant

Dr. Ernst Mach

#12 Ernst Mach to Wilhelm Jerusalem Haar b. München, Nov. 22, 1915

Honored Friend:

Finally today, I can answer you and your last two October letters, and I thank you for your good intention to write a review of my book [*Kultur und Mechanik* (1915)]. First of all, your remarks about Carl Stumpf's [hostile?] attempts [to review the book?][57] have naturally interested me, for one will already have tried.

But my aphoristic manner of handling the material, the heterogeneous character of the individual chapters, as well as the times we are living in, are much less than ideal for writing any review. After finishing it I suffered my normal author's letdown; only on this occasion it was stronger than otherwise, but nothing would have delayed it longer.

Thank you for all of the other news, and since it has been becoming more quiet around me, your letters are doubly pleasing. Naturally you must finish *Soziologie der Erkenntnis* although compared to me you still have infinitely much time left. [Theodor] Beer sent me an article,[58] which you and Popper have also [probably] received. Hearty greetings to you and yours

Your True Servant

Dr. Ernst Mach

#13 Wilhelm Jerusalem to Ernst Mach Vienna, December 14, 1915

Highly Honored Friend:

Your last letter for which I thank you dearly, has pleased me a good deal, because I gathered from it that your health has improved so much that you can write yourself [and not have your son Ludwig do it] and that you enjoy working again. Popper for some weeks now has fortunately been doing well, is in good spirits, writing a lot, and seems interested in everything.

On saturdays during the last few weeks, some discussion sessions of the university of Vienna teachers have taken place, which have been about the future political shape of Austria. I was there once and noticed that the old are liberal-minded [*freisinnig*] and the young conservative. It also benefits me to have to report to you about the scientific books which I have been reading, because by doing that it brings old reading once again to conscious attention.

Eduard Spranger[59] has sent me a treatise called *Lebensformen* which is from the commemorative volume [*Festschrift*] for Alois Riehl.[60] He tries to portray the living environment of science, economics, society, state, art, and religion as part of our mental environment and thereby to help delineate different types of human beings. But on the whole he falls well short of his aim.

For my book *Soziologie des Erkennens*, I have begun reading Edmund Husserl's *Phenomenologie* again, as well as articles by his follower Max Scheler.[61] I get the impression ever more strongly that there is a lot that is arbitrary and willful here and that what might be called hostility to life tends to prevail. In the latest issue of *Zeitschrift für Philosophie und Philosophische Kritik* (Fichte's old journal) a very instructive essay appears on the philosophical significance of the principle of relativity, but I have just started reading it.

In the American journal which I am still receiving during the war a book *Scientific Method in Philosophy* by [Bertrand] Russell is being presented by [slow?] installments, which is appearing now with the help of [Paul] Carus[62] Has Carus perhaps sent the book to you? I would very much like to read the whole thing immediately but I don't know how I can do it. Should one write to Paul Carus through *Open Court*, or do you know of a more exact address? As soon as I can get more stuff [reading material?], I will write to you again. In old and friendly respect

Your True Servant

W. Jerusalem

NOTES

1 The footnotes have been added by the editor.

2 Edmund Husserl was mostly bent on criticizing the contents of Mach's book: *Beiträge zur Analyse der Empfindungen* (1886). Earlier Wilhelm Jerusalem had criticized Brentano and crossed swords with Edmund Husserl. Between 1900 and 1905 the latter appears to have made a point of "reviewing" in print every book and many articles he came across which he suspected of being "guilty" of psychologism. There was no attempt at fairness or even of general informativeness about his target. Like Vladimir Lenin in 1909 his game was verbal intimidation, but he made an exception with Mach by being reasonably respectful, perhaps because Mach was well-known by 1900 and he himself had been influenced by Mach early in the 1890's. See also: Edmund Husserl, a review of Mach's article "Über die Prinzip der Vergleichung in der Physik" in *Archiv für systematische Philosophie*, **3** (1897) 241-244.

3 Edmund Husserl, *Logical Investigations, Volume One*, translated by J.N. Findlay [with some changes] , Humanities Press: New York, 1970 [1900] selected short parts from Sections #55 & #56, pp. 204-211.

4 Husserl's use of the word *Erkenntnistheorie* [epistemology] bears no relation to how the term is normally employed. The reader should also keep in mind that almost all of his basic terminology has been given idiosyncratic meaning distant from usual philosophical usage. He clearly hoped to catch and persuade young readers before they had learned the traditional definitions. Were we to try to clarify what he probably means for all of his basic terms here, the footnotes would have to become longer than the whole text. But for those readers determined to understand Husserl's re-definitions accurately in our short passages, please start off with the index to *The Cambridge Companion to Husserl*, edited by Barry Smith and David Woodruff Smith, Cambridge University Press: Cambridge, 1996.

5 The word "ideal" in philosophy can mean what has being but not existence, the product of mental construction, what is good or perfect, a simplified or idealized version of what is real, or treating something false or unreal as if it were true or real. Since Husserl seems to have believed in universals or Platonic Ideas, what was ideal or obeyed the rules of logic, had "being-in-itself", and could be understood by what he called "intuition". But when logical positivists and other philosophers of science took over his belief that logic was completely independent of psychology, they often meant by "ideal" the fictional product of mental construction and *not* something which was genuinely real in itself, though Sir Karl Popper did seem to incorporate aspects of Husserl's Platonism into his "World 3".

6 According to the traditional representative theory of truth, it is reality which determines truth, which makes truth an ontological or "metaphysical" problem, but Husserl like many philosophers who reject the representative theory (his was intuition-based), treated truth as if it were an epistemological question, but again, what he means by "epistemology" is odd. He seems to apply the term primarily to "ideal" entities and not like most philosophers to what one means, knows, or thinks one knows about the external physical world.

7 Many philosophers have held the view that our deepest presuppositions constitiute our real philosophy and that these presuppositions can, should, or do provide a foundation for science, but for Husserl the deepest presuppositions are logical or formal in character and yet they can be known by "intuition" as if they could be made conscious and as if such knowing could render philosophy "presuppositionless". By the word "science" he often appeared to mean these intuitable basic ideas, but he often contrasted "ideal science" with mere "empirical science", as if his science [of philosophy] was more certain.

8 The original book review appeared in *Archiv für systematische Philosophie*, **9** (1903) 523-543. A more accessible source has been "A Report on German Writings in Logic from the Years 1895 to 1899 - Fourth Article" in *Husserliana*, **XXII**, pp. 216-235, and in the English translation of this, pp. 261-269. The book which Husserl reviewed was: Wilhelm Jerusalem, *Die Urteilsfunktion. Eine psychologische und erkenntnistheoretische Untersuchung*, Wien und Leipzig, 1895.

9 Husserl's focus on "intuition" and the ideal world of logic, which seemed psychological to Mach and Jerusalem would completely ignore both the physical world and explanation. Thus, in his writings we will see ideal description praised and explanation, especially causal explanation, rejected. Husserl would also allege that intuition can "illuminate", a process which he seems to have regarded as superior to explanation. Mach also favored description within science, but his Darwinian and practical views included explanation.

10 Wilhelm Jerusalem seems to have been an early representative of Viennese linguistic philosophy, which was shared and expanded by Stöhr, Mauthner, Marty, Wittgenstein, and for a short time (1903-1905) even by Boltzmann. The Vienna Circle of the 1920's and 1930's would also pick it up and use it for their own ends, which would include to some extent concealing or minimizing their own epistemological and ontological assumptions.

11 This sentence perhaps best indicates the probability that Husserl's main intent or end in reviewing books on logic and psychology, largely from 1897 to 1905 was more to attack his opponents and defend his own philosophy than it was to present to his readers the views of other philosophers, logicians, and psychologists in a reasonable and fair-minded way. But perhaps we ourselves have become one-sided in criticizing Husserl's ruthless tactics. He may not have always been as arrogant and overbearing as the passage above indicates. But one thing is clear, one will almost never discover what the main contents of a book are by reading Edmund Husserl's review of it. Books are judged with a few salient points praised or abused with little or no serious attempt to describe or understand the contents, and when judged it is almost exclusively in terms of how close or distant the author's perspective is to Husserl's own philosophy. But not all of his reviews were as vindictive as those he directed against Jerusalem

12 Wilhelm Jerusalem, *Der kritische Idealismus und die reine Logik - Ein Ruf im Streite*, Braumüller: Wien und Leipzig, 1905, selections from pages 91 to 133.

* Jerusalem gives the reference to Edmund Husserl's *Logische Untersuchungen*, Band I, Leipzig, 1900, p. 39.

13 Husserl often attacks "relativism and scepticism" without clarifying what he means. In the real world there is a vast difference between relative truth and relative certainty and between scepticism which doubts everything and that which cleverly leaves "islands of indubitability" like Descartes and Hume. Given Husserl's "intuitive" theory of truth and his reduction of evidence to what is consciously "evident", he probably understands by relativism, relative certainty or if he is thinking in terms of relations, he may think that history, historical understanding or both, are somehow relative. But he does not mean relative truth as non-correspondence theory of truth, since that includes his own, and by scepticism he does not mean what most Greek sceptics meant, namely refusing to claim or believe that anything is unconditionally certain. Why doesn't Husserl clarify what he means, and why doesn't he warn his reader when he uses language in an unusual way? Perhaps because he feels he needs rhetoric and re-definition to persuade people. Perhaps he realizes that everyone does not believe in "intuition", and that most reasonable people often require more evidence than what is "intuitively evident".

14 If logic and mathematics depend on definitions, conventions, axioms, or rules, then how can one claim unconditional certainty for these abstractions or allege that they have "truth-in-themselves"? The answer for Husserl presumably lay in his belief in intuition and Platonism, and indeed, there was a school of mathematical theorists led by L.E.J. Brouwer (1881-1966) which did think that we "know" the truths of logic and mathematics by "intuition". But Jerusalem is surely correct that few reasonable people can accept this kind of thing. All theories of reason are not reasonable.

15 It seems that Leibniz and Kant before Husserl had termed belief that the real world is beyond conscious limits as "Idealism" because of its resemblance to Plato's position, but Plato's position as revived during the Middle Ages and as widely understood later has traditionally been called [ontological] Realism, not Idealism. Furthermore, realism as an epistemological position became fashionable as a rejection of Berkeley's idea-lism, his identification of the physical world with sensations which he called "ideas". Thus, to avoid confusion Husserl, probably should have rejected the usage of Leibniz and Kant and adopted the more widespread definitions of realism and idealism

which are located in most dictionaries and encyclopedias of philosophy, then and now, and which Jerusalem correctly pointed out.

16 Jerusalem is naturally upset when Husserl calls things which are either mental or non-existent "objective". For most people, only what is physical and trans-conscious can be objective in a literal sense, though we often say that people who are fair and even-handed are "relatively objective". What Husserl has done is to revive an old definition which has an opposite meaning which will allow him to consider idealities and what does not exist as "objective", and of course without informing his readers about the change in meaning. From a historian's point of view this is not honorable procedure.

** Jerusalem does not give a reference for this quotation which is presented in section 23 of *Der kritische Idealismus und die reine Logik* (1905) on page 123. But judging from a prior quotation it may be from Volume Two, Section §7, of *Logische Untersuchungen* near page 19.

17 This quotation explicitly shows many faults in Husserl's understanding and morality. His refusal to consider the question of whether there is a physical world and if so whether it is immanent or transcendent and his denial that it is an epistemological question is so willful, when that is the normal use of the term, as to seem dishonest. By re-definition of the word epistemology, and that is what he has done, he has sought to abuse and slander all other positions which are usually considered to be epistemological, not by offering any serious evidence against them, but by willfully re-classifying them as "metaphysical". It will strike many readers as cheating. Jerusalem may not have been as deep a thinker as Husserl, but like Mach, he was surely more honest and open in the manner of presenting his ideas. But perhaps allowances should be made for Husserl's extended "youth".

18 Husserl's refusal to accept the standard definition of the expression "epistemology" and instead declaring that it is the study of "ideal essences" will appear arbitrary to most readers, and hence, unacceptable, but alas, for historians it is not just important for us to describe points of view in terms of neutral classification, but we must also attempt to understand what seems arbitrary, indefensible, and unfair. Originally, Husserl seems to have been influenced by Leibniz's distinction between necessary and contingent truths, but necessary truths are ideal and hence not factually necessary, hence he felt forced to upgrade what was meant by ideal, which he tried to do with Platonism and "intuition". From his perspective, the choice is between intuition of idealities and "mere inference" about non-idealities, and he seems to regard the former as yielding "authentic" and the latter "unauthentic" knowledge, hence his attempt to re-define epistemology as theory of knowledge to only allow "authentic" knowledge. But in the historical or for us the real world, that is not the option. The choice is between weakly-supported guessing, which he calls "intuition" and well-supported inference, or, if there is really no such process as intuition, which seems most probable, between less-supported, unconscious inference, which is falsely called "intuition" and better-supported open inference. Neither yields unconditional certainty but relative certainty or a particular degree of probability as determined by the supporting weight of evidence. It is Husserl's failure to understand this which finally makes his attempt to re-define theory of knowledge understandable to us, *but it does not justify it.* He seems to be forgetting that other people also have a right to develop a position about knowledge, and, if they are right and he is wrong, as appears to us to be overwhelmingly probable, then his re-definition by blinding people to what is right or true hurts philosophy and science.

The problem is not re-defining words per se, this can be quite legitimate provided that no one is misled, but in order to avoid deceiving people one should explain what one is doing, that the new definition is not the normal one and that it - at least in Husserl's case - excludes much that had previously been covered by the old and more familiar definition. The historical legacy of Husserl's arbitrary definition of "epistemology" would eventually combine with Mach's expanded conception of "metaphysics" to make all of the few remaining positions within epistemology seem "merely linguistically different from each other", as if epistemology itself were no longer important, and all this without either solving basic epistemological problems or admitting the extent to which most "linguistic philosophers" retained presentist or formalistic epistemological assumptions.

*** Wilhelm Jerusalem (*Der kritische Idealismus* ... page 124) quoting Husserl (*Logische Untersuchungen* [1901], Volume II, page 20). It may be moral to oppose a position or even a type of philosophy or perhaps even all philosophy, but to re-define a fundamental category which has a standard meaning as Husserl does in order to exclude a position he opposes i.e. indirect realism, or indeed any position about the external world, so that it cannot even be considered as relevant to the question of knowledge is not the procedure of an honorable person. It is like changing the rules of a game in order to guarantee fake victory. Nor does the fact that he was able to gain followers after this kind of behavior *or even because of it* speak well of the followers either. Yes, he was original, but so were Caligula and Nero in their own very different way. A model philosopher should be moral and sound in both his living and in his philosophizing.

19 Jerusalem is employing Kant's notion that the middle-way between dogmatism and scepticism is "the critical attitude" which Kant holds is to refuse to inquire beyond what is sensory unless we can prove that it is possible to do so. In fact, Kant's approach re-defines "dogmatism" from overbelief into insufficiently justified inquiry. But Jerusalem, unlike both Kant and Husserl, almost always tries to be fair to all points of view by employing neutral or traditional terminology and classification, habits which help explain the popularity of his *Introduction to Philosophy* in both Austria and America.

20 The extant letters #1 & #13 are now said to be in the Deutsches Museum in Munich, Germany. #2 is in the Husserl-Archive in Louvain. #3 through #12 are in the University of Jerusalem Library Archive. #1 and #2 have been published in the original German in Joachim Thiele (ed.), *Wissenschaftliche Kommunikation*, A Henn Verlag: Kastellaun, 1978, page 157. #3 through #12 were published in German in John Blackmore and Klaus Hentschel (eds.), *Ernst Mach als Aussenseiter*, Braumüller: Wien, 1985.

21 Joachim Thiele in his account of Husserl's letter to Mach points out that the older man possessed in his library publications from Edmund Husserl dated 1891 and 1894 as well as *Logische Untersuchungen* from 1900-1901, and that all but the first had hand-written inscriptions from the author, that is, as if at least some written communication had existed between Mach and Husserl for several years before 1901. See Thiele, Ibid, p. 153.

22 It was precisely because deductive, that is, formal logic gave us no new information, no "enlightenment" as such, that Francis Bacon denied it a primary place in modern science. To be sure, because human beings do not always draw deductions from premises as soon as the premises are presented, but when they are brought to our attention there often is "enlightenment". Almost everyone except Husserl and his followers would consider such enlightenment *psychological*. Why? Because it is not dependent on formal logic, but varies from circumstance to circumstance and individual person to individual person. Some people notice "what follows" more readily than others. Such enlightenment seems to result from noticing one or more relations, noticing which almost certainly has particular physical or mental causes. This process is commonly classified as or called *explanation*.

23 Husserl is probably right that the logical manner in which we interpret or relate ideas drawn from experience does not necessarily itself reduce to or derive from experience, but first, contrary to his views, it may depend on the structure and evolution of aspects of the brain or mind, and second, this logical manner of interpretation or relation may well be psychological. There is certainly no evidence that Husserl's universals are real or that if they could be real they could be "neither physical nor mental".

24 The German word is *Wesen*, which regrettably suggests both essence and being, which encourages the kind of Scholastic hypostatization which Husserl is so fond of and which has so little evidence to support it.

25 Husserl liked to think that his own approach even if different from or even inconsistent with empirical science was also scientific. He struggled to put forward philosophy as a "pure presuppositionless science" of idealities in consciousness which could be intuited in their "pure" nature. Needless to say, the German word for science which is *Wissenschaft* is somewhat more generous than our own, and Husserl's definition is even more generous than that. If Husserl's

Phenomenology is science, either in the version of 1900 or later, then one might suppose that the highpoint of his version of science ought to be Medieval Scholasticism or a slight modification of it, a position which Bacon, Galileo, Descartes, Newton, and other prominent leaders of the Scientific Revolution just might have felt somewhat uncomfortable with along with almost all experimentalists and theorists today.

26 Hans Cornelius (1863-1947) had come out with his *Psychologie als Erfahrungswissenschaft* in 1897 and an article "Gestaltqualität" in 1899 [*Zeitschrift für Psychologie und Physiologie der Sinnesorgane*, **22** (1899) 101-121]. His views had many similarities to those of Ernst Mach and the two men often sent publications to each other. Epistemology and aesthetics rseemed to have remained Cornelius's principal academic interests. Eleven letters from Mach to Cornelius can be found in the *Bayrische Stadtsbibliothek* in Munich. Ten letters to Ernst Mach were in the Ernst-Mach-Institute in Freiburg im Breisgau, but are now said to be in the Deutsches Museum in Munich. Joachim Thiele published five letters from Hans Cornelius to Mach in *Wissenschaftliche Kommunikation*, A. Henn Verlag: Kastellaun, 1978, pp. 214-216.

27 Friedrich Wilhelm Foerster (1869-1966) was a professor for philosophy and pedagogy at the University of Vienna in 1913 but accepted a call to Munich in the following year.

28 Hans Driesch (1867-1941) was best-known for his defence of vitalism in biology. There are six extant letters from Mach to Driesch in the Leipzig University library and four letters from Driesch to Mach are in the Deutsches Museum in Munich. The letters by Mach have been published by Dieter Hoffmann and Hubert Laitko in their book: *Ernst Mach - Studien und Dokumente zu Leben und Werk*, Berlin, 1991, pp. 397-403. The letters from Driesch to Mach appeared in Thiele, op. cit., pp. 117-119.

29 Rudolf Eisler (1873-1926) was an Austrian philosopher and editor best-known for his dictionaries of, lexicons of, and introductions to philosophy.

30 The first Japanese translation of Mach's work was apparently *Der Mechanik* in 1931.

31 She was born Ludovica Marussig (1845-1919) and married Mach in 1867. She bore five children and seems to have had artistic tastes and had artist-relatives in Italy.

32 Marie Mach (1844-1928) became a governess, editor, and writer in various parts of the Habsburg Empire. For a recent English translation of an article by her on both her own childhood and that of her brother see: Chapter One in *Ernst Mach a Deeper Look*, edited by John Blackmore Kluwer: Dordrecht, 1992, pp. 3-16.

33 This book was published in Stuttgart in 1915.

34 Anton Lampa (1868-1938) was an experimental physicist in Prague and a supporter of Mach's philosophy. He became an official in adult education in Vienna after The Great War.

35 Anton Marty (1847-1914) was a linguistic philosopher and close follower of Franz Brentano. He was also a colleague of Mach for many years in Prague. There are twenty letters from Marty to Mach in the Deutsches Museum in Munich, Mach's letters to Marty have not yet been located.

36 Josef Popper-Lynkeus (1838-1921) was Mach's oldest and best friend.

37 This Wolfgang Pauli was the father of the famous physicist of the same name.

38 Hugo Dingler (1881-1954) was a young Munich philosopher of science much praised by Mach. He was known as an extreme conventionalist and teacher of Carnap. Twenty one letters from Mach to Dingler can be found in the Hofbibliothek Aschaffenburg, and there are ten extant letters from Dingler to Mach. All of these letters have been published in English translation by John Blackmore in *Ernst Mach - A Deeper Look*, Kluwer: Dordrecht, 1992, pp. 92-110.

39 Johannes Kleinpeter (1869-1916) was a Gymnasium teacher in Gmunden Austria, and a strong follower of Mach. His best-known work was *Der Phänomenalismus* (1913). There are forty-one letters from Kleinpeter to Ernst Mach in the Deutsches Musem in Munich, but Mach's letters to him have not been located yet.

40 Peter Rosegger (1843-1918) was an Austrian writer and nature-lover

41 An article "Soziologie des Erkennen" was published in 1921, but its relation to the book that

Jerusalem was writing in 1915 is not clear.

42 Wilhelm Braumüller is an old publishing house often associated with the University of Vienna.

43 Josef Jacob (1859-1918) was a Gymnasium Professor of mathematics in Vienna. The Deutsches Museum is said to have four letters from Jacob to Mach.

44 It is not clear which congress Mach is referring to, except that it was apparently one where Husserl gave an address.

45 The better-known expression today for Husserl's conception is *Wesenschau*. Literally, the word means "the conscious manifestation of being or essence." But Husserl drifting into Greek called his abstract types of being *Eidos*, which was to be studied by a "science" called *Eidetik* or *Eidologie*.

46 Joseph Eisenmeier (1871-1926) had been a student of Anton Marty and was a member of Brentano's circle. See *Die Psychologie und ihre Zentrale Stellung in der Philosophie. Eine Einführung in der wissenschaftliche Philosophie*, Halle, 1914. There are four extant letters from Eisenmeier to Mach in the Deutsches Museum in Munich.

47 Wolfgang Pauli [Sr.] [born Wolf Pascheles] was born in Prague in 1869 and became a medical doctor in 1893,. He changed his name to "Pauli" in 1898. Later he taught at the University of Vienna. In Vienna he proofread at least one of Mach's books. There are four letters from Pauli Sr. to Ernst Mach in the Ernst-Mach-Institiute, all from 1913-1914. For information about his Prague background see Frantisek Smutny [edited by Charles P. Enz], "Ernst Mach und Wolfgang Pauli's Ancestors in Prague", *Gesnerus*, **46** (1989) 183-194.

48 Ernst Mach became the Godfather of Wolfgang Ernst Friedrich Pauli [Jr.] (1900-1958) at birth. The middle name of "Ernst" came from Ernst Mach. Young Pauli grew up in Vienna, studied under Sommerfeld in Munich, taught in Zürich from 1928, and won a Nobel Prize in physics in 1945. Though he was clearly under Ernst Mach's philosophical influence until the end of the 1920's, he later tried to minimize the extent and importance of that influence.

49 For more information on Heinrich Mach see: John T. Blackmore, *Ernst Mach - His Life, Work, and Influence*, University of California Press: Berkeley and Losa Angeles, 1972, pp. 148-150.

50 Wilhelm Jerusalem, "Psychologen und Philosophen", *Zukunft*, **18** (1914) Heft 7, pp. 89-97.

51 Johann Nepomuk Mach (1805-1879) attended the University of Prague with allegedly perfect marks and became a well-educated tutor of the children of Austrian noblemen such as Baron Brethon in Zlin and Baron Stadion in Vienna who was a leading minister in the government. But when the Revolution of 1848 was suppressed in Austria, he then concentrated on farming and educating his three children at home. During the 1860's he bought a farm in what is now Croatia and raised oakleaf-eating silkworms. The lack of a market put him into debt. After his death, his youngest daughter Marie sold the farm and began a career as private governess in some of the less Westernized parts of the Empire, which she later described in a book and in some of her stories, which were all published anonymously.

52 Theodor Ziehen (1862-1950) was an original philosopher who in the spirit of those early times began his career by teaching psychology for many years in the Netherlands. He transferred to Halle in 1917 where he published books in many different branches of philosophy. He introduced many original terms into philosophy such as "binomistisch", "intrapsychisch", "Gignomene", "Redukte", etc. but they have not endured well. Ernst Mach may have been referring to Ziehen's book: *Die Grundlagen der Psychologie*, Leipzig, 1915.

53 Richard Wahle (1857-1935) was an Ausrian philosopher who taught at the University of Czernowitz starting in 1895. During the psychology-is-replacing-philosophy period Wahle wrote *Der Ganze der Philosophie und ihr Ende* (1904). According to Johnston he was a follower of Mach. See Wlliam M. Johnston, *The Austrian Mind*, University of California Press: Berkeley and Los Angeles, 1972, pp. 201-203.

54 Julian Benda (1867-1956) had written a book *Le Bergsonisme ou la philosophie de la mobilite* in 1912 and he had sent a copy to Mach. Benda would become best-known for his *Traison des clercs* (1927) and later criticisms of irrationalist trends in French thought such as Existentialism, Expressionist Art, etc. Unfortunately, irrationalism among many philosophers and self-styled

"intellectuals" remains influential in France.

55 Husserl had tried hard to bring back *a priori* categories into philosophy again with the help of his theory of idealities and "intuition". Mach and Jerusalem found this hard to accept. The notion of "pre-established harmony" of course comes from Leibniz but also may be implied by the Christian notion of Providence. Benda as a rationalist may have sought a way to reconcile mathematical science with Bergson's tendency to define reality in terms of what is changing. But Husserl's "solution" seemed so backward and Scholastic that Jerusalem looked for a way to eliminate the notion of the *a priori* once and for all from philosophy. But given the reign of mathematics and mathematicians in the 20th century this has been hard to do. They still want an alternative to empiricism which treats idealized abstractions like numbers, laws, rules, and conventions as real.

57 Carl Stumpf (1848-1938) was a philosopher who contributed to music, acoustics, and other branches of psychology and who helped persuade many people from about 1890 to 1910 that psychology was replacing philosophy. Like other former students of Franz Brentano he was strongly opposed to Mach's philosophy. For an account of his criticisms of Mach and the latter's reply see the chapter in this book on Musil and Stumpf.

58 Theodor Beer was a Vienna physiologist who had written a "non-critical" book on Mach.

59 Eduard Spranger (1882-1963) studied under Wilhelm Dilthey and like him became a professor of philosophy in Berlin (1920). Both men were influenced by Hegel's approach and adopted a manner of fairminded, social and historical generalization which remained popular and suggestive without being detailed or truly scientific. The sub-title of the book *Lebensformen* which he sent Mach was *Geisteswissenschaftliche Psychologie und Ethik der Persönlichkeit,* a book which by 1906 had gone through nine editions. Some other of his publications would also go through a vast number of editions. In order to give the English-speaking reader a taste of Dilthey's prose we have included a passage from him in the Appendix. The honorable nature of these men shines through their writings even if their ideas are rarely detailed enough for readers to determine either their full meaning or their soundness. Their manner of being gentle and philosophical about things contrasts a bit sharply with the attitude of polemically-focused gladiators like the young Husserl who seemed to be forever attacking rivals..

60 Alois Riehl (1844-1924) was an Austrian and neo-Kantian philosopher who taught in Berlin from 1905 to 1917. The Deutsches Museum in Munich has three letters to Mach.

61 Max Scheler (1874-1928) would become a well-known philosopher in his own right, teaching philosophy in Cologne from 1919 to 1928. He applied the intuition-oriented approach of Husserl to values. See the chapter on Musil.

62 Paul Carus (1852-1919), editor of the Open Court Publishing Company and its two philosophical magazines, *The Open Court* and *The Monist.*, was largely responsiblefor the publication of English translations of most of Mach's works.

REFERENCES

Benda, Julian. *Le Bergsonisme ou la Philosophie de la Mobilite*. Paris, 1912.

Benda, Julian. *Traison des clercs*. Paris: 1927.

Blackmore, John T. *Ernst Mach - His Life, Work, and Influence*. Berkeley: University of California Press, 1972.

Blackmore, John & Klaus Hentschel (hrsg.). *Ernst Mach als Aussenseiter*. Wien: Braumüller, 1985.

Cornelius, Hans. *Psychologie als Erfahrungswissenschaft*. Leipzig: 1897.

Cornelius, Hans. "Gestaltqualität". *Zeitschrift für Psychologie und Physiologie der Sinnesorgane,* 22 (1899) 101-121.

Eisenmeier, Joseph. *Die Psychologie und ihre Zentrale Stellung in der Philosophie. Eine Einfführung in der wissenschaftliche Philosophie*. Halle, 1914.

Hoffmann, Dieter & Hubert Laitko (hrsg.). *Ernst Mach - Studien und Dokumente zu Leben und Werk*. Berlin: Deutscher Verlag der Wissenschaften, 1991.

Husserl, Edmund. "Über das Prinzip der Vergleichung in der Physik". in *Archiv für systematische Philosophie.* 3 (1897) 241-244. (A review of Ernst Mach's article of the same title above. See: Ernst Mach. *Popular Scientific Lectures.* La Salle: Open Court, 1943. pp. 236-258.)
Husserl, Edmund. *Logische Untersuchungen.* Band I & II. Leipzig, 1900 & 1901..
Husserl, Edmund. a book review on Wilhelm Jerusalem's *Die Urteilsfunktion. Eine psychologische und erkenntnistheoretische Untersuchung.* Wien & Leipzig: 1895. in *Archiv für systematische Philosophie,* 9 (1903) 523-543.
Husserl, Edmund. *Logical Investigations.* Volume One. translated by J.N. Findley. New York: Humanities Press, 1970. (1st German edition 1900)
Husserl, Edmund. *The Cambridge Companion to Husserl.* edited by Barry Smith & David Woodruff Smith. Cambridge: Cambridge University Press, 1996.
Jerusalem, Wilhelm. *Die Urteilsfunktion. Eine psychologische und erkenntnistheoretische Untersuchung.* Wien und Leipzig: 1895.
Jerusalem, Wilhelm. *Der kritische Idealismus und die reine Logik - Ein Ruf im Streite.* Wien und Leipzig: Braumüller, 1905.
Jerusalem, Wilhelm. "Soziologie des Erkannens". *Kölner Vierteljahrshefte für Sozialwissenschaften.* 1 (1921) 28-34. (The late date of this article suggests that a work which originally had the same title may have been published by Jerusalem several years earlier.)
Jerusalem, Wilhelm. "Psychologen und Philosophen". *Zukunft,* 18 (1914) Heft 7, pp. 89-97.
Kleinpeter, Hans. *Der Phänomenalismus.* Leipzig, 1913.
Mach, Ernst. *Beiträge zur Analyse der Empfindungen.* Jena: G. Fischer, 1886.
Smutny, Frantisek. "Ernst Mach and Wolfgang Pauli's Ancestors in Prague". edited by Charles P. Enz. *Gesnerus,* 46 (1989) 183-194.
Spranger, Eduard. *Lebensformen - Geisteswissenschaftliche Psychologie und Ethik der Persönlichkeit.* Leipzig: 1914.
Thiele, Joachim. *Wissenschaftliche Kommunikation.* Kastellaun: A. Henn Verlag, 1978.
Wahle, Richard. *Der Ganze der Philosophie und ihr Ende.* Vienna: 1894.
Ziehen, Theodor. *Die Grundlagen der Psychologie.* Leipzig, 1915.

Chapter Eleven

ALOIS HÖFLER - POLYMATH

by the editors

Section Guide
1. Höfler and Meinong
2. "Inexistent Objects"
3. Höfler's Philosophy
4. Höfler and "The Society"
5. Mach and Höfler

1. HÖFLER AND MEINONG

Alois Höfler was born in Kirchdorf in Upper Austria on April 6, 1853. Both his father and mother died while he was in his early and middle teens. He and his two younger sisters were then raised by a second mother, Amalie Böheim, with great concern and care. He attended both Volkschule and Gymnasium in Vienna.[1] It is interesting that Ernst Mach who also had two younger sisters like Höfler also became a physicist. In fact, Alfred Adler, the one-time follower of Freud, made much of the tendency of "big" brothers to go into science, perhaps partly on the basis of Höfler, Boltzmann, Einstein, and Mach, who were all the oldest children in their families.

In the Fall of 1871 Höfler entered the University of Vienna where he studied math and physics under Ludwig Boltzmann and Josef Stefan. He passed his state examination and began teaching these same subjects in the *Josefstädter Gymnasium* on May 15 1876, that is, at the age of 23. He taught mathematics for four years, then physics for four years, and after then a combination of subjects. During his early years he also taught at other gymnasiums in Vienna. Höfler was also a pedagogical pioneer by teaching "goniometric functions" by means of graphic representations some twenty-five years before Felix Klein was able to get the same type of math taught in German gymnasiums. Höfler was also active preparing textbooks in natural science for lower and upper classes of middle schools and gymnasiums especially during the 1880's but also later.

During the 1870's and 1880's Höfler attended evening classes taught by Brentano and Meinong.[2] And like many other students he was overwhelmed by Franz Brentano's dialectical skill, ability to make sharp distinctions, and by Brentano's tutorial assistant, Alexius Meinong, and the determination of both men to make psychology as reliable and informative as possible, or in the perspective of the period to make philosophy itself scientific by transforming it into scientific psychology.

Franz Brentano (1838-1917) taught philosophy in Vienna from 1874 to 1894, first as a professor and later as a *Privatdozent*. He had so many students who later became famous like Edmund Husserl, Sigmund Freud, Thomas Masaryk, and Alexius Meinong that it often escapes attention that Alois Höfler had also been his student. Alois Höfler had studied little philosophy while a student and that was largely of an idealist orientation such as in Kant, Fichte, and Schopenhauer, but when he began teaching during the middle

J. Blackmore, R. Itagaki, and S. Tanaka (eds.). Ernst Mach's Vienna 1895-1930. 237-276.

1870's he found himself drawn to the evening "colloquium" of the master.

> Brentano's first year as professor in Vienna (1874) was my last as a student (mathematics with Boltzmann and physics with Stefan) such that little time was left from preparation for my state examination to become a Gymnasium teacher. While I had heard him give lectures on Aristotelian and Thomistic philosophy, I still remained rather distant. It was my youthful friend Ernst Boeck (who is now a psychiatrist) who brought me to hear the evening discussion sessions of Franz Brentano in 1877. I soon realized why my friend, who was some years younger, had become so enthusiastic about him, though I was not able to be quite that taken.[3]

Brentano hoped to make psychology scientific and in 1874 had published his famous book on "Act Psychology" which together with Wilhelm Wundt's "Content Psychology" dominated the discipline until the rise of Gestalt, Behavioral, and Freudian psychology shortly after the turn-of-the century. The heart of Brentano's approach seemed to be the insight that mental phenomena seemed to mean, "intend", or refer to something beyond themselves and that scientific psychology had to study such action and relations.

This insight became the basis of the work of most of Franz Brentano's followers, including Alexius Meinong. This young man, who came from an aristocratic family and who was born in 1853, the same year as Höfler, soon became Alois Höfler's friend and increasingly his mature guide in both psychology and philosophy. Why Meinong more than Brentano?

Höfler claims that inspite of his fascination with Brentano's methods that his own reading in Kant and his attraction toward Richard Wagner's music helped cause a measure of emotional dissatisfaction with Brentano since the latter often criticized Kant and made unkind jokes about Wagner and his music "of the future".[4]

Alexius Meinong (1853-1920) would be Alois Höfler's guiding star until both men died four decades later. His theory of objects and assumptions became especially important. Meinong as Brentano's assistant tutored many students including Alois Höfler, but Meinong soon found more support among the students he tutored for his own modifications than for many of Brentano's own views. Indeed, his star rose quickly and Meinong became an associate professor in Graz in 1882.[5] Both Höfler and Meinong in their lust to make philosophy more scientific tried to carry out many psychological experiments, Höfler in his Gymnasium lab and Meinong in his university laboratory. But Höfler continued his education in philosophy by continuing to occasionally attend Brentano's evening colloquia in Vienna from time to time, but when given a five month's vacation in 1886-87 he went to Graz and learned what he could from Meinong's latest attempts to add rigor to philosophy by transforming it into scientific psychology.

Alois Höfler's early attraction toward German idealism was eventually modified by his strong support of physical science. If Josef Stefan and young Ludwig Boltzmann were correct, then something physical must exist beyond conscious impressions, which meant that Ernst Mach's phenomenalism and Franz Brentano's apparent claim that mental phenomena "intend", only what is conscious, had to be false. Alois Höfler in his book on logic in 1890 which included some collaboration with Meinong pointed this out clearly.[6] Four years later another former student of Brentano, Kasimir Twardowski, made the same point but in a way which seemed more acceptable to other scholars. But in fact, while Meinong, Husserl, Twardowski and others seemed to treat the difference between conscious content and independent object as a sign of "realism" and later as a rejection of "psychologism", Alois Höfler went well beyond that. But this step was predicated on sharply distinguishing physics from philosophy while reducing philosophy to scientific psychology on the one hand and to "metaphysics" on the other.

Alois Höfler received his doctor's degree in 1885 with the work *Einige Gesetze der Unverträglichkeiten zwischen Urteilen* (i.e.Some Laws on the Incompatibility between Judgments) and three years later helped to found the University of Vienna Philosophical Society in 1888 and was its first *Obmann* or leader.[7] (He became *Obmann* again from 1898 to 1903 and from 1912 until 1922.[8]) The high esteem in which he was held can be judged from the fact that in 1903 at the time of his departure for Prague he was made and retained the title of honorary president of the Society until his death. But his main role in it was apparently to lead most of the discussion sessions.

He married Auguste Doernhoeffer in the year 1892 and had four sons of whom three survived. Karl became a professor of botany, Otto a professor of German, and Wolfgang a writer. Höfler had his work "*Psychische Arbeit*" accepted by the University of Vienna in 1894 which allowed him to teach psychology and philosophy as a *Privatdozent* there. But most of his time was still spent teaching in various Gymnasiums in Vienna for several more years..

The breadth of Höfler's interests was quite remarkable. Most of his gymnasium teaching was in math and physics, but his increasing interest in philosophy and in making it more scientific led him to study and write increasingly on logic and psychology as well. Indeed, he published books in all of these fields, and while they may have appeared to many observers to be "merely" textbooks and introductory works in fact he put his own ideas into all of them.[9] Some of his many books were also written in collaboration with top figures such as Felix Klein in mathematics and Alexius Meinong in logic.

He considered himself to be a follower of Meinong in philosophy and psychology and did all he could to help spread his ideas, but as we shall see, there was much more to the story than that. He had his own philosophical position. Nor were his interests confined to science or philosophy. He also loved music, poetry, and the arts. He strongly opposed attempting to reduce their role in the schools. Indeed, Höfler's father-in-law had been born in Bayreuth, and the two men would often visit this "sacred repository of Wagner's music dramas". Furthermore, as a member of the Vienna Wagner-Verein since 1874 he would finally be appointed its honorary chairman in 1919.

As a teacher of physics, he tended to be most proud of two students whom he had guided all the way through his gymnasium, namely, Boltzmann's successor, Friedrich Hasenöhrl,[10] who is perhaps best-known for anticipating aspects of Albert Einstein's famous equation $E = mc^2$ and Marian von Smoluchowski[11] who like Einstein would also contribute to the theory of Brownian motion and become strong believers in the reality of atoms. Both physicists would also become members of Höfler's philosophical society.

The 1890's were a period in which many people thought that philosophy would be replaced by scientific psychology and Brentano, Meinong, and Höfler seemed to be a part of this movement. But in 1900 Edmund Husserl would attack it severely on the ground that formal logic as something ideal could not be reduced to psychology or to mental phenomena. Other people argued that epistemology was the heart of philosophy and that it too could not be reduced to mere psychology, whether "scientific" or not. Höfler partly agreed with the critics that formal logic did seem to be idealized, but that the important thing was to use sound, useful logic in one's thinking and that this use was psychological, something especially important in all types of education. As for epistemology, Höfler would defend it, especially against the attempt of the then leading German philosopher Leonard Nelson in his book *Über das sogenannte Erkenntnisproblem* (1908) to reject *Erkenntnistheorie* or what we call epistemology.

Alois Höfler's efforts to help popularize Meinong's ideas, especially in what Franz Brentano would call descriptive psychology, took five forms, first, a collaboration on publications, second referring to Meinong's views in his own articles and books such as his theory of objects, judgment, evidence, and knowledge, third by his way of organizing The University of Vienna Philosophical Society and its lectures and discussions, fourth by making space available for lengthy presentations of Alexius Meinong's philosophy in education journal special publications which he was co-editing,[12] and fifth, by gathering much of Meinong's work together into a *Gesamtausgabe* and publishing it as a gift or present to him.[13]

We cannot go into as much detail as we would prefer in this paper for space reasons and because the estimated 1500 letters between Höfler and Meinong remain unpublished for the most part and this author has not yet found the time or health to plow through them and detail their unpublished content, but we can at least mention these five forms of help or assistance and hope that future research will fill in the gaps and correct any mistakes.

Three of Alois Höfler's many major books were on logic, psychology, and physics respectively. He claimed that Meinong was a co-author of the first, which Meinong was reluctant to admit, but the second book while written by Höfler was so heavily influenced by Meinong's contributions to experimental and descriptive psychology as to make his ideas the main topic of the book.[14]

Alexius Meinong in his autobiography mentions how he tried to make his thinking independent of Brentano's overwhelming influence,[15] but Höfler seems to have done the opposite, doing all he could to clothe his ideas within the views of Kant, Klein, and Meinong, as if he doubted that he could find an audience among his peers for his own "metaphysical" ideas, that is, his attempt to reconcile mind and matter and scientific physics with scientific psychology. As for his textbooks and introductory works to many fields, he was generally cautious concerning philosophy and trended to emphasize a sharp difference between what he thought could be dirrectly experienced and what he called "hypothetical". By following a prudent approach he seems to have gained a favorable reputation as a good teacher, a loyal follower of Meinong, and a wise if rather critical commentator on philosophical disputes, which he frequently engaged in mostly in the form of book reviews.

In time, he would become best-known as an educator and through the many editions of his books on psychology, particularly as one of the most influential advocate of Gestalt psychology. On the other hand, Alois Höfler's willingness to attribute so much to Alexius Meinong as if he were merely his follower is manifest in many if not most of Höfler's non-pedagogical and non-physical articles as well. For example, in "Die unabhängigen Realitäten" which he published in *Kant-Studien* in 1907[16] he argued as if he were merely mediating between articles by Meinong and Meinong's student Oelzelt-Newin, while in fact favoring Meinong's new theory of extended reference and both real and unreal objects.

Alexius Meinong rarely mentions or praises Höfler's writings and almost never in detail, an absence of attention which may have been a factor in Höfler's unseemly late praise for his own works. But in fact Alexius Meinong gave significant academic and philosophical help. What was lacking was proper recognition and praise for his friend and follower. First, in addition to studying under Meinong in 1882, it seems that Höfler received his doctorate in Graz from Meinong.[17] Second, he studied further in Meinong's laboratory in the Winter semester of 1886/87.[18] Third, Höfler thought that since the only way to save philosophy was to transform as much as possible into scientific psychology (except for the "metaphysical" connecting part linking physics and psychology and mind and matter) that it was important to help Meinong's work in descriptive and experimental psychology. Fourth, by showing in psychological terms that one may be able to refer to what is not conscious or not even real Meinong was providing a foundation which might be able to justify Höfler's view of physics, especially Alois Höfler's feeling based on the work of Stefan, Loschmidt, and Boltzmann that it was possible, though "hypothetical", that the physical world could be referred to, partly understood, and was probably real. But if the whole physical world for us finite and fallible human beings was at least possible, then this suggested that atoms and molecules might also exist.

If Meinong resembles John Locke with the latter's focus on psychology and faint support of an external physical world, then Höfler at least with respect to methodology of physics was more like Sir Isaac Newton with a cautious probablism about many aspects of the external world replacing Descartes' alleged certainty and with a reluctance to let inference get too far ahead of observation. But Höfler also had an interest in Locke's theory of education and one can see this in how he educated his own children.

But was Höfler's faith in Alexius Meinong's psychology well-founded? Especially given the criticism by Bertrand Russell and others early in the 20th century? This is naturally a disputed point, but if we are to maintain that Meinong *really* helped Höfler, then we will have to answer the question in the affirmative. Nor is this easy if we stick fully to the German language and some of its limitations, and it is perhaps impossible if we keep Meinong's misleading terminology. As for the extraordinary complexity of Alexius Meinong's full approach toward reference, much of that becomes less central if we properly distinguish between different definitions of the words "real" and distinguish carefully between what merely has identity and what has both identity and reality.

Our basic starting point is the realization that Brentano's original insight that mental phenomena "intend" an object outside of themselves may not be sufficient to provide a basis for reference, epistemology, and ontology.[19] It blurs differences between reference, consciousness, perception, and both direct and indirect objects of perception. Meinong, who had read English philosophy and was familiar with the English term "reference", may have felt this when he allowed that people can refer to whatever they want. But while Meinong possibly encouraged by Höfler might have agreed that objects and their content have to be distinguished and that the object may be trans-conscious at least in physics, Meinong went further *and allowed that we can also refer to unreal objects*.

A number of recent defenders of Meinong's basic approach appear to be looking to reform mathematical logic from the straightjacket which Russell is said to have placed on it[20] when he tried to make reference and ontology co-extensive. But our concern is more to point out how making the scope of potential or "meaningful" reference much more extensive than any particular epistemology or ontology helped Höfler justify believing in the probable existence of a trans-conscious physical world, thereby seeming to allow him to reconcile psychology and physics as scientific, even though they utilized very different methods, and in ontological terms even though mind and matter seemed so disparate in characteristics that interaction between them for all of its likelihood still seemed hard even to conceive. Höfler's solution was to distinguish between philosophy as scientific psychology and philosophy as "metaphysics". The latter was needed when comparing and relating the different sciences. Furthermore, physics was a science and in his opinion to the extent it was reliable and informative did not require either form of philosophy, neither as scientific psychology nor as "metaphysics". Meinong's extension of reference beyond experience and beyond Brentano's ambiguous notiuon of conscious experience "intending objects" made Höfler's own philosophical system possible.

But if reference is more extensive than reality, then reference to non-reality must be possible, and this was apparently a concept that Russell and many others found difficult to accept, not just because Meinong tried to avoid taking an ontological position and it is unclear what "non-reality" is or means, but because Meinong kept seeming to refer to non-realities which were neither qualities nor relations as *objects* .

It was common in his day as well as now among presentists in epistemology to reject indirect and representist epistemology as "mere metaphysics" as if ontological and transcendent in character, and worse, as if "incompatible with science", a strange charge given the indirect epistemologies of Galileo, Descartes, Newton, Darwin, and many other great scientists. There are many type sof presentist epistemology, and Mach held one form "phenomenalism". But phenomenalism also has an ontological dimension. Mach accepted both the epistemological and ontological forms, at least in philosophy of science, but Bertrand Russell seems to have attempted to fully reject epistemological phenomenalism while accepting ontological phenomenalism about physical objects and processes.[21]

Nevertheless, in one of his theories of formal logic he seems to have tried to avoid "metaphysical" discussion by doing the reverse, that is, by *narrowing* the scope of potential reference such that all "genuine" reference was to "what exists" as if there were no contrary or alternative to existence, and hence, as if the topic of reality and hence ontology itself were superfluous and did not have to be discussed. But in his defense, he was initially receptive to Meinong's view that reference is more extensive than existence and even continued to include it in his work on other types of philosophy than formal logic., such that reference to non-existence was possible. But finally while doing logic he decided that he couldn't clearly understand what Meinong "really" meant by "non-existent objects", hence he reverted at least in theory of logic to what seemed like the simplest or most "economical" solution (to use a term from Ernst Mach) even if grossly impractical from almost all perspectives outside of formal logic and even some positions within. He reverted to reference implies the existence what is referred to as if there were no "non-existent objects".

Alexius Meinong's way of avoiding "metaphysical problems" *broadened* the scope of potential reference as if to include possible reference to all possible ontologies, without explicitly committing himself to any particular ontology.[22] This position is exactly what most fair-minded historians want, and it is interesting to recall that one of Meinong's earlier fields of interest had been history. But of course his stand was the opposite of what most "original" philosophers desire, men who consider it important to persuade as many people as possible, something incompatible with presenting a fair-minded and fully inclusive conception of referential, epistemological, and ontological classification which when noticed and accepted allows us to fairly understand many *other* points of view.

Meinong's reluctance to commit himself to a particular ontological position goes together with his talk about "non-existent objects" i.e. objects of reference which lack tthe perceptual, conscious, or causal components of what "really exists". But while his view is fully compatible with common sense in most basic respects and should not be hard to understand in itself, it is so garbed in scholastic categories, vague classification, and otiose terms like "inexistence" and "non-existent objects" as to make the heart of his philosophy seem unnecessarily dense and incomprehensible[23]

2. "INEXISTENT OBJECTS"

From a practical perspective, everything we refer to can be said to have identity or character, but only what has causal potential or energy is real. A fictional hero has a measure of identity and even a round-square can be characterized at least in part. In fact, we can even characterize "nothing" as a complete absence of characteristics, self-contradictory as that may sound. In short, reference can give identity or character but in the absence of causal qualities such as mass, weight, and energy what is referred to is normally held to be unreal or non-existent. Meinong of course did not hold a common sense position in its entirety. Like many philosophers, his conception of causes was not normally in terms of pressure and interaction, but closer to the constant or necessary conjunctions among appearances which the phenomenalist Hume and the idealist Kant preferred. nevertheless, Meinong did broaden the scope of reference to allow reference to anything which could be mentally characterized such as fictions, phantoms, past existence, forms, types, and other abstractions.

Meinong's views have been interpreted as allowing for "inexistence",[24] something between nothing and something, but there does not appear to be a genuine need for this odd locution, especially where practical problems are concerned, even though his basic insight about our ability to refer to what is neither sensory nor conscious appears to be valid, at least when understood in volitional, common sense terms. Furthermore, since such targets of trans-sensory and trans-conscious reference can be characterized at least in part, it is reasonable to allow them a measure of identity. Were they also to have causal characteristics such as mass, weight, or energy as common sense and much science allows, then they could be said to have both full identity *plus full reality*. But even if they lack causal qualities such that in causal terms they are literally nothing, still, because they can be referred to at least in volitional terms and partly characterized one cannot deprive them of all identity. If Meinong's term "inexistence" were legitimate,[25] then it could well be attributed to partial identities which lack existence in a causal sense.

But if Meinong's basic insight is sound, that the scope of possible reference extends beyond existence, his terminology and way of expressing this insight is not. His term "inexistence" seems superfluous at best and could be misleading because what merely has identity, because it can be characterized, is not real at all in a causal or practical sense, and hence is not truly half-way between nothing and something. In causal terms it remains nothing. It is nothing. Like forms, qualities, relations, and fictions, it can neither resist nor exert any force or pressure in the real or practical world, hence for all of its measure of identity and characterizability, it is non-causal and hence non-existent.

Since our argument may be difficult to follow, let us approach it from an opposite tack. If we *were* to mean by "something" that it has character or identity, then "nothing" would mean what cannot be characterized or has no identity, but we *can* characterize nothing. It *does* have some identity. The mere act of referring to nothing includes some characterization or attribution of identity. How? Because negative characterization, referring to an *absence* of something, *is still a type of characterization.* This means that we cannot plausibly or even coherently identify what is real merely with what can be referred to, characterized, or what has identity. Why not? Because we would have to allow that both something and nothing were real, such that we cannot even refer to unreality without making it real. And terms with no contrary, with no capacity to understand what their absence would mean without abolishing that absence in the act of referring to it would literally make the concept of reality superfluous at best and possibly even illogical and beyond comprehension, something practical understanding cannot endure. Practical life is not possible without an ability to influence our environment, that is, help cause things to happen. To identify reality with what can cause or be caused is part of common sense and most non-idealized science. To reject that identificastion as many presentists want, almost guarantees a loss in practical understanding and behavior, a loss incompatible with human progress.

Unfortunately, the word "real" may mean genuine *or* what exists, and the meanings are quite different and are easily confused. We speak of a genuine pearl and contrast it with a fake pearl, but both can exist. In the sense of genuine only one is real, but in the sense of existence both are real. Thus the expression "that is really real" instead of being redundant normally means what genuinely exists. "Really" means genuine and "real" normally means exist. If to be real is to have causal capacity, then we are using the word in the sense of existence, not as genuine. Some philosophers such as Plato have confused a similar double meaning in Greek as if to be real meant to be "good" rather than to possess causal power in the sense of pressure or influence, as if reality were a value term.

What exists has complete identity in that it has a full body of qualities and relations as well as subordinate parts. But non-causal referents while having some identity since we can refer to them and partly characterize them lack full identity because as incomplete characterizations based on mere definition or incomplete understanding of them there are questions about them which in principle cannot be answered. For example, we can refer to a unicorn and partly characterize his features, but no one knows if he has an appendix or backbone. Hence the characterization of him is not complete, since our understanding of him is not complete. A real animal, however, has a full complement of features even if we don't know what they are yet or don't understand them. Unlike unicorns, all questions about causal agents or powers are at least in principle capable of a true or valid answer. Again, while only what has causal potential has complete identity, much that is non-causal and hence unreal may have at least a measure of identity, which unfortunately, Meinoing misleadingly called "unreal objects" or what possessed "inexistence". The practical distinction is between realities which have identity and identities without reality. Both are "real" in the sense of having identity but only what has causal potential has both existence and identity.

We have developed this argument to help defend Meinong's main point, namely, that neither consciousness nor sensation can restrict the scope of volitional reference and hence of characterizability, understandability, or reality, while rejecting his main term "inexistence" as well as the names of many other distinctions he makes. The problem is apparently that he doesn't identify reality exclusively, and in many contexts at all, with what has causal potential. He is not practical in that sense. And partly as a result, he is inclined to consider what can be referred to but is neitheir conscious nor sensory *nor causal* as "partly real", which it is not, at least not in causal or practical terms at all, though if one believes that so-called "universals" or other abstractions are real, then perhaps one might think that somehow his position could be defended after all.

Meinong's failure to define "reality" properly also caused problems for Höfler who did try to think of reality in causal terms with unreality as what lacked causal potential. Höfler also used Meinong's view that reference extends further than mundane reality to accept various type of "metaphysical" views. Nevertheless, he did try to make sense of Meinong's notion of "inexistence", though without fully seeing that there is no need to call or classify what has a measure of identity or characterization as real or even "inexistent". If it lacks energy it is simply unreal. (Höfler was sympathetic to a type of "energeticism", even if as a student of Boltzmann he stopped short of accepting the views of Ostwald and Helm on this controversial topic.)

But different commentators understand Meinong's philosophy in very different ways. David Lindenfeld, for example, in his very interesting book[26] identifies Meinong's views with a change in the nature of Positivism, which seems strange to this writer because of the Logical Positivist attempt during the 1930's to restrict the scope of reference (or "meaningful" reference) precisely to block Meinong's expamded conception. If we think of this well-known offshoot of the Vienna Circle as basically continuing the presentist approach which has been so dominant in most German philosophy since Kant, then one would normally think that Meinong's liberalism about the scope of possible reference would better fit the representism of Russell and Lovejoy. That Meinong could not fully or openly repudiate presentism about the external physical world is quite understandable given the trends of his time in Germany and Austria. Höfler, however, came closer to liberating himself.

The reader may ask, why discuss Meinong's philosophy when our focus of attention should be on Höfler? The answer is his influence on Höfler. A second question might be. why not make Meinong the focus of this chapter instead of Höfler? The answer is that Höfler through his textbooks has probably been much more influential among the mass of educated Germans and Austrians of his time than Meinong, whose ideas are even difficult for professional philosophers to penetrate. Furthermore, there was much interaction between Mach and Höfler concerning education, physics, and philosophy than between Mach and Meinong. Nor should we forget Alois Höfler's role in organizing and helping to select lectures and discussions for the University of Vienna Philosophical Society.

3. HÖFLER'S PHILOSOPHY

Because so little has been written on Alois Höfler's own philosophy and he was such a prolific writer, one must regard the following description as preliminary and conditional. Parts of it will have to be qualified or revised in the future as more publications and correspondence are taken into account. Nevertheless, a pioneer work, especially in English, may help suggest the philosophical treasures which can be found, even if this first attempt of ours is surely to be superseded by many later investigators. We have based our understanding of his philosophy primarily on what Höfler wrote from 1906 to 1909.[27] After that time, his focus on problems in theory of education seemed to dominate most of his thought. We have also read many works from many other periods, but they do not go into as much detail concerning philosophy. In other words, the description we provide does not include enough about possible *evolution* in his thinking which may have occurred even if he wrote or published little about it. This caveat has to be added, especially because it was after 1909 that the idea that scientific psychology could replace philosophy more or less collapsed in central Europe and we have had trouble finding out how Meinong and Höfler reacted to this development.. Hopefully, subsequent research will correct this omission.

If we want to start with a thumbnail sketch of Höfler's philosophical *sources* we can turn to an anonymous account in volume II of the *Österreichische Biographisches Lexikon 1815-1950*, p. 353:

> Höfler took from Franz Brentano the conception of the essence of judgment, evidence as the criterion of truth, and his essential point of view in ethics. From Meinong he took his theory of objects and his view of emotional presentation. And from Christian von Ehrenfels he would develop his own ideas on Gestalt perception, especially the characteristic of transference in order in his last years to grasp the Gestalt concept of music.

But we have chosen to concentrate on Höfler's own publications, an approach which is difficult largely because of changes in terminology from that era to ours. First, let us distinguish between typical German and Anglo-American terminology on its most crucial sore point. Namely, that we often regard as epistemological what Germans and Austrians have traditionally considered to be ontological or metaphysical. In short, especially for us, there is a key difference between *direct* realism where one acknowledges that an object can be sensed or made conscious but exists independently and an *indirect* realism where the object is both independent of *and always beyond* what can be sensed or made conscious (to us).[28] The reason Austrian and German philosophers tend to ignore the epistemological distinction is because they have accepted some of Kant's misleading re-redefinitions of terms like "Idealismus", "Dogmatismus", and "Metaphysiik".[29] Unable to find definitions of "Wissenschaft", "Philosophie", and "Metaphysik" in Höfler, we must conjecture. One view is that he thought science was "value free" but by identifying value with feeling, perhaps following Schopenhauer, Brentano, or Meinong, he considered metaphysics and especially aesthetics *more important*, but this interpretation is not fully compatible with the idea that philosophy would eventually become scientific psychology.

Alois Höfler considered both physics and psychology not only to be *Wissenschaften* or sciences but empirical sciences. But he also allowed for the possibility of non-empirical sciences such as logic and mathematics, which he considered to be "a priori sciences". Indeed, some of Höfler's most interesting work took place in those two fields, and his book on logic which would be republished many times was revised by him shortly before his death to escape its status as a textbook and in the process ballooned to over 900 pages. We cannot claim to have mastered it in detail, but its mere existence meant that Höfler was faced with another problem. In addition to attempting to reconcile physical science with psychological science, or mind with matter., he also had to relate and try to incorporate "a priori science" into his insufficiently clarified understanding of science.

Some philosophers like Frege and Husserl would speak of math, logic, and their components as ideal or idealized, a view which naturally evolved into treating numbers and logical relations as universals, while Mach and Avenarius thought they could be derived from experience. Höfler, however, followed Kant by treating them as apriori categories of the mind, though he may have been tempted by Meinong's theory of unreal objects.

Nor was that the end of his "metaphysical" problem. He would have to stretch his "covering theory" to include a quasi-realm as well. Physics as a discipline was becoming increasingly mathematical and theoretical, or in his terminology it was becoming more "hypothetical". But how should one classify logic and mathematics as components of physics, that is, as cases of applied logic and applied mathematics? If one sticks to doctrines of logic and mathematics as aspects of thinking, as how we mentally use them, then they would have psychological characteristics, but if we reduce them to "propositions" and formal components of equations and scientific laws, then there is a split between meaning and form with the first empirical or hypothetical and the second a priori, which is not a very satisfactory solution. Höfler like Newton would have liked more to "deduce" physical understanding from sensory appearances than rely on hypotheses in physics, but unlike Newton, Höfler seemed unable to verbally confuse induction with deduction. He simply allowed that empirical and theoretical science were different, that the latter was gaining strength, and that more science than ever would probably have to be called "hypothetical" regardless of how likely its soundness or truth would be or become.[30] One may suspect that Höfler's pessimism about the mind-matter problem and difficulties in reconciling empirical and a priori science reflect the influence of Emil du Bois Reymond whose gloom and doom comments about "seven unsolvable problems in science" were quite influential during the 1870's and 1880's in German-speaking countries, but this would just be speculation. On the other hand, it is interesting that they still haven't been solved to this day.

Two other words employed very differently in English and in German-speaking lands were "evidence" and "*Evidenz*". Following Brentano, Höfler seemed to think that only what was evident could be *Evidenz*, whereas in England and America, evidence is a term used in law courts and has a very wide meaning, namely, anything which affects the *probability* something is true or false, right or wrong, or real or unreal. Thus, in the English-speaking world evidence includes much more than is evident. Höfler, however, did have a concept of probability and though he did not tie it with *Evidenz* he seemes to have allowed that the truth of argumentation could be more or less likely or probable, and as if argumentation could include much that the English-speaking world call "evidence".

Höfler was best at comparing different views in terms of their strong and weak points and then offering a provisional judgment. But he often becomes so involved in the ideas of other people and his own preferences are often so hidden, that it is often difficult to tell what his "real" views are. But in a sense this is the mark of the good pedagogue he was widely reputed to be, though it can be frustrating for we scholars writing almost a century later. Needless to say, he seems to be an excellent source for determining what the climate of Central European opinion was about physics, psychology, logic, and philosophy in his time.

Understandably, since the work of Russell and Whitehead on logic only occured a half-dozen or so years before his own death he was not in a position to fully appreciate it, but he tried, as his revision of his Logik whioh extended it to over 900 pages showed. As a mathematician he was competent on the technical side, but his interests were overwhelmingly toward logic as a guide to helping pupils, students, and teachers think in a more rational and reasonable way. It is in this sense and perhaps only in this sense that it would be fair to allege that he was guilty of "psychologism" with respect to logic. Nor should one expect anythiung else from an educator. One might argue that "a priori categories of the mind" also could be called "psychologistic", but this does not seem just since they are not psychological in an empirical sense, which is what Frege and Huserrl mainly seem to have had in mind when they authored the criticism in the first place..

Because human beings are fallible and physical theories come and go, Höfler tended in his more sceptical moods to think of the realm of physical entities as a middle stage between the world of appearances and "things-in-themselves". But in his more optimistic moments he tended to take a more tolerant attitude toward the "hypothetical" existence and character of atoms, molecules, and other physical "realities", perhaps because they seemed to have energy and causal influence, which are indispensable characteristics in the practical world as well as applied science Nor did he think that it would be possible to dispense with the whole range of disciplines which interested him from physics, epistemology, metaphysics, psychology, mathematics and logic. From our perspective, what he called "metaphysics" has to be held most important for him since it represented the only possibility for unifying all forms of empirical and a priori science as well as physics, psychology, logic, and mathematics. He would not have accepted our tendency to identify his philosophy with metaphysics, but that was because he thought that everything potentially scientific in philosophy would eventually become part of psychology. His view of metaphysics was not just that it was unscientific, but that it was both necessary and incapable of being made fully reliable in an objective sense.

His most original "metaphysical" insight seems to have been keeping physics and philosophy apart as if philosophy was evolving into scientific psychology and as if physics and scientific psychology could be both sound even if drastically different. This apparent discovery thereby seemed to legitimize metaphysics as something different from science but necessary if we are to understand how physics and psychology are related to each other and hence how mind and matter are connected. But his position alleging that it is possible for physics to do without philosophy introduces problems. One suspects that like his contemporary Wilhelm Ostwald that he somehow defined "hypothetical" in two different ways. He calls atoms and molecules "hypothetical" but metaphysics is also "hypothetical". For Ernst Mach, "hypothetical" was a kind of euphemism for unreal, but for Höfler scientific hypotheses like the reality of atoms and molecules seemed highly probable to him. Presumably, "metaphysical" hypotheses were less probable.

But Alois Höfler who seems to have identified his own philosophy with a number of principles and guidelines rather than a system never seems to have worked out a fully explicit system. The scientific disciplines of physics, psychology, and logic/mathematics remained isolated pillars. Metaphysics was supposed to add the roof (or if one will the foundation) of the temple of knowledge, but there is little evidence that he worked it out.

Perhaps, it was pessimism about its possibility which discouraged him and led to his surprise when shortly before his death (about 1920) he would be asked to include "his philosophy" in a major project of the time edited by Raymund Schmidt which would be titled: *Die Deutsche Philosophie der Gegenwart in Selbstdarstellungen.* (German Philosophy of the Present in Self-Representation). Höfler's contribution would be in Volume II, pp. 117-169. Other volumes would follow and it would be an especially valuable work for historians of philosophy up to the present. But Höfler should not have been surprised at being asked to contribute. Incomplete as it was, he had a philosophy and it helped to guide his many contributions to much educational theory and to how to teach several particular fields.

Höfler apparently feared that *if* physicists based methodological views once again on their epistemological and ontological assumptions as they had often seemed to do in the time of Descartes and Newton, that phenomenalist and idealist philosophers with a superior understanding of psychology as Berkeley, Hume, and Kant have often been alleged to have had would again undermine or circumscribe sound physical science. While he himself leaned toward a much qualified Newtonian indirect realism and mind-matter dualism in ontology but not in philosophy (as he understood the term), he felt that physicists would have to understand psychology much better than they did if they were to successfully protect physics from the "attempts at reform" authored by phenomenalist scientist-philosophers like Mach and Ostwald, that is, to subordinate physics to their restrictive conceptions of epistemology and philosophy of science.

Nor should we overlook the fact that while the leading philosophers of science of the period like Mach, Duhem, and Poincaré would seem to make grievous mistakes by opposing atomism and relativity theory, that Höfler by becoming more tolerant toward "hypotheses" in physics and thus being less doctrinaire was able to shift views on most controversial subjects at least in physics as the weight of evidence accumulated, though perhaps because of his caution he does not seem to have been a pathbreaker in physics and was even rather slow to accept the work of Einstein and Minkowski. (He liked to make jokes about Einstein as "ein Stein" i.e. as one stone, while wondering whether it was better to be hit on the head by one stone, a Newtonian apple, or Machian elements.)

In 1906 he still seemed to treat the ether theory as a possible option in physics, but by 1915 he was giving lectures and leading discussions on early versions of Einstein's general theory of relativity,[31] but in his normal way without fully accepting it. But such flexibility in one's outlook if it allows for shifts in line with the *Zeitgeist* is perhaps advantageous for a leading educator who would remain discussion leader of his philosophical society for almost three decades. But this does not mean that he ever became a relativist about truth. He was not, but he did use the term "hypothetical" in a rather broad sense.

Concerning his interesting theory of energy, he seems to have felt almost certain that *judging* and *desiring* required some form of mental energy in addition to measurable physiological or physical expenditure of energy. Indeed, Höfler showed considerable interest in the philosophies of Arthur Schopenhauer, Friedrich Nietzsche, and Eduard von Hartmann, the former two men concerning the Will and the latter for some aspects of his little-known philosophy of science. But while Höfler's psychology-based causal approach seemed foreign to the various types of energy employed in physics and chemistry and quite unlike Ostwald's reductive Energeticism, in fact, Höfler's identification of energy with an abstract *capacity* for work was compatible with both mental and physical conceptions of energy without specifying an ontological status for energy. But what clever thinking gained, may well have been lost though insufficient informativeness. What did he mean by "capacity"? Can what is merely conditional or dispositional ever be fully clear or real?

Höfler also criticized the notion of "substance". One might suspect that this reflects the influence of Kant, Meinong, or even Mach. But a much simpler explanation would be that everything empirical and physical is changing. It evolves, and is more process than literally particle or object. Also, he liked to say that "substance" referred to a category and had a formal rather than a factual reference.

Höfler's views seem to imply that there are distinct types and gradations of energy or causal "capacity" with mental energy being the most rare and important, and that tests to measure one type or gradation will commonly not measure the others, which also suggests that we need a larger vocabulary instead of calling them all "energy". He was not happy with the word, but does not seem to have suggested the substitutes needed.

For many years, from at least 1890 to 1907, Höfler attempted to help philosophy become more scientific by carrying out psychological experiments and by promoting lectures and discussions of a psychological and physiological character in his philosophical society.

His favorite slogan concerning physical science was: "We need a physics without substantiality but with causality".[32] There is clearly a Kantian touch here, since in German "Kausalität" means law-determined while a cause as a force or pressure is closer to "Ursache" in that language. If everything is changing and possesses energy, then the notion of substance as if itwere something unchanging would be inappropriate. Höfler followed Kant in accepting "the causal law", i.e. the notion that everything has a cause and that like-causes will have like-effects, and he even identified complete or full causes i.e. "deterministic" causes with laws and necessity, which also sounds very Kantian, but his main concern was apparently with what he called *Teilursachen* or partial causes. This is understandable, since in the real world we are not aware of all of the relevant variables, that is, all of the causal factors or links in the causal chain. For Höfler, these partial causes possess energy and when combined with all other relevant *Teilursachen* "necessarily" produce a particular result or effect. Thus, from his point of view, by making action the primary reality and energy its primary "capacity" (abstract as it was), he was apparently trying to connect common sense causal explanation with the very different causal understanding of many physicists.

Physical objects have often been thought of as a combination of qualities and relations, but without the quality of energy they seem to remain at least physically unreal, but if energy itself is merely a capacity for work, then physical reality is still not a substance, even if it possesses causal potential or existence, unlike "bundles" of qualities and relations with no energy which possess merely non-causal properties. One can see in Höfler's apparent reasoning a certain dependence on Alexius Meinong's ideas, especially concerning "subsistent relations" and "unreal objects".

Höfler as a very busy, overworked person through much of his life and though fair to most other points of view, tended to spoil his writing with hasty asides and insufficient elucidation of his own perspective. But for all that, he was a responsible thinker of careful judgment, and if he erred in his understanding it was almost always on the side of caution, but this is a fault which almost always wears well over time, hence, it is quite appropriate for philosophers, scientists, and historians to try to learn more about him, his ideas, and his influence.

Was Alois Höfler - who, given the extent of his writing on other subjects, actually philosophized a great deal - frequently *too* cautious in "hiding his own light under a bushel"? He selflessly promoted the ideas of his friend and mentor when he had a philosophy of his own. His reticence is hard to explain merely on the ground of modesty or shyness or even caution. It may well have been the case that he actually believed that he was deeply indebted to Meinong's philosophy as scientific psychology, and that it was right to consider him merely Meinong's disciple. But as a physicist, logician, and even as a mathematician it is clear that he held many ideas which escape that designation.[33] The next section will mention more about his philosophical principles.

4. HÖFLER AND "THE SOCIETY"

While there are different accounts of the original motivation for founding the University of Vienna Philosophical Society with most of the participants emphasizing the influence of Franz Brentano's evening colloquia[34] there can be no doubt about who was the first leader and how he put his imprint on the Vienna Society. The first members, who were almost all current or former students of Brentano, nominated Alois Höfler. The time was late 1887 and early 1888. Alois Höfler, who had not even obtained an *Habilitation* or permission to lecture at the University at that time, alleged he was too busy as a full-time teacher at a Viennese Gymnasium to accept, but his friends insisted, and he finally with considerable reluctance accepted the role of *Obmann* or leader. In organizing the Society he opted for monthly meetings, starting in October and usually ending in June, but most important the meetings not only would represent many different points of view within philosophy but about half of them or more would not be primarily in philosophy at all but in a wide variety of fields ranging from math and physics to physiology, medicine, history, music, and art.

We do not know exactly why Höfler was chosen, but one presumes that his broad interests, skill as a moderator, and friendship with many of the participants were factors. One must also note that most of his philosophical articles were prefaced by a careful consideration of other points of view before putting forward his own. Perhaps even his unwillingness to be dominated by Brentano entered the picture. For all of his attempt to consider himself a follower of Meinong, he was clearly his own man. Concerning Brentano, that dynamic leader would couple his departure for Italy in 1894 with a somewhat critical publication: "Last Wishes for Austria".

Alois Höfler's broad interests helped make him an ideal leader of the University of Vienna Philosophical Society. His major goal was to combine diversity of subject matter and point of view with enough opportunity for audience participation to help as many people as possible philosophize and become increasingly interested in philosophy. It is true that because of his belief that philosophy could be transformed into scientific psychology that he may have given more emphasis to experimental psychology on the one hand and to physics and philosophy of physics on the other than is normal in philosophical societies. But in the process he also helped lay the groundwork for the tremendous development of psychology and philosophy of science which would take place in Vienna in the first decades of the 20th century.

But in retrospect, maybe the connection with Meinong was primary. The leading members of the Society who occupied all of the important posts for many years were almost all closer to Meinong than to Franz Brentano, both personally and philosophically. And since Höfler was regarded as Meinong's closest follower and Meinong was in Graz and not in Vienna, then Höfler in spite of not possessing a university appointment was the "logical" choice. But this is conjecture.

Who were Höfler's fellow officeholders in the Society at the start or for the longest periods of time, that is, for assistant leader, secretary, treasurer, and librarian? They were Edward Fechtner, Franz Hillebrand, Josef Klemens Kreibig, Hans Schmidkunz, Kasimir Twardowski, and Christian Freiherr von Ehrenfels,[35] with Kreibig, Twardowski, and Ehrenfels as Höfler's strongest support, though almost all of these men would eventually have appointments outside of Vienna.

After three semesters, Alois Höfler gave up his position as leader. It had simply been too time-consuming. Since the only full professor of philosophy in Vienna at that time was Robert Zimmermann, this elderly professor became *Obmann* until his death in 1898[36] when the position was returned to Höfler, who by this time had acquired a Habilitation in the same year (1894) that Brentano had left Vienna for retirement in Italy. Höfler received his right to teach at the University of Vienna in education and philosophy. This honor probably made it easier to elect him as assistant leader in 1895 and ready to take over if something happened to Zimmermann..[37] But it seems that from the earliest days of the Society that Höfler was and would remain the most frequent leader of discussions, which may well have been the most important practical position in the Society.

He also gave numerous speeches commemorating great philosophers and dead members. He also gave many commemorative speeches, published numerous obituaries, and discussed the ideas of numerous thinkers, living and dead. He commemorated Immanuel Kant and Gustav Fechner (twice), Theodor Meynert and Friedrich Nietzsche, and in the Society lectured or led discusions on the ideas of Albert Einstein, Felix Klein, Alexius Meinong, and Robert Reininger (his successor). He was, so to say, a "people-person" much more than one would expect from a mathematician and physicist, and his articles on physics, psychology, logic, mathematics, cosmology, and philosophy often refer to people as much as to ideas.

This is probably connected to his identification of people with "partial causes", as if one must understand a person's psychology to comprehend either his ideas or his actions in order to grasp why things influenced by people actually happen. In fact, one of the singular and most successful policies of the Society had been and continued to be until Höfler's death in 1922 the introduction of non-lecture sessions where after a few introductory remarks by a speaker presenting a controversial thesis the whole evening was opened up for free discussion. It seems that Ehrenfels along with Alois Höfler was the first to introduce this approach in 1892. In fact, it soon became so popular, that one or two additional evenings would often have to be opened up to additional informal debates on the same topic. Ehrenfels, Kreibig, and Höfler all seem to have been skilled at thinking up controversial topics which would arouse audience participation. Eventually, the number of discussion sessions would equal almost half of the number of lecture sessions, of which many would also be followed by discussions. In short, everybody could be or at least play the role of philosopher. Höfler would be accused of making the discussion sessions last longer than necessary,[38] but his apparent insistence on allowing everyone to speak and contribute naturally made them stretch out in time.

Christian von Ehrenfels would also help interest Höfler in Gestalt theory in psychology, which Höfler's lectures, books, and textbooks on psychology would help disseminate. It is not clear how the lecture and discussion topics were chosen in the Society or by whom before 1922, but one can trace about one-third of them to a particular interest which Höfler had, but given his breadth and range of interests this by no means represented a restriction of serious proportions.

In addition to most of the Society's lectures and discussions on math, physics, education, psychology, and philosophy of science one can detect his hand on the lectures and discussions by followers of Meinong and those about Meinong's ideas. In addition to Alois Höfler's own lectures and the discussion sessions devoted to Meinong's philosophy, other students of Meinong such as Kreibig would lead at least eighteen discussion sessions, Ehrenfels would give at least seven lectures and lead seven discussions, and the following friends of Alexius Meinong would all give one or more lectures: H. Pichler, K. Zindler, E. Martinak, and E. Mally.[39] Oddly, Alexius Meinong himself never lectured or led discussions in the Society but he did along with several followers from Graz attend the 25th year celebration on November 13, 1913 and perhaps unbeknownst to us attend one or more of the sessions.[40] Mach remained a *regular* member until at least 1913 as did Sir Karl Popper's father. Both Husserl and Meinong remained *corresponding* members for as long as we have membership lists, which alas, is only until 1913. Edmund Husserl who would open up his own philosophical society in Göttingen, like Alexius Meinong does not seem to have lectured at all in the Vienna Society. Unlike Meinong, it seems that there were no lectures or discussions on Husserl's ideas there until after Höfler's death. Sir Karl never mentions the Society. Did he know that his father had been a member?

Franz Brentano also missed out. Though he could well have inspired the Society in the first place including Meinong, and he did give six lectures before he departed for Italy in 1894, he appears to have declined to be or remain a corresponding member. Given the interest of Brentano, Meinong, and Höfler in both experimental and descriptive psychology, the numerous lectures on these topics given under Höfler's three reigns could be said to partly represent his influence, but after Franz Brentano left there were no lectures or discussions on Franz Brentano's own philosophy or of his close followers such as Anton Marty and Oskar Kraus. Only well after Alois Höfler's death did the Society finally recognize Brentano again by lecturing on his ideas.[41]

In 1903, Alois Höfler was finally offered a professorship in a university. It was in education but at the German University in Prague. When he left Vienna, the Society in gratitude made him honorary president for life. The real and honorary promotions, however, seem to have spurred him to write more, not just on pedagogical subjects but also on philosophy. Meinong was also quite active at this time, and Höfler's philosophical articles while ostensibly rather descriptive or reconciliationist between Meinong's views and those of other commentators also presented his own views of that period. For example, let us quote Alois Höfler from "Die unabhängigen Realitäten" which as already mentioned was published in 1907. The first four theses in this paper openly represent the views of Meinong and Oelzelt-Newin, but theses five and six are his own. His use of quotation marks for his two theses probably represent the kind of caution for which he became so well-known in Vienna.

> Thesis Five: "The physical external world, represented as cause of our physical phenomena".

He then adds in a footnote: "I was once asked, whether I didn't mean 'psychical'? No, in no way do I identify all mental phenomena as 'causes'...." [But as mentioned above Höfler did regard Judging and desiring as having causal energy and influence.]

> Thesis Six: "There is intercourse (*Kommerzium*) between the Ego and metaphysical realities which are independent of it, but where the I or self is in turn dependent on those physical phenomena a kind of causal relation is produced in which we have thought to replace the time component by a metaphysical trans-temporal relation."

But to return to history of the Society during the absence of Höfler from 1903 to 1907, his replacement as *Obmann* turned out to be a literal disaster. The University of Vienna Philosophical Society picked Friedrich Jodl, who had taken Zimmermann's chair at the University after the old man had died in 1898. But Friedrich Jodl had several axes to grind. First, he wanted to make the Society exclusively philosophical in its character. He was tired of physicists and psychologists running matters. Mach and Boltzmann, the two leading physicists in Vienna, both seemed to dabble in philosophy as had Alois Höfler. Their influence had to be reduced. Second Jodl was an active, anti-clerical humanist strongly opposed to theological contributions to the Society, and in his way of thinking even though Brentano had left the Catholic Church and was no longer a priest he still represented in his mind "Medieval obscurantism" and his influence had to be rooted out.

But Jodl like many outsiders had apparently failed to notice the split within Brentano's supporters between the close followers of the master and independent spirits like Husserl, Meinong, and Höfler. In short, he failed to see that the Society - if it could be said to have any particular orientation - was not dominated by Brentano or by religion but by the ideas of Meinong and Höfler who strongly favored science. Brentano's direct influence seemed to be as fully excluded as that of Husserl.

And third, back to Jodl, who was also the leader of an ethics society in Vienna, he tended to be adept at using high moral principles to justify low behavior or at least strong criticism which hurt the feelings of other people. The main event under his leadership took place in January 1905 when Ludwig Boltzmann who had taken over Mach's philosophical lectures in the University of Vienna gave a devastating attack in the Society on the ideas of Arthur Schopenhauer.[42] In the discussion afterwards Jodl valiantly defended the great pessimist and regretted that Bolzmann had been hired to teach philosophy, since it had been under the false impression that he would bring physics and philosophy closer together.[43] In fact, Jodl argued that if life is as painful as Schopenhauer maintained then suicide was legitimate as was using philosophy to justify such a negative act. Boltzmann replied that if Jodl wanted to commit suicide then he should do it, but that he could never prove it was rational.[44] As is well-known, Boltzmann himself committed suicide the next year, that is, in September 1906.

Less-known is that Jodl never gave another lecture in the Society after that year and that Höfler who was still in Prague seems to have been almost immediately made discussion leader again. But instead of being an informal position as it had been previously, it was now apparently recognized as an official one in the Society, as if Friedrich Jodl were now excluded from conducting sessions. If the Society's documents are correct, Höfler resumed this task still in 1906,[45] that is even before he was officially invited back by the University of Vienna to become Professor of Education in 1907. Jodl, however, refused to resign as *Obmann*, and when he became Dean in the University of Vienna he even refused to allow Höfler to give lectures on the philosophies of Mach and Boltzmann as Höfler had planned to do.[46] Later deans permitted Alois Höfler to give philosophy lectures, but the bitterness between the two men remained. When Jodl finally resigned as *Obmann* of the Society in 1912 it was offered once again to Höfler who then assumed his typical triple role of *Obmann*, discussion leader, and honorary president of The University of Vienna Philosophical Society which he had helped found. He seems to have continued to exercise all three positions including his university professorship until his death in 1922.[47]

Höfler's primary focus slowly began to shift with his professorship in education in 1907. He increasingly tried to make clear that his first scientific loyalty was to education in physics and psychology and not to philosophy. This accorded well with Mach's view of him as primarily an educator and most people in Vienna increasingly honored him as a successful writer of textbooks for *Realschule* and *Gymnasium* students. His contributions to pedagogical theory, demonstrated teaching ability, renowned tolerance in the Society, and his attempt to reconcile diverse opinions both verbally and in writing also won him honor and praise, even if he still felt sensitive over the behavior of Jodl as well as a former principal of a Gymnasium who declined to reduce his teaching load in order to allow him to finish some of his books. But Höfler had other problems, especially when his view that philosophy was being reduced to scientific psychology was increasingly challenged, an endangered position shared by Wilhelm Jerusalem and in part by Ernst Mach himself.

The final blow came when many psychologists themselves accepted behaviorism and rejected "introspective psychology". The new approach by ignoring or denying the existence of consciousness tended to presuppose a peculiarly naive form of direct realism. Did Höfler shift with the period or did he continue to support the independence of physics from philosophy primarily on the ground that scientific philosophy was confined to scientific psychology? Our main source from 1920, his autobiography, suggests that he *probably* retained his earlier view, but he certainly ceased to give it prominence. Furthermore, his strong support for Gestalt psychology seemed at least in part to pre-empt the old controversy.

But Höfler-research and scholarship is still in its infancy, and we are not sufficiently acquainted with the contents of all of his books and textbooks, a monumental task, nor have we read enough of his vast correspondence to make this article as definitive as a book on his ideas will eventually be. We have assumed that Alois Höfler and Alexius Meinong retained their basic beliefs in psychology as science even this late, and Höfler does bluntly say in 1920 that "I see no reason to take anything back from what I wrote in 1897" (p. 136) but did this also apply to his view about philosophy and identifying it with scientific psychology?.

In answer, it does *seem* that Höfler did retain his definition of scientific philosophy if we have properly understood his footnote on page 147 in his autobiography, where he mentions that he repeated it in a major revision and enlargement of his *Logik*, which had not yet been published when Höfler probably wrote this passage (1920). It is also interesting that he has admitted deviating from Meinong on this point, since the Graz philosopher according to Höfler seems to have been reluctant to give any definition of philosophy at all. Still, we have discovered nothing yet which would be inconsistent with his 1906 position on psychology and metaphysics, merely a shift in emphasis in the types of lectures and discussions given in the Society. His views on physics were somewhat affected by Einstein's theories, which he treated in a rather cautious way. But while he was willing to extend the scope of "hypotheses" in physics, this mostly meant treating some views as more probable such as the reality of atoms and less probable such as the existence of the ether. Indeed, he never ceased to remind readers in his book reviews that the ether theory is a theory and like other theories should not be treated as a fact, since doing this would tempt us to exclude the possibility of other theories being true.

It is of course ironic that it was very likely the popularity of the Society which from 1906 to the First World War hovered around 200 paying internal and external members[48] and apparently many more non-paying listeners[49] which later helped to add so many followers to the positivist Vienna Circle of the 1920's which was opposed to Höfler's modest "middle position" in methodology of physics and even more to his underlying indirect realism and mind-matter dualism in "metaphysics".

Höfler gives no sign in any of the writings which we have read that he was aware that he and Meinong were being accused of "psychologism" and tolerating metaphysics, as if from a positivist view that they were somehow "out of date". In fact, by the year 1920 he was accusing Ernst Mach with his phenomenalism and opposition to atoms and relativity theory of being out of date. But something *did* seem to sour his attitude, though possibly connected with the war years and their aftermath. Several long footnotes in Alois Höfler's autobiography written in sections from 1916 but mostly in 1920 were full of untypical complaints, pessimism, and grim innuendoes as well as asides. Indeed, the autobiography combines so much praise of his own accomplishments and criticism of people who hindered his academic career as to leave a most unwelcome taste in our mouths. *Surely this could not have been the same man who had obviously been such a popular leader of the University of Vienna Philosophyical Society!*

Höfler had long placed great emphasis on writing fast and publishing as much as he could, but gradually he began to realize that this might well be counterproductive, Apparently, by his middle sixties he felt that the end was near and that his contributions were not being properly recognized, and he wanted to set the record straight. His friends like Kreibig and Meinong were either dead or would die before his autobiography was published. Alois Höfler seems to have doubted in those first miserable postwar years that any one else would remember him long, and for much of Germany and Austria he was correct, though the hundredth anniversary of his birth would be commemorated conspicuously in Vienna in 1953.

The Vienna Circle of the 1920's apart from Neurath rarely mentioned him, even though it was made up very largely of philosophically inclined mathematicians and physicists. He may have remained in the philosophical shadow of Meinong too long. But one must also keep in mind that positivists and many young left-wing people *still* thought of almost everyone who had ever studied under Brentano[50] as "reactionary" and "anti-science", which was not the case, and was very misleading given Höfler's lifelong dedication and contributions to several sciences.

Alois Höfler may have felt he was slipping away.[51] While the ostensible date of completion of his autobiography was April 19, 1920, which was a few months before Meinong died, he kept inserting additional remarks into the autobiography until shortly before his own death on February 26, 1922. This same year would see Hans Hahn, a mathematician, former member of the Society and lectuer therin, and one of the more radical Machists anywhere, invite Moritz Schlick to Vienna. This former student of Max Planck would soon start his Circle and (not always willingly) help revive epistemological Positivism. Unfortunately, Höfler's autobiography, which barely touches on the details of his own philosophy, is the only one of his writings which is still relatively well-known. Höfler does, however, in his autobiography repeat at least the following outline of his late philosophical credo:

> Natural science is completely independent from philosophy; in many ways philosophy is dependent on natural science.
>
> I. I believe that next to the physical knowledge (*Erkennen*) of *real* objects that there is also mathematical knowledge of *ideal* objects.
>
> II. I believe, that physical knowledge is *empirical* and that mathema-

> tical knowledge is *a priori.*
>
> III. Both types of leading concepts: *real - ideal* and *empirical - a priori* which all processes of knowing make use of and which is examined by epistemology, take us beyond the physical-mathematical realm into what is more generally philosophical, namely to the doubled characteristics of object and method of the most general philosophical disciplines: *metaphysics* and *theory of objects (Gegenstandtheorie).*[52]

But within these quotations one *may* detect three basic positions which Alois Höfler had long defended so tenaciously. First by denying that physics had anything to do with philosophy he was *probably* still identifying the field of philosophy according to one of his definitions with an empirical *Wissenschaft* which became genuinely scientific in his opinion only to the extent that it became scientific psychology. Second, while he defended the reality and importance of "metaphysics", he tried not to let it influence his understanding of either physics, psychology, mathematics, or logic. And third, he apparently let Emil du Bois-Reymond's despair that *science* could ever solve "the seven puzzles"[52.1] including mind-matter duality, empirical-apriori duality, and consciousness-unconsciousness duality, discourage him from seriously trying to solve them by means which seemed to be outside of what he understood as science. Even if the only hope for a rational solution was "metaphysical" in the sense of transcendent, this failed to motivate him to make the required "metaphysical" effort, hence perhaps his acceptance of the fact that he was not really a philosopher, if one were to mean by that designation not a scientific psychologist but a "metaphysician". As for the numerous monistic approaches which he thought ducked the basic problems proposed by du Bois-Reymond, like Ernst Mach's "elementism" and that of his followers, he seems to have had little use for their apparent escapism at all. He preferred to live with honest problems rather than with fake solutions.

Nevertheless, he was an admirer of Gustav Fechner[53] who had helped to originate "elementism" as well as the so-called Weber-Fechner law which seemed to begin to make psycho-physics a science, but he was well aware of its shortcomings and of the need, contrary to Mach's opinion, to still make "metaphysical" assumptions when performing psycho-physics. He sems to have remained optimistic about the hope of psychology becoming a reliable science, and to the best of our understanding even after the *Zeitgeist* turned against it, he still saw the future of philosophy as evolving into scientific psychology à la Meinong. How he could maintain any distinction between philosophy and "metaphysics" may seems obscure, but apparently he still wanted hard to believe that there was a difference between scientific and unscientific philosophy. Edmund Husserl was of course beginning to think of Phenomenology as a science, but Höfler preferred Meinong's views to those of Husserl, and few people in any recognized science took Husserl's claim seriously anyhow. So-called "Intuition" was not science. Philosophy might still be called a *Wissenschaft,* a broad term, but it was clearly not an *exakte Wissenschaft,* except perhaps some aspects of formal logic. Husserl, who at one time had focused on mathematics and logic seems to have lost his grasp that what made science trustworthy was its *public* verifiability and practical utility, not private "intuition" or talk about "intersubjectivity".

Strange as it may seem in light of Höfler's intense opposition to phenomenalism and epistemological positivism, was the fact that he almost certainly contributed significantly to their revival by stressing the importance of physics and by denying that it had anything at all to do with philosophy. Indeed, the common denominator in the opinions of all three of the major Viennese philosophers of science of the period, Mach, Boltzmann, and Alois Höfler was that all three men were physicists by profession and all three maintained that one could do good work in physics without being concerned with philosophy. This attitude naturally helped influence younger men, especially students of mathematics and physics in Vienna and beyond, that the proper goal was to develop a methodology of physics as if it could apply to all of the sciences and as if it did not need epistemological and ontological assumptions. To be sure, there was both much ignorance of their own assumptions and wishful thinking about "presuppositionless" methodology.

Indeed, even some opponents of phenomenalism such as Moritz Schlick before he came to Vienna and Viktor Kraft in Vienna[54] would be caught up by the new physics-first movement in philosophy of science, which had clearly been stimulated by the new developments in theoretical physics led by Planck and Einstein earlier, and during the 1920's by Bohr, Heisenberg, Schrödinger, Pauli, and many others. Surely, identifying philosophy with methodology of physics seemed timely with the hope that it might also strengthen the methodological approaches of other sciences as well. Such hope did not spring eternal, but at least it lasted for several decades.

But neither Mach, Boltzmann, nor Höfler could have imagined how far the new iconoclasts would go. What Höfler's progeny i.e. Hahn,[55] Frank,[56] and Neurath[57] would do, especially under the influence of Wittgenstein and Carnap, would be, first, to reject what they all called "metaphysics" (which Höfler had defended as necessary given du Bois-Reymond's scientifically unsolvable problems), and second, they rejected his notion of philosophy as scientific psychology as being "psychologistic", an expansion of what young Husserl had been insisting upon for twenty years (even though the same criticism could be directed at Husserl's own approach, as indeed it was by Ernst Mach and especially by his follower Wilhelm Jerusalem), and third, the new positivists tried to restrict philosophy of science to methodology of physics *while still calling it philosophy of science as if there were no kind of "meaningful" or "rational" philosophy outside of methodology of physics*. These radical positivists as led by Rudolf Carnap were able to popularize this very narrow view by denying that it needed any indirect epistemological and ontological assumptions, first by claiming that all such distinctions were "merely linguistic", and second, by rejecting indirect epistemology as itself being "metaphysical".

But most conspicuous of all, many members of the Vienna Circle and later Logical Positivism tried to shrink the scope of "meaningful" reference, which Meinong and Höfler had tried so hard to open. The new generation of Machists or as they preferred to be called "empiricists" labeled what could not be verified or falsified in direct sensory terms, "meaningless", and while later withdrawing the charge still tried to make the scope of reference almost entirely subordinate to rules and definitions, as if reference were not a real process and relation whose actual character could only be determined by extensive investigation into the psychological facts.

To avoid giving the erroneous impression that the Philosophical Society completely collapsed after Höfler's death in 1922, let us add that while the Society seems not to have published any membership figures after the First World War that the new *Obmann*, Robert Reininger, long the assistant leader under both Jodl and Höfler, decided to re-orient the Society more toward professional, largely idealist philosophy. Indeed, while Höfler had written on Kant's philosophy and has sometimes been considered to be a "neo-Kantian himself, Reininger went further and made the Society an *Ortgruppe* of the *Kant-Gesellschaft* in the late 1920's. One might add that the journal of the Kant Society once gave the figures of its Vienna branch as being over 300 members, which suggests they were counting people who attended sessions but didn't pay dues. Nevertheless, even without Höfler and his discussion sessions it still seems to have been a very popular society. Interestingly, even Harvard University's philosophical society was listed as an *Ortgruppe* of the *Kant-Gesellschaft* at the time.

Perhaps because of Robert Reininger's emphasis on formal lectures by local and professional German philosophers and reduction in the number of discussion sessions to virtually zero, many of the Society's former members who liked to talk like Hahn, Frank, and Neurath and even the less talkative like Viktor Kraft gradually sought to organize their own discussion circle. Schlick's willingness to chair such a group then led to the Vienna Circle.

5. MACH AND HÖFLER

Mach's path often crossed Alois Höfler's not because they agreed on very much in the field of philosophy, which they did not, but because they respected each other as educators and both Austrians had broad interests which included physics, psychology, education, music, and philosophy. Mach became a conspicuous figure in educational circles with a lecture before the *Realschulmännerverein* in Dortmund on April 16, 1886. He said what his audience wanted to hear, namely, that more physical science was needed in the schools and less Greek and Latin.[58] During the next year, an educational journal for physical science, *Zeitschrift für den physikalischen und chemischen Unterricht*, was founded in Berlin with Dr. F. Poske as editor and Mach and Dr. B. Schwalbe as co-editors. In that same year Mach co-authored a textbook for the lower classes of middle schools[59] and four years later alone for the upper classes,[60] that is, books for what were called in German *Realschulen, Realgymnasien,* and*Humanistischegymnasien*. He also published two physics textbooks on the university level in 1891.[61] Thiele, has listed eight textbooks partly or wholly prepared by Ernst Mach which totaled over 30 editions by 1917.[62] Mach also wrote thirteen articles for Poske's journal from 1888 to 1896,[63] mostly it seems on teaching techniques and laboratory instruments, but almost all were quite short.

Alois Höfler, who had already been preparing a textbook for classes in physics and would do the same for logic, psychology, and other subjects began to correspond with Mach. Adolf Hohenester in his article *Ernst Mach als Didaktiker* (1988) has quoted from several of the 35 extant letters from Mach to Höfler from 1887 to 1891.[64] Both men, while quite enthusiastic about the prospects for more physics in middle schools were depressed by the apparent reluctance of Austrian authorities to approve their textbooks, But the grass was greener in Germany and after a decade or so the textbooks of both men became widely employed in the Austrian half of the Habsburg Empire as well, that is, in Bohemia and Moravia as well as Austria per se. It is not clear if they or translations of them were also widely used in Austrian Poland, that is, in Galicia, though Mach's philosophical ideas were already becoming known in some university circles in Cracow as well as in Prague. After Poland became independent again in 1918, one of Höfler's textbooks would be translated into that language.

We have already discussed in the introduction to this book how there was an apparent split between German and non-German physics textbooks from mid-nineteenth century until about 1920. Whereas those outside of Germany, especially those by Adolphe Ganot were translated worldwide, and seem to have taken the reality of atoms and molecules for granted, those inside Germany, especially textbooks by Mach, Höfler, and Pfaundler tended to regard atoms as merely "hypothetical". Part of the problem was apparently over the term "hypothetical" itself. Whereas outside of Germany and Austria, what seemed very probable was usually treated as real, in Austria, at least as Alois Höfler allowed in the first textbook he edited (1881), *everything* probable was "hypothetical" as if experience as the only alternative to what was hypothetical was absolutely certain.[65] It would be very difficult to say how widespread Alois Höfler's textbook definition was in the German and Austrian empires as a whole, but if it was repeated or assumed in most of Alois Höfler's textbooks, then it could have been very widespread indeed.

Though Ernst Mach may have published a few more editions of physics textbooks than Höfler (though this is not certain because of discrepancies in how we count editions) Höfler published more initial versions in physics and many more textbooks in terms of both versions and editions outside of physics. But with respect to apparent popularity they may have been roughly equivalent in physics, though this is also difficult to determine because of the absence of figures on total books sold..[66]

On the other hand, because Höfler largely revised his own textbooks and Mach did not, it is possible that Höfler's philosophical *assumptions* may have had a larger overall impact, at least in disciplines outside of physics per se. What is conspicuously lacking, apart from Alois Höfler's one translation into Polish, is *verified* translations into other languages. Nevertheless, Table #1 is a list of those textbooks of theirs which went through at least one further edition. But unfortunately, while we have data on Ernst Mach's books which were prepared for *different types of schools*, we have no equivalent breakdown for Höfler's textbooks but only an overall number of editions. In short, if there were also separate editions of Höfler's books for different types of schools, then the number of his editions may be double or triple the figures below. Table #2 is a list of textbooks which seem *not* to have been republished. Table #3 is a list of Alois Höfler's "didactic" books to help teachers teach better. In a late letter to Höfler (January 27, 1916) Ernst Mach mentions a textbook presumably by Höfler titled *Didaktik des physikalischen Unterrichts*, but we have not yet found a reference to it elsewhere. If it existed it would complete a triad of books for teachers which Höfler had been working on since he became a professor of education in Vienna in 1907.

Textbook Table #1

Ernst Mach[67]	Alois Höfler[68]
1) *Grundriß der Naturlehre für die unteren Classen der Mittelschulen.* Prag, 1887 for Gymnasien 7 editions to 1910. for Realschulen 7 editions to 1910 (authors: E. Mach & J. Odstrcil) (reviser: K. Habart from 1896)	1) *Grundlehren der Logik* (I. Theil) Wien & Prag, 1890. 8 editions to 1926. (author: A. Höfler) (reviser: A. Höfler)
2) *Leitfaden der Physik für Studierende.* Prag & Wien, both books 1891. for university & teachers, 2 editions (authors E. Mach & G. Jaumann 1st ed.) (reviser: E. Mach, 2nd edition 1891)	2) *Naturlehre für die unteren Classen der Mittelschulen.* Wien, 1893. 5 editions to 1915 (authors: A. Höfler & E. Maiß) (reviser: G. Schilling 1915)
3) *Grundriß der Naturlehre für die oberen Classen der Mittelschulen,* Prag, 1891. For Gymnasien 1 edition; Realschulen 1 ed. (author E. Mach)	3) *Grundlehren der Psychologie.* Wien, Prag, & Leipzig, 1897. 10 editions to 1930 (author: A. Höfler) (rev.: Alois Wenzel, 1930) XV + 642

4) *Grundriß der Physik für die höheren Schulen des Deutschen Reiches*, Leipzig, 1893. 1. Theil: 6 editions to 1910
2. Theil: 6 editions to 1917.
(author: E. Mach)
(revisers: F. Harbordt & M. Fischer from 1893)

5) *Grundriß der Naturlehre für Mädchen-lyzeen*, Wien, 1913.
3 editions, all 1913.
(authors: E. Mach & K. Habart)
(reviser: Gallus Wenzel, 1913)

4) *Hundert psychologische Schulversuche*. Leipzig, 1903
4 editions to 1918.
authors: A. Höfler & S. Witasek)

5) *Grundlehren der Logik und* Psychologie Wien, 1903.
8 editions to 1921.
(author: A. Höfler)

6) *Naturlehre für die Oberstufe der Gymnasien*, Braunschweig und Wien, 1903. 2 editions to 1910.
(aut: A. Höfler, E. Maiß, & F. Poske)

Textbook Table #2

Other textbooks by Höfler which do not seem to have been republished or at least not with Höfler as author, editor, or reviser.

1) *Schabus' Anfangsgründe der Naturlehre - Zum Gebrauche an den unteren Classen der Mittelschulen*, Vierzehnte Auflage, neu bearbeitet und mit Übungen von Alois Höfler, Carl Gerold's Sohn: Wien, 1881 (erste Auflage 1854). VIII + 256 pages.

2) Alois Höfler unter Mitwirkung von Alexius Meinong, *Philosophische Propädeutik. Logik*. F. Tempsky: Prag & G. Freytag: Leipzig, 1890. VIII + 244 pages.

3) Alois Höfler unter Mitwirkung von Alexius Meinong, *Philosophische Propädeutik. Zehn Lesestücke aus philosophischen Klassikern*, F. Tempsky: Wien & G. Fretag: Leipzig, 1890. VI + 40 pages.

4) Alois Höfler, Eduard Maiß, Friedrich Poske, *Hilfsbuch zur Naturlehre für die Oberstufe der Gymnasien, Realschulen und verwandter Lehranstalten, enthaltend Zusätze aus der angewandten Mathematik, der Logik und der Psychologie und 80 Leitaufgaben*, F.Wieweg und Sohn: Braunschweig & K. Gerold's Sohn: Wien, 1904. 93 pages.

5) Alois Höfler, Eduard Maiß, Friedrich Poske, *Physik mit Zusätzen aus der angewandten Mathematik aus der Logik und Psychologie und mit 230 physikalischen Leitaufgaben*, F. Vieweg und Sohn: Braunschweig, 1904. XXXI + 966 pages.

Table #3

"Didactic" books to help teachers teach better:

1) Alois Höfler, "Zur gegenwärtigen Naturphilosophie", *Abhandlungen zur Didaktik und Philosophie der Naturwissenschaft*, edited by E. Poske, A, Höfler, and E. Grimsehl, erster Band, Julius Springer: Berlin, 1906, pp. 63-196.

2) Alois Höfler, *Drei Vorträge zur Mittelschulreform*, W. Braumüller: Wien & Leipzig, 1908. V + 167.

3) Alois Höfler, *Didaktik des mathematischen Unterrichts*. B.G. Teubner: Leipzig & Berlin, 1910. XVIII + 509 pages.

4) Alois Höfler, *Didaktik der Himmelskunde und der Astronomischen Geographie*, B.G. Teubner: Leipzig & Berlin, 1913. XII + 414 + 26 pages.

As mentioned earlier, when Mach called atoms "hypothetical" this seems to have meant that he thought their existence was improbable or unreal, but in Höfler's textbooks he appears to have allowed that "hypothetical" meant that the existence of atoms and molecules might be probable, even very probable. In short, Even when asked to revise Schabus' textbook (1881) he tried to introduce topics in physics in an empirical and *anschauliche* way like Mach would later do in his textbooks, but unlike Mach he would at the end of each part or chapter shift from sensory description to causal explanation, often in terms of atoms and molecules, but which he regarded as "hypothetical" regardless of how probable the existence of these invisible particles were. But he also admitted that causal explanation was lacking for many physical phenomena.

His "hypothetical" approach also seemed to be rather prudent given the widespread hostility to the reality of atoms and molecules by so many main German-speaking thinkers including Kant, Hegel, Schopenhauer, and even Nietzsche. But the more extensive treatment which Alois Höfler gave to the "atomic hypothesis" compared to what Mach would do in his textbooks would support our tentative conclusion that Höfler was distinctly more "atomism-friendly" than Mach, at least from the 1880's on.

Another difference between Mach and Höfler over textbooks apart from the fact that Höfler wrote them as well as "didactic" works for teachers on many disciplines outside of physics, was that by the middle of the 1890's Ernst Mach was handing new editions over to other writers to continue what he had started. Höfler, on the other hand, almost always continued to prepare new editions himself, though there was some revising and reworking of material, especially in books published after his death.[69] We should add that the three lists above are not complete, partly because it is not clear whether some books are textbooks or not. For example, he was working on a huge book on logic shortly before his death, which presumably he hoped would be treated as more than a textbook, though it is often listed as a second edition of a textbook originally written in 1890 which was only about a fourth as long.

Höfler, who by the 1890's had still not yet received a professorship in a university himself, would contribute much to Mach becoming known in Vienna even before Mach was invited to occupy "Franz Brentano's Chair" in the Austrian Capital in Fall 1895.[70] Mach had given a talk "The Principle of Comparison in Physics" at a convention of natural scientists in September 1894.[71] Höfler found it so stimulating that on December 13 of that same year he led a discussion session in The Philosophical Society on Mach's lecture which he continued on January 24, 1895 and still again (under J.K. Kreibig) on March 14, 1895. These events seem to have influenced several faculty members of the University of Vienna to recommend that Ernst Mach fill Brentano's Chair which had been left open for many years. Since Brentano - who is alleged to have jealously discouraged other applicants - had just left Vienna for Italy, it was felt that the right time had come to appoint a new philosopher. So Mach was chosen.

But Mach as a professor only taught in Vienna for three years because of a stroke which paralysed half his body in 1898 and which forced him to abandon teaching. He asked to be relieved from his university duties and go into retirement in 1901. In that same year he also began criticizing some of the physics of his colleagues in education, namely F. Poske and A. Höfler, that is, at roughly the same time he resigned as co-editor of *Zeitschrift für den physikalischen und chemischen Unterricht*. First, in Mach's resignation notice he recommended Alois Höfler as his replacement as co-editor as if to counter-balance criticism of his physics, a kind gesture which Alois Höfler would remember and appreciate. Second, speculation might suggest that Mach resigned either because of his stroke and the paralysis of half his body or because of the criticisms he had made or was about to make concerning the ideas of his colleagues. And third, F. Poske himself wrote many years later (1916) that "Mach resigned in late 1901 . . . [because] he believed that his activity in science was taking ever more time such that he could not devote as much attention to the journal as it deserved . . . " We have suggested several possible explanations. Future research, however, should narrow this number down.

Höfler had published a work in 1900 titled *Studien zur gegenwärtigen Philosophie der mathematischen Mechanik* ("Studies about the Current Philosophy of Mathematical Mechanics").[72] The earliest reply from Mach that we have been able to pin down is the English-language 1902 edition of the 1901 *Die Mechanik* (translated as "The Science of Mechanics"). The criticisms of Höfler's physics, however, would be included in all of the later English editions of Mach's book up to the present. Unfortunately, many of the criticisms overlook that Alois Höfler appears to have frequently meant something other than what Mach thought he meant, which is a natural result of the two scientists having very different epistemological and ontological assumptions. Höfler's view that physics has nothing to do with philosophy was clearly false, at least about basic assumptions and particularly in so far as communication between physicists was concerned. Let us quote five criticisms while adding a few comments.

> It is not correct as Alois Höfler asserts that [my] definition [of mass] tacitly assumes *one and the same force* acting on both masses. It does not even assume the notion of force, since the latter is built up subsequently upon the notion of mass, and gives them the principle of action and reaction independently and without falling into Newton's error.[73]

The natural response to Mach's claim is that if the notion of force is not even assumed, then we are dealing with kinematics and not dynamics, that is, with description and neither causes nor causal explanation, which seems to trivialize his "interactive" definition of mass.

> Our instinctive craving for concepts involving quantities of things, to which Höfler is doubtless seeking to give expression, and which amply suffices for every-day purposes, is something that no one will think of denying. But a scientific concept of "quantity of matter" should properly be deduced from the proportionality of the single physical quantities mentioned, instead of contrariwise, building up the concept of mass upon "quantity of matter".[74]

If the concepts of matter and mass are to be distinguished, then if energy or potential force is attributed to matter, then it is quite appropriate to define mass as "quantity of matter". But Ernst Mach in phenomenalist fashion is of course unwilling to attribute such "metaphysical" things as potential force or energy to matter, hence he believes that defining mass as mere "quantity of matter" is empty, circular, and should be replaced, that is, even if Newton had given another definition but one still within Newton's basic views.

> Poske and Höfler are unable to give their assent to my [Mach's] interpretation of this passage, for the reason that Galileo does not expressly refer to the limiting case of transition from the inclined to the horizontal plane; although Poske grants that the consideration of limiting cases was frequently employed by Galileo, and although Höfler admits having actually tested the educational efficacy of this device with students.[75]

This passage does not include enough information to settle this dispute between the three physicists or even to know what it was about, but it does indicate that Mach the phenomenalist was capable of employing and relying on inference more than the "metaphysical realists" Poske and Höfler, a result which epistemologically is not supposed to happen.

> Höfler is of the opinion that the existence of "absolute motion" is denied, because it is held to be "inconceivable". But it is a fact of "more painstaking self-observation" that conceptions of absolute motion do exist. Conceivability and knowledge of absolute motion are not to be confounded.[76]

The problem seems to be that the English-language terms "concept" and "conceive" are sometimes used to help refer to what can be *imagined* by us and sometimes to what can be *understood*. Alois Höfler apparently correctly means that absolute motion cannot be literally imagined in the strict sense of imaged and Mach seems to correctly reply that it can be understood. Hence, given this interpretation, both physicists were right, and Ernst Mach merely misunderstood Höfler's use of language.

Since Ernst Mach could read English and usually checked English editions of *Die Mechanik* we are assuming that the words "concept" and "conceive" are bote reasonably accurate substitutes for the original German words and contained a similar basic ambiguity. But there is also an interesting philosophical point, namely, that Mach seems to allow for the possibility of understanding some things such as "absolute motion" which could not be literally observed and presumably were not real. This suggests that he was capable of employing a liberal theory of meaning and reference, much more liberal than that of many of his followers who supported the Logical Positivist's so-called "empirical criterion of meaning".

> Finally, when Alois Höfler brings forward the argument that "in every relative motion one at least of the bodies moving with reference to each other must be affected with absolute motion", - I can only say that for the person who considers absolute motion as meaningless in physics, this argument has no force whatever. But I have no further concern here with philosophical questions. To go into details as Höfler has in some

places would serve no purpose before an understanding has been reached on the main question.[77]

This last quote is the most interesting and for several different reasons. First, it confirms that Höfler *was* distinguishing between inability to literally imagine or image "absolute motion" with a distinct ability, not only to partly understand it but even to accept it as real. Second, It makes clear that while Mach allowed - *unlike* most Logical Positivists - that "absolute motion" was understandable that very much *like* later Logical Positivists he also held that it was "meaningless" in physics. Now it may strike some observers that there is a contradiction between absolute motion being both understandable and meaningless. But the real solution at least in this editor's opinion is that absolute motion is imaginable, understandable, *and even real* when the medium is influenced by the moving object, whether it is air, water, or an "energy plenum" as in general relativity. Human problems in measuring such motion cannot refute its reality or make it "meaningless".

And third, there is the very strong likelihood that what Höfler meant by "absolute motion" was quite different from what Mach "understood". Like most "realists" Höfler probably thought that even if there were no people that the physical universe and the particular spaces in it would continue to exist and hence in that sense that space as a combination of all particular spaces was real and could be called "absolute space". Furthermore, given such absolute space, then motion which could be measured at least in principle from the "edge" of the universe could be called "absolute motion". Clearly, such a conception and use of terminology was foreign to Mach, who both called such motion "relative" and tended to think of physical space as consisting of relations between sensations and of motions as relative in the sense of being related to sensations such that any motions without such relations were in his extreme terminology "meaningless". In short, epistemological and ontological differences made it difficult for the two physicists to understand the same things by the words they used, thus causing miscommunication.

In 1903 recommendations from several scholars including Mach helped Höfler finally receive a university chair.[78] However, it was not in philosophy which Höfler was very much engaged in at that time but in education in the German University in Prague. Mach's recommendation was repeated in 1907 when Alois Höfler was interested in returning to Vienna for a chair also in education. This would allow Alois Höfler once again to take charge of most of the discussion sessions in the University of Vienna Philosophical Society. Furthermore, Poske seems to have finally followed Mach's suggestion and made Höfler co-editor of his education journal.[79]

After Boltzmann's death in 1906 and Alois Höfler's return to Vienna in 1907 the new Professor of Science Education wanted to give a lecture course which compared the philosophies of Ernst Mach and Ludwig Boltzmann, the two best-known Viennese physicists. But unfortunately as we have already mentioned, Friedrich Jodl, who had been appointed head of The Philosophy Society during Höfler's absence in Prague and who was also a prominent Dean in the University hierarchy denied Alois Höfler's request, alleging that Höfler's new position in Vienna was in education and not in philosophy,[80] and presumably because Höfler had lost his earlier position as a *Privatdozent* in philosophy by leaving Vienna for Prague. But as we also mentioned, later deans of the University permitted Alois Höfler to lecture on the philosophies of Mach and Boltzmann and when Friedrich Jodl resigned his position as *Obmann* of The Society Höfler was immediately elected or rather re-elected to replace him and remained leader and honorary president until his death in 1922, though it was presumably his ability to handle discussion sessions fairly and wisely (which Jodl seems to have had trouble doing) which was probably paramount.

Nor should we forget that after Alois Höfler received his Vienna chair his interests naturally focused much more on problems in education and the teaching of science. He obviously still considered philosophy to be important as his top post in the Society and his continued assistance to Meinong who was becoming increasingly blind indicated, but Alois Höfler's philosophical concern now seemed to be much more on how teachers of physical science should handle philosophical opinions and controversies than in extending his own philosophical principles and guidelines or in attempting to link them together into a system. He often criticized the very idea of a philosophical system, but the acute reader will already have noticed that Höfler's various writings do suggest the outline of one.

His fellow textbook writer and editor Leopold Pfaundler had also been somewhat troubled by philosophical controversies which seemed to affect physics. In September 1905 in Volume I of the textbook he had been editing for thirty years and which had gone through ten editions, he added a long footnote to the first page of text.[81] The wish he expressed was that physics textbooks and physicists should abstain from philosophical matters altogether, but the words he employed surely bewildered many readers. The first half of his footnote was a mixture favoring both indirect realism in epistemology and avoiding philosophical issues. But the last few lines came close to direct realism and even phenomenalism while still urging physicists to ignore philosophy. In short, Pfaundler's warning failed, but another writer of textbooks took up the cause.

Höfler wrote a long article on Mach's philosophy which was published in 1910.[82] It was ostensibly on Mach's *Erhaltung der Arbeit* (1872) which had just been re-published (1909). He suspected that the re-publication of Mach's first philosophy book which was without any revisions except for a new preface was somehow connected with the *Polemik* which was then taking place between Mach and Planck. This realization changed his article into a sharp debate with himself about whether it was appropriate to mention the polemics, or the philosophical disagreements which underlay it, in physics textbooks. But while his long discussion with himself wandered as widely as had Leopold Pfaundler's relatively short note, Höfler finally did approach a coherent conclusion.

Alois Höfler's effort was aided by his belief which went back to Franz Brentano that philosophy would gradually evolve into scientific psychology and his own ideal that physics should have nothing to do with philosophy. He also seems to have retained his view that what he called "metaphysics" or "unscientific philosophy" still had value as a way to relate scientific physics with scientific psychology, or if one will, "mind" with "matter". He even saw that the increasing emphasis on theory and mathematics in physics was not compatible with Ernst Mach's sense-based experimental approach which was based on description of the appearances, a view that theories were transient, and deep distrust of all hypotheses. But since Höfler felt that physics could be "philosophy-free"[83] he apparently thought that he ought to be able to be truly neutral between the philosophies of Mach and his opponents. As for methodology of science, like Ludwig Boltzmann he tended to identify it with science and not philosophy, which in the real world created a major problem, since the distinction between philosophy and "right method" in science is not always clearly drawn and it may be seriously questioned if methodology without philosophical assumptions is possible. Nevertheless, Höfler would do his best to be both neutral between philosophical factions and advise teachers of physics on how to carry this out.

Max Planck had rejected Ernst Mach's phenomenalism and his theory of economy in a 1908 lecture and in a published version in 1909.[84] Mach had answered by re-publishing *Erhaltung der Arbeit* (which in spite of the title "*Conservation of Work*" was mostly a philosophy book) which included a short reply in the Foreword to Planck. But just a few months later Mach replied again and this time much more forcefully, ridiculing Max Planck's understanding of philosophy and repeating his opposition to the reality of atoms.[85] Planck was so angry and infuriated that he wrote a more severe criticism of Mach's ideas which was written in late 1910 and came out in early 1911.[86] Mach's response was to encourage two active followers Friedrich Adler and Joseph Petzoldt to continue the debate against Planck. This they did and in 1917 Philipp Frank defended Mach once more against Planck.[87]

It is doubtful if Höfler was familiar with Planck's second article or even with Mach's second reply at the time he intervened into the controversy with his "neutral" article for educators (January 1910), but he tried to set the following guidelines for what to include and what to leave out concerning this and like disputes. First, it was clear that physics was at a turning point where more theory and more explanation were being emphasized and that textbooks would have to come to terms with this development. Second, Ernst Mach's notion that textbooks should start with what was experiencable and seemed concrete and easily understandable was appropriate, especially for beginning students. It was very important to distinguish between what was hypothetical and what could be experienced. The approach of starting with hypotheses and then deducing "propositions" about particulars, regardless of how useful in theory-focused "research", was unsatisfactory as a teaching technique. To this extent Mach seemed surely correct (though it is not clear if Höfler still held an extremely broad view of hypotheses which he seems to have done twenty-five years before, i.e. as if everything probable or even very probable should still be interpreted as being merely "hypothetical").

On the other hand, however, many of the physical phenomena predicted by theory which were neither sensory nor experiencable, such as atoms and electrons, were being supported so strongly by objective instruments and measurement, that it might be unreasonable to neglect or oppose their reality and probable characteristics. One should not let excessive caution prevent textbooks from keeping up to date, especially given the major changes which were then happening in physical science.

In other words, as theoretical physics was appearing to become more important than experimental physics, especially with the work of Maxwell, Boltzmann, Planck, and Einstein - though the matter seemed far from settled by 1910 - Höfler was trying to adjust. But this required overcoming an old prejudice, namely, that experience and observation were certain and indirect methods were not. In the real world both were conditional and were subject to different degrees of probability or relative certainty. Höfler was beginning to approach the current view that objective measurement by machine which can be checked by many observers should generally be treated as more reliable than any single person's conscious or sensory impressions, even though instrument measurement is more indirect from the perspective of the conscious inquirer. Public evidence must be preferred to private evidence in science, whether one is discussing experimental or theoretical physics. In fact, private evidence unsupported elsewhere is merely hearsay for *other* people, regardless of how direct or certain it seems to the particular person who experiences the private evidence. Many current philosophers often talk about "intersubjective evidence" but literally there is no such thing. Nor is meaning attributed to language or to other symbols literally objective. But instrument-based evidence is objective and if confirmed by many people is the best we can do, especially if in so-called Popperian terms it can survive serious attempts to refute it. Also, allowances must be made as Boltzmann suggested for the idealized character of most theoretical physics, idealization which makes exact correspondence with reality as it exists in particularized, historical terms impossible.

Höfler seems to have decided it was unwise to discuss the Mach-Planck polemic as such in textbooks. Also, one should not even discuss the issue of "science versus metaphysics". Neither Planck's "defense of metaphysics" nor Mach's "anti-metaphysics" had a place in science textbooks. *There should be no mention of metaphysics at all.* In short, Alois Höfler praised Mach the educator but increasingly rejected Mach the epistemologist, philosopher, and physicist. Scientists may well have epistemological and ontological views and assumptions like everyone else and these can influence their work and what they mean by scientific terms and definitions, but arguing about them is more likely to waste time and embitter relations between scientists than to solve fundamental differences, desirable as the latter surely is. Ultimately, science would determine which epistemology was correct, if it had not already done so. Nor is discussing epistemological controversies the most suitable use of teaching-time available for physics teachers in middle schools or in undergraduate university classrooms. To sum up, Höfler in his article gave reasonably coherent advice to future physics textbook writers, much more coherent than Pfaundler who had had roughly the same intentions four years previously, though without any help or "guidance" from the Mach-Planck fireworks.

Höfler wrote another article on Mach in 1916 but as an obituary. By this time while still respecting Mach's experimental work in physics and his ideas on how to teach beginning pupils and students, he increasingly felt that the future of physics probably lay more with theoretical physics and belief in the reality of atoms and molecules than with Ernst Mach's sensation-focused methodology and philosophy of science.[88] But he preferred in his article to emphasize their shared views on how to teach physics.

He concluded with regret that he was forced to share the opinion that Mach's philosophy had gone much too far. He had permitted his love of "phenomena" to become a phenomenalism which fully excluded "noumena", that is, the real world.

He then quoted from a late letter Mach sent, apparently to the journal editor F. Poske, which transcended physical and philosophical differences.

Dear Honored Sir: Haar near Munich, January 27, 1916

Thank you for what you sent me [*Didaktik des physikalischen Unterrichts*][89] from which I can determine how deeply you have mastered the material. That we differ on so much, seems to me to be nothing less than bad luck, for even in these things there are different paths which can be found to reach what is reliable and practical. I was never persuaded of my own infallibility and have passed through many a polemical controversy to remind me of it. On this occasion I would like to emphasize that your [educational] journal has given me much pleasure over the last years, because I noticed again and again how instruction and methodology have crystalized into something new and better. But you are right that a new age is coming, and it pleases me that it will find you at the top of your capacity . . . Please live well and let yourself be heartily greeted from your servant[90]

Dr. Ernst Mach[91]

NOTES

1 Alois Höfler, "Lebenslauf des Alois Höflers", in *Die Deutsche Philosophie der Gegenwart in Selbstdarstellungen*, Zweiter Band, Hrsg. Raymund Schmidt, Felix Meiner: Leipzig, 1921, p. 118. The key publications for all research on Alois Höfler are: *Verzeichnis der Veröffentlichungen von Alois Höfler*, edited by Karl, Otto, and Wolfgang Höfler, *Wiener Zeitschrift für Philosophie, Psychologie, Pädagogik*, Heft 3, Band 4 (1953) 165-181 and even more basic: Ulf Höfer, *Verzeichnis des Nachlasses von Alois Höfler (1853-1922), Teil I*, Forschungstelle und Dokumentationszentrum für österreichischen Philosophy, Graz, 1994. 250 pages.

2 Ibid. Höfler 1921, pp. 120-121.

3 Alois Höfler, "Franz Brentano in Wien", *Süddeutsche Monatschrift*, 4 (1917), 320. See also: Alexius Meinong, "Leben und Streben", in *Die Philosophie der Gegenwart in Selbstdarstellungen*, edited by Raymund Schmidt, second edition, Verlag von Felix Meiner:
Leipzig, 1923, pp. 102-103.

4 Höfler 1921, op. cit., p. 121.

5 Meinong became a full professor in Graz in 1889, but like Husserl in Germany had had to wait longer for the government to approve his appointment than he liked. Indeed, bitterness about delayed appointments was a common phenomenon of the period, and Höfler who had to wait the longest of the three became perhaps most bitter, at least it seems so on the basis of his autobiography. Apparently it was difficult even for leading philosophers to be philosophical about such matters. Maybe only independent scholars fully outside of the academic "rat race" could become truly philosophical in the old Hellenistic sense of peace of mind or *ataraxia*.

6 Alois Höfler, *Logik*, 1. Auflage, Tempsky: Wien, 1890.

7 (Alois Höfler), *Rückblick auf die ersten fünfundzwanzig Vereinsjahre der Philosophischen Gesellschaft an der Universität zu Wien mit dem 26. Bericht für das Jahr 1912/13*. In Selbstverlage der Philosophisches Gesellschaft: Wien, 1913, p. 4.

8 Richard Meister, "Die Geschichte der Philosophischen Gesellschaft an der Universität Wien 1888-1938" in Robert Reininger (Hrsg.), *50 Jahre Philosophische Gesellschaft an der Universität Wien 1888-1938*, Verlag Philosophischen Gesellschaft an der Universität Wien: Wien, 1938, p. 13.

9 See **1)** *Logik*, first edition, Tempsky: Wien, 1890, 244 pages, A major revision occurred in 1921 with 936 pages; **2)** *Psychologie*, first edition 1897, with 600 pages; The original text went through many editions. **3)** *Physik*, with 966 pages, published by Vieweg in 1904. Höfler also prepared much shorter versions as textbooks for *Gymnasium* and *Realschule* students..

10 Alois Höfler, "Die Philosophie des Alois Höfler", *Die Deutsche Philosophie der Gegenwart in Selbstdarstellungen*, herausgegeben von Dr. Raymund Schmidt, zweiter Band, Felix Meiner: Leipzig, 1921, p. 121. [This and all subsequent references to Alois Höfler 1921 are to this article which in spite of the title is more autobiographical than philosophical and includes the *specifically* autobiographical remarks in footnote #1. Subsequent references are *not* to the second edition of his *Logik* which also came out in 1921.]

11 Ibid., p. 124. See also (Höfler), 1913, p. 47.

12 See Alexius Meinong, "Über die Erfahrungsgrundlagen unseres Wissens", *Abhandlung zur Didaktik und Philosophie der Naturwissenschaften*, herausgegeben von Poske, Höfler, Grimsehl, Springer: Berlin, 1906, Band I, Heft 6, 113 Seiten.

13 Alois Höfler in order to honor Meinong's 60th birthday gathered together Meinong's previous publications and with the help of other former students published them in 1913-14 in the form of a two-volume *Gesamtausgabe*. For Meinong's reaction see the following letter: Alexius Meinong to Edith Landmann-Kalischer, Graz, November 8, 1912 in Rudolf Kindinger (Hrsg.), *Philosophenbriefe aus der Wissenschaftlichen Korrespondenz von Alexius Meinong*, Akademische Druck- u. Verlagsanstalt: Graz, 1965, pp. 187-188.

14 Alois Höfler, *Psychologie*, first edition 1897, with 600 pages.

15 Meinong 1923, op. cit., p. 104. (See footnote #17.)

16 Alois Höfler, "Die Unabhängigen Realitäten,"*Kant-Studien*, **12** (1907) 361-392.

17 Alexius Meinong, "Leben und Streben", in *Die Philosophie der Gegenwart in Selbstdarstellungen*, edited by Raymond Schmidt, second edition, Verlag von Felix Meiner: Leipzig, 1923, p. 105.

18 Höfler 1921, op. cit., pp. 122-123. (See footnote #10.)

19 Unfortunately, the German language seems to have no well-known *noun* form for the English term "reference" in the sense of what people think *about* or talk *about*. Franz Brentano's rersurrection of the Scholastic term "Intention" in German has a very different meaning than the English word, is not a natural German expression, and is hard to find in all but the largest non-specialized German dictionaries, such that there is an artificial air about his theory, especially when he talks about mental experience having an "Intention" rather than people intending something. In particular, it seems to objectify a mental characteristic as if it were independent of human choice and aim, which it is not or need not be.

The *English* word "intention" means *Absicht* in German, and "to intend" in English means *beabsichtigen* in German, But Brentano rarely uses these normal German words. To "refer to" in English is often translated in German with *beziehen auf* or *sich beziehen auf*, but the German terms are closer to "relate to" than "refer to", and only what is living can *literally* do the latter. In English, reference is normally understood as a mental action as well as a relation, and the doer of the action is the person who does the referring. Reference is most commonly *volitional* in English. We can *choose* what to refer to, whether the intended target exists or not. The best German equivalents are probably *bezug auf* and *Bezugnahme*.

Largely because of these differences and the problems they create it is hard to handle the question of "theory of reference" in German, but it is fundamental in Anglo-American philosophy, hence we will emphasize it here both for its own value and as a way to help straighten out and clarify some of the problems in German philosophy.

20 Terence Parsons, *Non-Existent Objects*, Yale University Press: New Haven, 1980. More recently see: Dale Jacquette, *Meinongian Logic - The Semantics of Existence and Non-Existence*, De Gruyter: New York, 1996.

21 Bertrand Russell, "Hard and Soft Data" in *Perception and the External World*, edited by R.J. Hirst, Macmillan: New York, 1965.

22 Meinong 1923, op. cit., p. 112-114. (See footnote #17.)

23 This is from "The Walrus and the Carpenter", *Alice in Wonderland*, by Lewis Carroll.

24 Meinong 1923, op. cit. p. 116. (See footnote #17.)

25 For a more detailed and formal study of "inexistence" see Dale Jacquette, *Meinongian Logic - The Semantics of Existence and Non-Existence*, De Gruyter: New York, 1996 (See footnote #61).

26 David F. Lindenfeld, *The Transformation of Positivism - Alexius Meinong and European Thought, 1880-1920*, University of California Press: Berkeley, Los Angeles, and London, late 1970's.

27 For basic details of Höfler's philosophy see his long article "Zur gegenwärtigen Naturphilosophie", *Abhandlungen zur Didaktik und Philosophie der Naturwissenschaft*, Verlag von Julius Springer: Berlin 1906, Band I, Heft 2, pp. 3-136; "Die Unabhängigen Realitäten, "*Kant-Studien*, **12** (1907) 361-392; and "Erkenntnisprobleme und Erkenntnistheorie", *Zeitschrift für Philosophie und philosophische Kritik*, **137** (1909), 1-61.

28 The distinction was made quite clear in American philosophy when the "Critical Realist" manifesto of 1920 which was supported by Arthur Lovejoy, George Santayana, and many others defended indirect realism and criticized the "New Realism" of Ralph Barton Perry and others circa 1910 which had alleged that we could have a direct awareness of the external physical world. See the Appendix in this book.

29 The tendency in German philosophy to identify indirect and what Americans at that time called "critical realism" with metaphysics made it easier for Rudolf Carnap and his anti-metaphysical followers to argue during the 1930's that the two main *direct* epistemological positions,

phenomenalism and direct realism, were only "linguistically different from each other", that is, as if *all* epistemological positions were direct in sensory or conscious terms. In short, failure to accept indirect realism as at least an epistemological alternative left all options consciously or empirically *direct* which seemed to encourage linguistic reduction (as if direct epistemology exhausted *all* epistemology), a reduction which became a hallmark of Logical Positivism and distinguished it from earlier types of Positivism, such as Ernst Mach had held.

30 Alois Höfler, "Zur gegenwärtigen Naturphilosophie", *Abhandlungen zur Didaktik und Philosophie der Naturwissenschaft*, Band I, Heft 2, Julius Springer: Berlin, 1906, pp. 94-114. Whereas most positivists had contrasted science and metaphysics as if they were incompatible with each other, Höfler bravely tried to make clear that in terms of our current stage of understanding that we could not relate the physical and psychological realms without it. Physics studied one and scientific philosophy the other, but there was no science as yet of the connection, which for Höfler with his dynamic theory of partial-causes was interactionist.

31 Ibid., p. 53, footnote 57 for the reference to his 1894 support of the energy principle as primary. The first half of the 1906 article is a criticism of Ostwald's Energeticism. See also Reininger 1938, op. cit. p. 33 for mention of his lecture and discussion session (with Philipp Frank) on Einstein's theory of relativity. For Ernst Mach's criticisms of Höfler's views on mechanics, see Section IV of this article.

32 Alois Höfler, "Die Unabhängigen Realitäten", *Kant-Studien*, **12** (1907), 365; and Höfler 1906, op. cit., p. 44. Höfler's rejection of substantiality in the sense of substances, as philosophers have understood them, does not seem to have been related to Ernst Mach's attempt to reduce everything physical to sensations or to relations between them but to John Locke's inability to understand what a substance could be in addition to its qualities (1906, p. 44). But as a physicist he was also aware that physical phenomena were actually processes and not entities and that these processes were related to other processes, hence they could not be substances in the sense of being independent of everything else. But in the physical or non-philosophical sense he preferred, talk about substances was avoided by identifying energy with capacity for work, partial-causes with what has energy, and processes with what uses energy. Höfler relates his defense of causality to a book by Edmund König titled *Die Entwicklung des Kausalproblems*, Band II, 1890, p. 423.

33 Some of the following rather obscure articles by Alois Höfler also include a few of his philosophical ideas which are otherwise difficult to locate: 1) "Erkenntnisprobleme und Erkenntnistheorie", *Zeitschrift für Philosophie und philosophische Kritik*", Band **137**, J.A. Barth: Leipzig, 1909, 61 pages. 2) "Die Ideale des Realismus", *Zeitschrift für den physikalischen und chemischen Unterricht*, **30** (1917) 1-12. 3) "Relativitäts und Erkenntnistheorie", *Zeitschrift für Physikalischen und Chemischen Unterricht*, **35** (1922) 88-90. 4) "Theorie und Hypothese im physikalischen Unterricht", *Zeitschrift für den physikalischen und chemischen Unterricht*, **22** (1909) 50-52. 5) "Über dogmatisches und kritisches Kant-Studium . . . , *Zeitschrift für Hochschulpädagogik*, Heft 1, Jg. **5** (1914) 16 pages. 6) "Relativitätstheorie und Erkenntnistheorie". *Neues Wiener Abendblatt*, Nr. 38. **55**. Jg., February 8, 1921, page 3, Vienna. 7) "Welt Denken und Weltanschauen", *Süddeutsche Monatshefte*, Leipzig/München, May 1921, pp. 133-139. and 8) "Naturwissenschaft und Philosophie". *Sitzingsberichte der Akademie der Wissenschaften in Wien*, Philosophisch-historische Klasse, **196** Band, 1. Abhandlung, A. Hölder, Wien, 150 pages.

34 (Höfler) 1913, op. cit., pp. 3-4 and Meister 1938, op. cit., p. 4. (See Footnotes 7 & 8.)

35 Ibid. (See footnote # 7)

36 Ibid., p. 6.

37 Ibid., p. 36.

38 The criticism came from Otto Weininger in one of his letters to Hermann Swoboda which has been published in a book by Hannelore Rodlauer titled *Otto Weininger - Eros und Psyche Studien und briefe 1899-1902*, Verlag der Österreichischen Akademie der Wissenschaften, Wien, 1990.

39 Robert Reininger, *50 Jahre Philosophische Gesellschaft an der Universität Wien 1888-1938*, Verlag Philosophischen Gesellschaft an der Universität Wien: Wien, 1938, pp. 21-37.

40 (Höfler) 1913, op. cit., p. 10.

41 From 1934 to 1938 Alfred Kastil gave four lectures on Brentano and his philosophy before the University of Vienna Philosophical Society. (See Reininger 1938, op. cit., pp. 44-45.)

42 The Lecture was given on January 21, 1905. It was published both by the Philosophy Society itself and by Ludwig Boltzmann in his book *Populäre Schriften*, Leipzig, 1905, pp. 385-402. The Society, however, in its version added remarks by Jodl and Boltzmann which had taken place after the lecture. See *Wissenschaftliche Beilage achtzehnten Bericht der Philosophischen Gesellschaft an der Universität zu Wien*, Selbstverlag: Wien, 1905, pp. 60-62. (For an English translation of the discussion remarks by Jodl and Boltzmann, see John Blackmore, *Ludwig Boltzmann His Later Life and Philosophy 1900-1906*, Book Two: *The Philosopher*, Kluwer: Dordrecht, 1996, pp. 309-312.)

43 Ibid., pp. 60-61.

44 Ibid., P. 62.

45 (Höfler) 1913, p. 36.

46 Höfler 1921, pp. 125 & 141.

47 Richard Meister, "Die Geschichte der Philosophischen Gesellschaft an der Universität Wien 1888-1938" in Robert Reininger (Hrsg.), *50 Jahre Philosophische Gesellschaft an der Universität Wien 1888-1938*, Verlag der Philosophischen Gesellschaft an der Universität Wien: Wien, 1938, p. 13.

48 (Höfler) 1913, op. cit., p. (38) "Übersicht über die Mitgliederzahl". (See also Blackmore 1995 Book Two, p. 300.)

49 Both Otto Weininger and Friedrich Jodl mentioned the high number of non-paying visitors who would attend sessions of the Society. It was known that Laurenz Müllner, the Professor of Religious Philosophy, declined to become a member of the highly science-focused group. What is less-known is that he nevertheless sometimes attended on a non-paying basis anyhow (See Hannelore Rodlauer, *Otto Weininger - Eros und Psyche Studien und briefe 1899-1902*, Verlag der Österreichischen Akademie der Wissenschaften: Wien, 1990.)

50 It is true that Franz Brentano and some of his close supporters did oppose Darwin's theory of evolution by natural selection and did oppose Einstein's special theory of relativity, but in general Brentano and almost all of his students *strongly* favored science, especially psychology. It is not yet clear where Höfler stood with respect to Darwin, and his views on relativity theory seem to have gradually evolved from initial deep scepticism to conditional acceptance, which was rather common among physicists of the time. But Höfler was certainly more sympathetic to belief in the reality of atoms and molecules than almost all epistemological positivists were then and now. It is positivists and other epistemological presentists who have tended to be "reactionary" on this matter, indeed, often preferring the euphemism "theoretical entities".

51 Höfler 1921, op. cit, p. 44.

52 Ibid., p. 24.

52.1 Emil du Bois-Reymond, "Die sieben Welträtsel" (1880) in *Vorträge über Philosophie und Gesellschaft*, Felix Meiner Verlag: Hamburg, 1974, pp. 159-187. See also his article "Über die Grenzen des Naturerkennens" (1872), pp. 54-77.

53 Höfler gave two lectures in the Society on Fechner. Translated into English their titles were: "Remembering Fechner's Hundredth Birthday", April 24, 1901 and "Remembering the 30th Anniversary of Gustav Fechner's Death", December 20, 1917. [For a complete list of the Society's lectures from 1888 to 1922 in English please see Blackmore 1995 Book Two, op. cit., pp. 277-288.]

54 Viktor Kraft who gave five lectures and discussion sessions in the Society from 1911 to 1917 expressed his opposition to epistemological positivism in a lecture in the Society titled in English "Toward a Critique of Epistemological Positivism", December 6, 1911. The subject was so interesting or controversial that he added two discussion sessions on the same topic, November 20 and December 4, 1911. Kraft is best-known today, however, for his sympathetic history of the Vienna Circle published almost forty years later.

55 Hans Hahn gave lectures or led discussions three times in the Society from 1906 to 1922. All were on mathematics. The first was on geometry and the last two were on Bolzano's paradoxes of infinity. [One might add that the Society after Höfler discovered many unpublished manuscripts by Bolzano in Professor Zimmermann's *Nachlass* in 1898 made it a project to organize and publish them, but it has taken almost a century and it is not clear if the task has been fully completed yet.]

56 Philipp Frank gave lectures or led discussion sessions four times in the Society from 1907 to 1915. The first in English "Hans Driesch's Philosophy of Nature and its Relation to Vitalism" was given on December 4, 1907. The second was titled: "Is There an Absolute Motion?", December 4, 1909. The last two were discussion sessions with Höfler on "The Problem of Physical Relativity", March 22 and April 22, 1915. [Philipp Frank was also an official secretary of the Society for 1911-1912.]

57 For the German titles of Otto Neurath's seventeen lectures and discussion sessions in the Society (not counting those given after Höfler's death) see Reininger 1938, op. cit., pp. 21-36. For English titles see Blackmore 1995 Book Two, op. cit., pp. 277-287.

58 Ernst Mach, *Der relative Bildungswert der philologischen und der mathematisch-naturwissenschaftlichen Unterrichtsfächer der höheren Schulen*, Tempsky: Leipzig, 1886, 29 Seiten. (For an English translation see "On Instruction in the Classics and the Mathematico-Physical Sciences" in *Popular Scientific Lectures*, fifth edition, Open Court: La Salle, 1943, pp. 358-374.)

59 Ernst Mach and Johannes Odstrcil, *Grundriss der Naturlehre für die unteren Classen der Mittelschulen - Ausgabe für Gymnasien*, Tempsky: Prague, 1887. (The seventh edition appeared in Vienna in 1910.)

60 Ernst Mach, *Grundriss der Naturlehre für die oberen Classen der Mittelschulen - Ausgabe für Gymnasien*, Prague, 1891. (His book with the same title except for *Realschulen* appeared in 1892.)

61 Ernst Mach and Gustav Jaumann, *Leitfaden für Studierende*, Tempsky: Vienna and Freytag: Leipzig, both 1891. (A second edition appeared the same year but without the contribution by Jaumann.)

62 Joachim Thiele, "Ernst Mach-Bibliographie", *Centaurus*, **8** (1963) 224-225.

63 Ibid., pp. 208-212.

64 Adolf Hohenester, "Ernst Mach als Didaktiker, Lehrbuch- und Lehrplanverfasser", in *Ernst Mach - Werk und Wirkung*, edited by R. Haller and F. Stadler, Hölder-Pichler-Tempsky: Wien, 1988, pp. 150-152 & 162. (The Deutsches Museum only lists three letters from Höfler to Mach (1894-1914) and those from a period when there are no extant letters from Mach to Höfler. In short, a large number of letters on both sides seem to have disappeared.)

65 Alois Höfler (ed.), "Einleitung", *Schabus' Anfangsgründe der Naturlehre.* Vierzehnte Auflage, Carl Gerold's Sohn: Wien, 1881.

66 See Ulf Höfer, "Handexemplare und Separatabdrucke von Alois Höflers publizierten Schriften", *Verzeichnis des Nachlasses von Alois Höfler (1853-1922), Teil I*, Forschungstelle und Dokumentationszentrum für österreicische Philosophie, Graz, 1994, pp. 1-34. For publications on Mach's didactic work in addition to Thiele see: D. Hoffmann & W. Manthei, "Ernst Mach als Schulphysiker, Didaktiker, und Bildungstheoretiker", in *Ernst Mach - Studien und Dokumente zu Leben und Werk*, edited by Dieter Hoffmann und H. Laitko, Deutscher Verlag der Wissenschaften: Berlin, 1991, pp. 279-315; and Adolf Hohenester, *"Ernst Mach als Didaktiker, Lehrbuch- und Lehrplanverfasser", in Ernst Mach - Werk und Wirkung*, edited by Rudolf Haller and Friedrich Stadler, Verlag Hölder-Pichler-Tempsky: Vienna, 1988, pp. 138-163.

67 Thiele 1963, op. cit., pp. 224-225.

68 Höfer 1994, op. cit. pp. 1-34. (See footnote #1.)

69 See Alois Höfler, *Didaktik des mathematischen Unterrichts* (Didaktischer Handbücher für den realistischen Unterricht an höheren Schulen, Band I), B.G. Teubner: Leipzig und Berlin, 1910, 18 + 509 pages. And *Didaktik der Himmelskunde und der astronomischen Geographie* (Didaktischen Handbücher für den realistischen Unterricht an höheren Schulen, Band II), B.G. Teubner: Leipzig/Berlin, 1913. 7 + 414 + 32 pages.

70 Franz Brentano as an ex-priest was not allowed to marry in Austria, so he went to Germany to tie the knot, but on returning to Austria lost his professorship, but remained as a very influential *Privatdozent* until 1894.

71 Ernst Mach, "Über das Princip der Vergleichung in der Physik", *Verhandlungen der Gesellschaft Deutscher Naturforscher und Ärzte* (Leipzig), Theil I, 1894, , S. 154-156.

72 Höfler's article was a *Nachwort* (afterword) of 168 pages tacked onto the end of a 101 page republication of a book by Immanuel Kant: *Metaphysische Anfangsgründe der Naturwissenschaft* (1786). The combination book was published by J.A. Barth of Leipzig as Band **III** in the University of Vienna Philosophical Society's series of books which Höfler had started when he became *Obmann* again of the Society in 1898.

73 Ernst Mach, "In Answer to Criticisms of the Views Expressed by the Author in Chapters I and II", *The Science of Mechanics*, Open Court: La Salle, 1960, p. 328; Höfler *Nachwort* 1900, op. cit., p. 77.

74 Ibid., Mach 1960, p. 329; Ibid., Höfler 1900, p. 72.

75 Ibid., p. 334; Ibid., pp. 111-112.

76 Ibid., p. 337; Ibid., pp. 120 & 164. But contrast this quotation with the next one.

77 Ibid., p. 338; Ibid., pp. 124-126 & 133.

78 Alois Höfler, "Ernst Mach", Sonderabdruck aus der *Zeitschrift für den physikalischen und chemischen Unterricht*, **29** (1916) 57, footnote 1. [Footnote 1 was added by the editor of the journal, F. Poske.]

79 Friedrich Poske, "Alois Höfler" in *Zeitschrift für den Physikalischen und Chemischen Unterricht*, **35** (1922), 51. [The change in journal heading is responsible for the difference in capitalization of words between this and the 1916 edition.]

80 Alois Höfler, "Die Philosophie des Alois Höfler", *Die Deutsche Philosophie der Gegenwart in Selbstdarstellungen*, Hrsg. von Raymund Schmidt, zweiter Band, Felix Meiner Verlag: Leipzig, 1921, p. 125.

81 Leopold Pfaundler, "Einleitung", *Mechanik und Akustik*, Band I (Müller-Pouillets Lehrbuch der Physik und Meteorologie, zehnte umgearbeitete und vermehrte Auflage in vier Bänden herausgegeben von Leopold Pfaundler). Friedrich Vieweg und Sohn: Braunschweig, 1906, p. 1. An earlier edition of Pfaundler's textbook reflected the influence of Emil du Bois-Reymond's philosophy.

82 Alois Höfler, "Zur Geschichte und Wurzel der Machschen Philosophie", Sonderabdruck aus der *Zeitschrift für den physikalischen und chemischen Unterricht*, **23** (1910), 1-16.

83 Ibid., p. 16, footnote 19.

84 Max Planck, "Die Einheit des physikalischen Weltbildes", *Physikalische Zeitschrift*, **10** (1909) 62-75.

85 Ernst Mach, "Die Leitgedanken meiner naturwissenschaftlichen Erkenntnislehre und ihre Aufnahme durch die Zeitgenossen", *Physikalische Zeitschrift*, **11** (1910) 599-606.

86 Max Planck, "Zur Machschen Theorie der physikalischen Erkenntnis: Eine Erwiderung", *Physikalische Zeitschrift*, **11** (1910) 1186-1190.

87 Friedrich W. Adler, "Die Einheit des physikalischen Weltbildes", *Naturwissenschaftliche Wochenschrift*, **8** (1909), no. 52; Joseph Petzoldt, "Die vitalistische Reaktion auf die Unzulänglichkeit der mechanischen Natursicht", *Zeitschrift für allgemeine Physiologie*, **10** (1910) 69-119. Philipp Frank, "Die Bedeutung der physikalischen Erkenntnistheorie Machs für das Geistesleben der gegenwart", *Naturwissenschaften*, **5** (1917) 65-72.

88 Alois Höfler, "Ernst Mach", Sonderabdruck aus der *Zeitschrift für den physikalischen und chemischen Unterricht*, Zweites heft, **29** (1916) 61.

89 Mach, Höfler, or Poske inserted the title of the book in brackets, but no book with this title has been found which suggests that the reference is actually to Poske's revision of Höfler's textbook *Unterstufe der Naturlehre. Ausgabe A. (Physik nebst Astronomie und Chemie)*. 5. verbesserte Auflage, F. Vieweg und Sohn: Braunschweig, 1916.

90 Höfler (Mach Obituary 1916), op. cit., p. 63.

91 This letter is not to be confused with one which Mach sent to Höfler on the same day (January 27, 1916).

REFERENCES

Adler, Friedrich. "Die Einheit des physikalischen Weltbildes". *Naturwissenschaftlichen Wochenschrift*, 8 (1909) no. 52.

Blackmore, John. *Ludwig Boltzmann His Later Life and Philosophy 1900-1906*. Book Two. Dordrecht: Kluwer, 1995.

Boltzmann, Ludwig. *Populäre Schriften*. Leipzig: J.A. Barth, 1905

Du Bois-Reymond, Emil. "Die sieben Welträtsel (1880)". in *Vorträge über Philosophie und Gesellschaft*. Hamburg: Felix Meiner Verlag, 1974, 159-187. (See also his article "Über die Grenzen des Naturerkennens (1872)" in the same book, pp. 54-77.

Frank, Philipp. "Die Bedeutung der physikalische Erkenntnistheorie Machs fUur das geistesleben der Gegenwart". *Naturwissenschaften.*, 5 (1917) 65-72.

Höfer, Ulf. *Verzeichnis des Nachlasses von Alois Höfler (1853-1922) - Teil I*. Forschungsstelle und Dokumentationszenrum für österreichische Philosophie. Graz: 1994.

Höfler, Alois. "Einleitung". in *Schabus' Anfangsgründe der Naturlehre*. 14. Auflage. A. Höfler (Hrsg.) Wien: Carl Gerold's Sohn, 1881.

Höfler, Alois. *Einige Gesetze der Unverträglichkeiten zwischen Urteilen*. Doctoral dissertation, Vienna, 1885.

Höfler, Alois. *Logik*. 1. Auflage. Wien: Tempsky, 1890. 244 pages. The 1921 edition was expanded to 936 pages.

Höfler, Alois. *Psychologie*. 1. Auflage . Wien & Prag: Tempsky, 1897. 604 pages.

Höfler, Alois, Eduard Maiss, & Friedrich Poske. *Physik mit Zusätzen aus der angewandten Mathematik aus der Logik und Psychologie* . . . Braunschweig: F. Vieweg und Sohn, 1904. 966 pages.

Höfler, Alois. "Zur gegenwärtigen Naturphilosophie". *Abhandlungen zur Didaktik und Philosophie der Naturwissenschaft*. Band I. Berlin: Springer, 1906, Heft 2, pp. 3-136.

Höfler, Alois. "Die unabhängigen Realitäten". *Kant-Studien*. 12 (1907) 361-392.

Höfler, Alois. "Erkenntnisproblem und Erkenntnistheorie". *Zeitschrift für Philosophie und philosophische Kritik*. 137 (1909) 1-61.

Höfler, Alois. "Theorie und Hypothese im physikalischen Unterricht". *Zeitschrift für den physikalischen und chemischen Unterricht*. 22 (1909) 50 -52.

Höfler, Alois. *Didaktik des mathematischen Unterrichts*. Leipzig und Berlin: B.G. Teubner, 1910. 500 pages.

Höfler, Alois. "Zur Geschichte und Wurzel der Machschen Philosophie" . *Sonderabdruck aus der Zeitschrift für physikalischen und chemischen Unterricht*. 23 (1910) 1-16.

(Höfler, Alois). *Rückblick auf die ersten funfundzwanzeig Vereinsjahre der Philosophischen Gesellschaft an der Universität zu Wien mit dem 26. Bericht für das Jahr 1912/13*. Wien: In Selbstverlage der Philosophischen Gesellschaft, 1913, pp. 1-47.

Höfler, Alois. *Didaktik der Himmelskunde und der astronomischen Geographie*. Leipzig & Berlin: B.G. Teubner, 1913. 453 pages.

Höfler, Alois. "Über dogmatisches und kritisiches Kant-Studium". *Zeitschrift für Hochschul Pedagogik*. 5 (1914) Heft 1, 16 pages.

Höfler, Alois. "Ernst Mach". Sonderabdruck aus der Zeitschrift für den Physikalischen und Chemischen Unterricht. 29 (1916) 57.

Höfler, Alois. "Franz Brentano in Wien". *Süddeutsche Monatsschrift*, 4 (1917) 319-325.

Höfler, Alois. "Die Ideale des Realismus". *Zeitschrift für den physikalischen und Chemischen Unterricht*, 30 (1917) 1-12.

Höfler, Alois. "Lebenslauf des Alois Höfler". in *Die Deutsche Philosophie der Gegenwart in Selbstdarstellungen*. edited by Raymund Schmidt. Volume II. Leipzig: Felix Meiner, 1921,

pp. 117-169.
Höfler, Alois. "Welt Denken und Weltanschauen". *Südeutsche Monatshefte*. Leipzig/München: May 1921, pp. 133-139.
Höfler, Alois. "Naturwissenschaft und Philosophie". *Sitzungsberichte der Akademie der Wissenschaften in Wien*. Philosophische-historische Klasse. 1 Abhandlung. 196 Band 1. Wien: A. Hölder. 150 pages. (Not to be confused with a 122 page article with the same title and in the same transactions, 191 Band 3, 1920.)
Höfler, Alois. "Relativitäts und Erkenntnistheorie". *Zeitschrift für Physicalischen und Chemischen Unterricht*. 35 (1922) 88-90.
Höfler, Alois. For a list of his textbooks see pp. 259-261 in this book and in Ulf Höfer, *Verzeichnis des Nachlasses von Alois Höfler (1853-1922) - Teil I.* Forschungsstelle und Dokumentationszentrum für österreichische Philosophie, Graz: 1994.
Höfler, Karl, Otto, & Wolfgang. "Verzeichnis der Veröffentlichungen von Alois Höfler". in *Wiener Zeitschrift für Philosophie, Psychologie, Pädagogik*. Band 4 (1953) Heft 3, 165-181.
Hoffmann, Dieter & W, Manthei. "Ernst Mach als Schulphysiker, Didaktiker, und Bildungstheoretiker". in *Ernst Mach - Studien und Dokumente zu Leben und Werk*. edited by Dieter Hoffmann & Hubert Laitko. Berlin: Deutsche Verlag der Wissenschaften, 1991, 279-315.
Hohenester, Adolf. "Ernst Mach als Didaktiker". in *Ernst Mach - Werk und Wirkung*. Hrsg. Rudolf Haller & Friedrich Stadler. Wien: Verlag Hölder-Pichler-Tempsky, 1988, pp. 138-163.
Jacquette, Dale. *Meinongian Logic - The Semantics of Existence and Non-Existence*. New York: De Gruter, 1996.
Kindinger, Rudolf (Hrsg.). *Philosophenbriefe aus der wissenschaftlichen Korrespondenz von Alexius Meinong*. Graz: Akademische Druck- u. Verlagsanstalt, 1965.
König, Edmund. *Die Entwicklung des Kausalproblems*. Band II. 1890.
Lindenfeld, David F. *The Transformation of Positivism - Alexius Meinong and European Thought, 1880-1920*. Berkeley, Los Angeles, & London: University of California Press, late 1970's.
Mach, Ernst. *Die Geschichte und Wurzel des Satzes von der Erhaltung der Arbeit:* Prag: Calve, 1872.
Mach, Ernst. *Der relative Bildungswert der philologischen und der mathematisch-naturwissenschaftlichen Unterrichtsfächer der höheren Schulen*. Leipzig: Tempsky, 1886. 29 pages. (See English translation below.)
Mach, Ernst & Gustav Jaumann. *Leitfaden für Studierende*. Vienna & Leipzig: Tempsky & Freytag (respectively), 1891. (A second edition appeared in the same year but without the contribution by Jaumann.)
Mach, Ernst. Die Leitgedanken meiner naturwissenschaftlichen Erkenntnislehre und ihre Aufnahme durch die Zeitgenossen". *Physikalische Zeitschrift*. 11 (1910) 599-606.
Mach, Ernst. "On Instruction in the Classics and the Mathematico-Physical Sciences". in *Popular Scientific Lectures*. fifth edition. La Salle: Open Court, 1943, 358-374.
Mach, Ernst. "Über das Prinzip der Vergleichung in der Physik". *Verhandlung der Gesellschaft Deutscher Naturforscher und Ärzte* (Leipzig). Theil I. 1894, pp. 154-156. (For an English version see his *Popular Scientific Lectures*. La Salle: Open Court, 1943, pp. 236-258.)
Mach, Ernst. For a list of Mach's textbooks see pp. 259-260 in this book and also: Joachim Thiele, "Ernst Mach-Bibliographie". *Centaurus*, 8 (1963) 224-225.
Meinong, Alexius. "Über die Erfahrungsgrundlagen unseres Wissens". *Abhandlung zur Didaktik und Philosophie der Naturwissenschaften*. Band I. edited by Poske, Höfler,, Grimsehl. Berlin: Springer, 1906, Heft 6, 113 pages.
Meinong, Alexius. "Leben und Streben". in *Die Philosophie der Gegenwart in Selbstdarstellungen*. edited by Raymund Schmidt. second edition. Leipzig: Verlag von Felix Meiner, 1923, pp. 101-158.
Meister, Richard. "Die Geschichte der Philosophischen Gesellschaft an der Univerzität Wien 1888-1938". in Robert Reininger (Hrsg.). *50 Jahre Philosophische Gesellschaft an der Universität Wien 1888-1938*. Wien: Verlag Philosophischen Gesellschaft an der Universität Wien, 1938, pp. 3-20.
Parsons, Terrence. *Non-Existent Objects*. New Haven: Yale University Press, 1980.

Petzoldt, Joseph. "Die vitalistische reaktion auf die Unzulänglichkeit der mechanischen Natursicht". *Zeitschrift für allgemeine Physiologie.* 10 (1910) 69-119.

Pfaundler, Leopold. "Einleitung". *Mechanik und Akustik.* Braunschweig: Friedrich Vieweg, 1906.

Planck, Max. "Die Einheit des physikalishen Weltbildes". *Physikalische Zeitschrift.* 10 (1909) 62-75.

Planck, Max. "Zur Machschen Theorie der physikalischen Erkenntnis - Eine Erwiderung". *Physikalische Zeitschrift.* 11 (1910) 1186-1190.

Poske, Friedrich. "Alois Höfler". *Zeitschrift für den Physikalischen und Chemischen Unterricht.* 35 (1922) 51.

Reininger, Robert. (Hrsg.). *50 Jahre Philosophische Gesellschaft and der Universität Wien 1888-1938.* Wien: Verlag der Philosophischen Gesellschaft an der Universität Wien, 1938. 43 pages.

Rodlauer, Hannelore. *Otto Weininger - Erose und Psyche Studien und Briefe 1899-1902.* Wien: Verlag der Österreichischen Akademie der Wissenschaften, 1990.

Russell, Bertrand. "Hard and Soft Data". in *Perception and the External World.* edited by R.J. Hirst. New York: Macmillan, 1965.

Chapter Twelve

THE UNIVERSITY OF VIENNA PHILOSOPHICAL SOCIETY*

by thc cditors

Section Guide

1. Alois Höfler's Society 1888-1922
2. Robert Reininger's Society 1922-1938
3. Mode One: Lectures & Discussions
4. Mode Two: Mach's Probable Influence

1. ALOIS HÖFLER'S SOCIETY 1888-1922

In a sense clubs and societies were to many late 19th century Europeans what television and computers have become for most denizens of our fast developing electronic age. There were scores of intellectual societies, but comparatively few of them were formal, few kept full records, and few have been anything like as influential in their day or later as the Philosophische Gesellschaft an der Universität Wien.[1] Indeed, for thirty-four years, that is, from 1888 to 1922, the centerpiece of reflective Vienna would remain the University of Vienna Philosophical Society, nor did it finally expire until 1938 and the Hitler *Anschluss*.

If one looks back on the best-known European thinkers of the twentieth century almost all were connected with the city of Vienna. *Brentano* taught there; *Husserl* listened; *Mach* wrote and retired in the ring city; *Boltzmann's* attacks on philosophy echoed down its halls; *Wittgenstein* as a Viennese youth may well have heard about him at home or later in Linz, *Schlick* had a chair and organized his circle; *Neurath* spoke wherever he could; *Carnap* "abandoned phenomenalism" by using "realist expressions"; *Popper* learned from Karl Polanyí; and *Paul Feyerabend* heard Ehrenhaft on physics. Seven of these sons or stepsons of high Vienna either lectured in the Philosophical Society, or were lectured about. And the remaining three had teachers from the Society.

We do not intend to disparage or minimize other cities. Both of the Cambridges also contributed a lot to philosophy, Moore, Russell, and Wittgenstein in England and James, Santayana, and Quine at Harvard, but Wittgenstein was Viennese and Quine was surely influenced by Carnap and the Vienna Circle. In short, Viennese influence seems to have been greater, even if Logical Positivism and Behaviorism have since lost their glitter and fields like psychology once again view consciousness and hence epistemology itself as important and even as if "psychologism" and causal realism had returned.

Politically, Vienna like the Austrian Empire itself did not prosper and the defeat of the Central Powers in 1918 saw the latter fall apart. It was not victorious in diplomatic or military terms, and much of its presentist and phenomenalist philosophy has often been considered either utopian or escapist, anything but realistic or practical. But in the midst of ever rising opposition and hostile prejudice a cultural Renaissance mostly - but by no

J. Blackmore, R. Itagaki, and S. Tanaka (eds.). Ernst Mach's Vienna 1895-1930. 277-314.

means entirely - led by educated Jews of a secular orientation had taken place in Vienna starting in the last quarter of the 19th century, and its influence and fame has been such that we historians of science, culture, and philosophy cannot fairly disregard it, but we can and must investigate its nature and roots more extensively, especially in order to discover how it arose and could appeal so strongly to intellectually and scientifically-minded thinkers in economically and politically more fortunate countries. Vienna was a cultural center in music, art, economic theory, and above all in philosophy, and no amount of causal "impotence", "decadence", or political "defeatism" of that period or later can obscure that remarkable cultural development. Much of our current academic philosophy still reflects ideas developed in or first actively propagated from Vienna.

Coming back to Höfler's Society, according to the membership list for 1912/13, it included philosophically-oriented economists and physiologists as well. *Menger* and *Schumpeter* belonged, as did *Josef Breuer, Siegmund Exner, and Heinrich Obersteiner*. Among psychologists *Alfred Adler* had lectured earlier. *The Bühlers* did so many times later. *Freud,* however, had been scheduled to lecture but became ill and cancelled. Numerous physicists belonged to the Society in 1912/13 including *Marian Smoluchowski, Arthur Haas, Karl Przibram, Stefan Meyer, Egon v. Schweidler, Philipp Frank,* and *Ernst Mach* himself (but Schrödinger, Thirring and Meitner though also from Vienna and interested in philosophy were not on the list for that year and Wolfgang Pauli Junior, though ranked by his Machist father as a genius, was still too young).

Ludwig Wittgenstein's early hero, *Otto Weininger,* had attended sessions of the Society religiously until shortly before his suicide, and it seems that "Ludi" learned much of his linguistic anti-philosophy from reading the books of Stöhr, Boltzmann, or both, or from hearing their lectures or hearing about their arguments against philosophy in the Society. But it will still take much detailed and extensive research to pin down *exactly how* Wittgenstein was influenced by earlier Viennese linguistic philosophers.

Also, when members, including the philosophers, left Vienna they would almost always remain "corresponding members" even for decades like Husserl, Ehrenfels, and Meinong or for somewhat lesser periods like Hahn, Frank, Wahle, and Mach.[2] They knew where the action was, and seem to have felt reluctant to lose all contact with it.

Why has the existence and importance of the Society been so neglected since World War II? Perhaps because efforts to popularize the Vienna Circle and Logical Positivism have appeared reluctant to mention it and because most documents about its activities and members were almost exclusively in the hands of aged members of the Society who gradually died off. Almost no libraries or archives outside of the city of Vienna have relevant material, and even inside Vienna the documents are incomplete and have been difficult to obtain, though now one hopes that some will become more accessible.

What made a lot of sessions of the University of Vienna Philosophical Society so popular? Almost certainly the controversial nature of so many of the lectures and the continuing discussions which were allowed during subsequent weeks on the same topic. The discussion sessions appear to have continued until almost everyone had his chance to participate, regardless of field or background. This drew people from numerous fields outside philosophy who nevertheless had philosophical views.

How large was the Society? It began with 58 paying members in 1887/88 and grew to 217 as a maximum in 1907/08,[3] where the number leveled off until the First World War, but almost from the start they seem to have been outnumbered by non-paying visitors. The largest auditorium held about six hundred people, and it was often filled, especially when Ludwig Boltzmann lectured against philosophy or was a leading critic, for example, when he demolished Wilhelm Ostwald in 1904[4] and Friedrich Jodl in 1905.

The first officers of the Society were largely former or then current students of Franz Brentano who mostly agreed with him that scientific psychology was well along in the process of replacing philosophy, which may partly explain why so many of the early lectures were on research discoveries in psychology and in physiology.[5] Prominent Brentanoids included Alois Höfler, Christian von Ehrenfels, F. Hillebrand, J.K. Kreibig, and K. Twardowski, who tended to dominate the organizational side of the Society until their deaths during or shortly after World War I. Though Robert Zimmermann as the other major professor of philosophy in Vienna became *Obmann* or leader of the Society from 1890 to 1898 he was already elderly and few of his own students seemed to participate as actively as Franz Brentano's followers. The physicist Ludwig Boltzmann seemed to be the most influential member who had been a close student of Zimmermann's ideas,[6] and like his mentor was quite interested in aesthetics* but Boltzmann\s return to Vienna in 1894 was cut short twice, once to teach in Leipzig (1900-1902) and of course by his suicide in 1906.

In order to clarify Alois Höfler's dominant role from 1888 to 1922, let us set up a table which shows his parallel activity when other people were at least officially head of the Society.

LEADER (*Obmann*)	HÖFLER'S OTHER POSITIONS
1888-89 Alois Höfler	
	1894 Becomes *Privatdozent*
1890-98 R. Zimmermann	
	1895-1898 Assistant Leader
1898-03 Alois Höfler	
	1903-1907 Prague Professor 1903 Honorary President of the Society for Life
1903-12 Friedrich Jodl	
	1906-12 Discussion Leader 1907-22 Professor in Vienna
1912-22 Alois Höfler	

From 1889 until rather late in the First World War, the Philosophical Society published extensively: first, yearly reports [*Berichte*],[7] which described the activities, officers, and members; second, a yearbook [*Jahrbuch*] from or before 1902[8] until 1916-17; third, a series of books by both members and non-members,[9] and fourth, two commemorative volumes.[10] But it seems the various publications were only sent to members, and no library or archive even in Vienna has a complete collection.[11] The yearbook published about half of the lectures given in a particular year, including some discussion sessions and after-lecture-remarks which had been taken in shorthand.[12] This usually meant four articles per yearbook which could range over a hundred pages per issue.[13] While the Society published the *Berichte* and both the 25th and 50th year anniversary volumes,[14] which have become so important for scholars, the bulky yearbooks were published by Johann Ambrosius Barth in Leipzig. But limiting circulation as if "only for its members" has unwittingly delayed research. Some publications are still unlocated at least in the public domain and may be extinct.

From about 1907 to the 1920's, Otto Neurath played the role of *enfant terrible* in the Society. He was strongly seconded by Hans Hahn the mathematically-minded super-phenomenalist, and the young conventionalist Philipp Frank who seems to have tried his hardest to eliminate causes altogether. Rudolf Haller has honored them as "The First Vienna Circle" (circa 1907-1910), but in fact the context of the University of Vienna Philosophical Society, the lectures they gave in it, and the offices which they held there, probably contributed more to their intellectual development than the informal weekly discussions they held with each other at night, but admittedly it is difficult to be certain about such matters.

The Society continued to be quite active during the Great War (1914-1918), in fact, they held more sessions than usual and often on controversial moral issues related to the conflict, but it seems that all of their publications including *Berichte* stopped (or were stopped) by 1917 and would not be resumed until almost a decade later (1926). Neurath continued to be a leading speaker and together with Kraft they both lectured on Mach's ideas.

The Highest Number of Discussion Sessions were
Devoted to the Following Topics and Problems:

1894 Does Ethics Depend on Metaphysics?	Four Sessions
1894-95 On Mach's Lecture: On Comparison in Physics	Three Sessions
1895 Is Everyday Scepticism about Ethics Refutable?	Three sessions
1895-96 on Hertz's Book *Introudction to Mechanics*	Three sessions
1899 Are Mental Pictures More than Sensory Images?	Three sessions
1899-00 Is Vitalism Plausible?	Four Sessions
1900-01 On Nietzsche's Transvaluation of Values	Four sessions
1901-02 On the Crisis of Darwinism	Seven Sessions
1902-03 What are Concepts?	Four Sessions
1905 How Should Space Be Represented	Four Sessions
1906-1909 Jodl's Restrictions on Discussion Sessions	
1910-11 On Intuition	Three Sessions
1911 On Immortality	Three Sessions
1911-12 Must Epistemological Idealism Be Solipsistic?	Four Sessions
1912 Are There New Proofs for the Existence of God?	Four Sessions
1912-13 What Is Philosophy?	Three Sessions
1913 What Is Metaphysics?	Three Sessions
1914 What Are Platonic Ideas?	Three Sessions
1914 What Criteria Should determine Value?	Three Sessions
1914-15 On Function and Structure	Four Sessions
1915 The Problem of Physical Relativity	Three sessions
1915 Is Ethics Also Valid for Peoples & National States?	Three Sessions
1916 On Generation, Degeneration, and Destruction	Three Sessions
1916-17 On Mach's Inertial & Philosophical Views	Four Sessions
1916-17 Monads, Atoms, Electrons, and Quanta	Four Sessions
1918 On the Proper Relation of Science to Philosophy	Four Sessions
1918-19 What Is Personalism?	Four Sessions

1918-19 Is Causation Law-Based or Conditional?	Six Sessions
1920 On Schopenhauer's "Blind Will"	Six Sessions
1920 On Spengler's *Decline of the West*	Four Sessions
1920-22 On the Notion of Infinity in Mathematics	Four Sessions
1921 On Alexius Meinong's Philosophy	Three Sessions

2. ROBERT REININGER'S SOCIETY 1922-1938

Robert Reininger (1869-1955) who had been Assistant Leader from 1906 to 1912 under Friedrich Jodl and who had presumably stayed in that same position from 1912 to 1922 under Alois Höfler, first, became full professor of Philosophy in Vienna, and second, at about the same time finally became *Obmann* or boss of the Society himself from 1922 until 1938. He is known as a Neo-Kantian in philosophy and rather stoic in personality. He drastically changed the Philosophical Society, not at once, but increasingly through the years. Together with Richard Meister, he put the main emphasis on historical lectures, on inviting German philosophers to give talks, and he virtually abolished the discussion sessions. Indeed, by 1927 the Society had become simply part of the *Kant-Gesellschaft*. Information on activities, officers, and membership was apparently no longer published, and while the Society did restore publication of some parts of some lectures from 1926 to about 1937[15] it was on poor paper and in no way comparable to the amount and quality of what had been published prior to 1914.

In statistical terms, of the 600 lectures and discussions given from 1888 until 1938, **447** were given during what might be called the Höfler era, that is, until May 27 1922, and **153** during Reininger's reign from June 11, 1922 until June 10, 1938. But since the Höfler era of 34 years was about twice the Reininger reign of 16 years we can determine that an average Höfler year included **13** lectures and discussions and an average Reininger year about **9.25** sessions. But the difference is almost completely explained by the presence of discussion sessions during Höfler years and their absence in the Reininger years. Or to be more exact, the Höfler years as a total included **339** lectures and **108** discussion sessions, while the shorter Reininger period contained **148** lectures but only **5** discussion sessions. In this respect Robert Reininger may have been following Jodl's only briefly successful attempt to cut back on the number of discussion sessions and thereby give professional philosophers greater control of the Society.

Why are these numbers so very important? Because having become accustomed to discussion sessions under the long Höfler era, their comparative absence which had started as early as 1921 was followed only a few years later by the development of a large number of separate discussion groups centering around a particular leader or professor *which met outside the Society*.

The best-known of these small societies has surely been Schlick's Circle, which was later called the Vienna Circle. Hans Hahn who had lectured in the Philosophical Society before the war was called back to the University as a professor of mathematics in 1921. Moritz Schlick came in 1922, A circle began soon after and was firmly established by 1925. Rudolf Carnap had also arrived that year and the Circle's interest in Wittgenstein and Russell's work in mathematics and logic gave them a subject matter to discuss as well as Otto Neurath's "physicalism" and other contributions. Otto Neurath had not only contributed much to the Philosophical Society but starting in 1908 had become its most prolific lecturer. His final lecture there in 1924 was followed by attendance at Schlick's Circle. He clearly loved discussion. Viktor Kraft and R. von Mises also joined the trek.

Other discussion groups were those of Heinrich Gomperz[16] and Robert Reininger himself. Apparently, Prof. Reininger and some of his other colleagues didn't like "free" discussions since it could involve criticism or even ridicule of professional philosophers themselves, especially if they were not alert or argumentatively agile. For a small circle, however, where each professor was almost a patron-saint, he could invite the people he wanted, and he could expect high respect and civility as well as many shared basic assumptions . Discussion within these congenial confines offered a lot to timid souls, and apparently even to some of the less timid like Neurath. According to popular legend, the notion of a patron-saint, total-control circle would be perfected by the often autocratic Ludwig Wittgenstein in Cambridge. Most Austrians, however, were more polite.

Sir Karl Popper, who as a student was far too argumentative, outspoken, and critical to be invited to Schlick's Circle in Vienna or later to endure Herr Wittgenstein's arrogance has also mentioned in his autobiography that there were other philosophical circles in Vienna during the 1920's, that is, during the immediate post-Höfler and post-discussion years of the Philosophical Society. But because the leaders of the other circles were sometimes asked by Schlick to visit his "*Klatsch*", their independence has been doubted and membership of the Vienna Circle inflated. Here are Sir Karl Popper's words:

> The circle itself was, so I understood, Prof. Schlick's private seminar, meeting every Thursday evening. Members were simply those whom Schlick invited to join. I was never invited to join. I was never invited, and never even fished for an invitation. But there were other groups meeting in Victor Kraft's or Edgar Zilsel's apartments, and at other places; and furthermore there was also Karl Menger's famous "*mathematisches Colloquium*". Several of these groups of whose existence I had never even suspected, invited me to present my criticisms of the central doctrines of the Vienna Circle. It was there in Edgar Zilsel's apartment, in a most crowded room, that I read my first paper. I still remember the stage fright.[17]

Most members of the Vienna Circle were either strongly influenced by Ernst Mach like Philipp Frank, Otto Neurath, and young Rudolf Carnap or at least shared his aversion to "metaphysics" while focusing on methodology of science, hence there is value in examining the education of these people as well as members of some of the other circles in post-Höfler Vienna. For example, all of the people who Popper mentions in the quotation above had corresponded with Mach (except Menger but whose father the economist had done so) and had also been either members of Höfler's Society or given lectures in it. As for Sir Karl Popper, his father had also been a longtime member of the Höfler-led University of Vienna Philosophical Society.[18] An oddity about Sir Karl is that while he alleges that he thought of himself as an "orthodox Kantian" at that time that he almost never refers to Robert Reininger, who was the acknowledged leader of Vienna Neo-Kantians as well as *Obmann* of the Philosophical Society during most of the 1920's and later.

When the Nazi's took over Austria in 1938 it is said Reininger was asked to expel all of the Jewish members. He refused, and instead closed down the University of Vienna Philosophical Society itself.[19] It was started again after World War II, but in the absence of easily accessible documents about its proceedings it is difficult to say much about it. Robert Reininger and Richard Meister are to be praised for putting together the fiftieth year anniversary volume which included a history of the Society and what they regarded as a complete list of lectures and discussions from 1888 to 1938 and not least for daring to publish it (albeit for members only) during the first few months of Nazi domination of Austria. Unfortunately the list has not been easy for most scholars to locate, especially in the English-speaking world, hence, we have provided an English version. People listed in **bold** type corresponded with Mach or had some special relation with him or with the ***Vienna Circle*** which combines **bold** and *italic* type.

SECTION 3. **MACHIST MODE ONE**

LECTURES AND DISCUSSIONS OF THE UNIVERSITY OF VIENNA PHILOSOPHICAL SOCIETY IN CHRONOLOGICAL ORDER 1888 - 1938 [20]

( ) = discussion leader, [ ] = whom about, [[ ]] = commemorated

1888

April 27: **F. Brentano,** Methods of Historical Investigation in Philosophical Fields
May 29: **S. Exner,** Some New Studies on Sensations of Movement (with Examples)
T. Gomperz, Report on the New Archive for the History of Philosophy
June 22: **A. Höfler,** On Paradoxes in Teaching How to Determine Distance and Size
E. Löwy, The Significance of Philosophy for Students of Law
Nov. 27: T. Meynert, On the Brain and Human Culture

1889

Jan. 15: F. Raab, On the Organization of Mental Life
Feb. 13: R. Zimmermann, On the Beginnings of Mathematical Psychology in Vienna
March 27: **F. Brentano,** On the Concept of Truth
June 26: **W. Jerusalem,** On the Education of Laura Bridgemann
E. Leisching, On Scherer's Theory of Poetry
Nov. 16: T. Meynert, On the Mechanism of Melancholy
Dec. 17: **F. Brentano,** Schelling's Philosophy

1890

Jan. 18: **E. Kulke,** On Both Basic Problems of Beauty
Feb. 24: A. Berger, On the Dramatic Expression of Psychological Phenomena
April 26: V. Knauer, On Humor
May 31: E. Fechtner, Locke's Relations to Newton
E. Leisching, On the Discourses of Sir Joshua Reynolds
Nov. 25: T. Meynert, On the Coordinated Activity of Parts of the Brain
Dec. 16: **F. Brentano,** Modern Errors in our Understanding of the Laws of Deduction

1891

Jan. 31: J. Kolbe, On Number, Space, and Some of the Relations Between Them
March 10: **G. Adler,** On Current Physical Views of the Constitution of Matter
May 5: T. Meynert, The Mechanics of Physiognomy
June 11: **A. Höfler,** Seven Theses in A. Frick's "The Steady Filling of Space by Mass"
Nov. 28: A. Berger, On the Production of Drama
Dec. 12: V. Knauer, R. Hammerling vs. the Pessimism of Schopenhauer & Hartmann

1892

Jan. 9: **(A. Höfler)**, Relating Political to Philosophical Education (the Rector's Address with Special Consideration Given to the Coming Reform of the Study of Law)

Jan. 30: **C. Ehrenfels,** About Unprovable Knowledge
Feb. 9: **(C. Ehrenfels),** Richet's Book on Thought-Transfer and so-called Bright-Sight
Feb. 24: **W. Jerusalem,** Wundt's Institute for Experimental Psychology in Leipzig
March 8: **(C. Ehrenfels),** Continuation of the Discussion of February 9th
March 22: **F. Brentano,** On the Future of Philosophy
May 24: **A. Stöhr, Wiesner**'s Theory of Plasma
June 10: **A. Höfler,** Remembering Theodor Meynert & His Relations to the Society
Nov. 16: **(C. Ehrenfels),** Discussing Some Theses about Moral Assessment
Dec. 7: **(C. Ehrenfels),** Continuing the Discussion of November 16th
Dec. 17: A. Berger, **Friedrich Nietzsche**

1893

Jan. 29: **F. Brentano,** The Phenomenal Green
Feb. 22: R. Zimmermann, Semper as Aesthetician
April 27: **(C. Ehrenfels),** Probability Grounds Against Monism
May 10: **(C. Ehrenfels),** Continuing the Discussion of April 27th
May 20: H. Schmidkunz, Terminological Confusion
June 27: **R. Wahle,** The Discovery of the Godhead
Nov. 18: **F. Hillebrand,** The Basic Laws of Depth Perception
Dec. 2: (K. Twardowski), C. Lange's Linking the Emotions and Vasomotor Processes

1894

Jan. 20: (K. Twardowski), Is There Sense in Conceiving the Physical & Psychical as "Two Sides" of the Same Reality?
Jan. 27: **G. Jäger,** On the Divisibility of Matter
Feb. 26: **J. Breuer,** The Static Mind
March 5: **J. Breuer,** Continuing the Discussion of February 26th
April 21: (I. Himmelbaur), Do Ethical Principles depend on Metaphysical Factors?
May 10: **C. Ehrenfels,** On the Influence of Metaphysical Views on Ethical Education
May 22: Discussion on the Lecture of May 10th
June 2: V. Urbantschitsch, On the Value of Hearing Exercizes on the Sense of Hearing
June 10: Continuing the Discussion on the Lecture of May 10th
Oct. 31: **L. Müllner,** Commemorating the 70th Birthday of Robert Zimmermann
Dec. 7: M. Straszewsky, How Investigation into Oriental Philosophy May Help Us Better Understand Philosophy in General
21 Dec. 13: **(A. Höfler), Mach's Lecture on the Principle of Comparison in Physics** (a s Given at the 66th Convention of Natural Scientists in September in Vienna)

1895

Jan. 17: **L. Boltzmann,** On a Physical Application of Probability Statistics
Jan. 24: **(A. Höfler), Continuing the December 13th Discussion of Mach's Lecture**
Feb. 15: H. Schenker, On the Spirit of Musical Technique
March 14: **(J. K. Kreibig), Further Discussion on Mach's Lecture**
March 18: H. Schenker, Continuing the February 15th Lecture
March 28: **W. Jerusalem,** On the history of the Problem of Judgment
April 25: **(J. K. Kreibig),** How Can Everyday Scepticism about Ethics Be Refuted?
May 9: **(J. K. Kreibig)** Continuation of the Discussion of April 25th
May 16: R. Zimmermann, Spinoza as Politician
May 30: **(J.K. Kreibig),** Further Continuation of the Discusssion of April 25th
June 15: L. Gegenbauer, Unsolvable Problems
Oct. 30: **S. Exner,** On Sensations of Direction

Nov. 12: E. Schweidler, On the World Model of Mechanics (Followed by Discusssion on **Heinrich Hertz**'s Book *Einleitung zur Mechanik*)
Nov. 26: **E. Kulke,** On Winckelmann's Theory of Beauty
Dec. 3: (E. Schweidler) Continuing the Discussion of November 12th on Hertz's Book
Dec. 17: E. Graf Lamezan, On the *principium individuationis*
1896
Jan. 28: (E. Schweidler) Further Continuation of the Discussion of November 12
March 2: A. Berger, Aristotle's Catharsis Theory in Light of Modern Nerve Data
March 17: E. Czuber, Critical Remarks on the Basic Concepts of Probability Statistics
April 21: S. Mekler, On the Genesis of the Cosmopolitan Doctrine in Ancient Times
April 28: (K. Zindler), Meinong's Theory of Evident Suppositions plus Relating Probability Statistics to Theory of Judgment (Related to March 17th Talk)
May 16: E. Reich, Ethical Education and Social Problems
May 20: H. Obersteiner, On the Material Foundation of Consciousness
June 23: **(C. Ehrenfels),** The Problem of Psychological Egoism
Nov. 17: I. Himmelbaur, On Our Ability to Reform Criminals
Nov. 25: **F. Jodl,** Some Thoughts on the Causal Problem
Dec. 16: R. Kralik, Socrates and His Philosophy
1897
Jan 20: F. Wickhoff: On Simmel's Problems Concerning the Philosophy of History
Feb. 24: R. Sterneck, On an Infinite Quantity of Points
May 31: **R. Wlassak,** Richard Avenarius's Empiriocriticism
June 15: **R. Wlassak,** On Richard Avenarius's "Empirical Concept of the World"
Oct. 26: H.S. Chamberlain, On Indian Thought
Nov. 5: R. Sterneck, Psychological Analysis of the Concepts of Density & Intensity
Nov. 22: K. Zindler, On the Spatial Picture of the Color Perception Continuum
Dec. 7: [R. Sterneck], Discusssion of the Lecture and Topic of November 5th
Dec. 14: B. Hatschek, On the Sense Organs of Lower Animals
1898
Jan. 14: R. Kralik, On Philosophy of History
Feb. 17: **(T. Gomperz),** Philosophy of History
April 25: **R. Wlassak,** Physiological and Mentalistic Views about Theory of Feeling
May 2: **[R. Wlassak],** Discusssing the Lecture and Topic of April 25th
May 9: E. Schweidler, Some Experiments Bearing on Helmholtz's Color Theory
May 18: **J. Wiesner,** On the Law of Unity in the Inner Structure of Organisms
June 20: E. Seidler, On the Philosophical Foundation of Economic Theory
June 22: (R. Kralik), On the Decadence and Regeneration of Modern Art
[E. Seidler], Discussing the Topic of June 20th
Nov. 3: H. Obersteiner, On the Psychology of the Sense of Smell
Nov. 21: **M. Smoluchowski,** Is the Presupposition of Logical Simplicity in Ultimate Natural Laws Justified? (Followed by Discussion)
Dec. 5: **[M. Smoluchowski],** Continuing the Discussion of November 21th
Dec. 16: H.S. Chamberlain, Richard Wagner's Philosophy
1899
Jan. 16: **S. Exner,** Studies on the Periphery of Localized Vision
Jan. 30: (L. Linsmayer), Are There Mental Pictures [*Vorstellungsgebilde*] which are More than the Mere Sum of the Contents of Perception and Memory?
Feb. 18: **F. Jodl,** Natural Law and Chance

March 6: (L. Linsmayer), Continuing the Discussion of January 30th
March 15: (L. Linsmayer), Further Continuation of the Discussion of January 30th
May 4: R. Kralik, Experimental Investigations into Mental Phantoms by C. Ruth
May 25: E. Leisching, Inner Forms of Art and Other Matters of Aesthetic Controversy
May 30: [E. Leisching], Discussion of the May 25th Lecture

June 8: [R. Kralik], Discussion of the Lecture of May 4th plus a Historical Report on the Capacity for Expression in Music
Nov. 13: A. Berger, The poetic Development of Hermann Sudermann. footnote
Nov. 22: **(H. Swoboda),** Are there Strongly Proven Grounds Against Solipsism? 22
Nov. 29: **(H. Swoboda),** Continuing the Discussion of November 22nd
Dec. 14: M. Kassowitz, The Old and the New Vitalism

1900

Jan. 15: [M. Kassowitz], Discussion of the Lecture of Dec. 14 (introduced by Lectures by L. Linsmayer and **J. Breuer**)
Jan. 29: [M. Kassewitz, L. Linsmayer, & **J. Breuer**], More on December 14th Lecture
Feb. 8: [M. Kassewitz, L. Linsmayer, & **J. Breuer]** Still More on December 14th Talk
Feb. 17: **F. Jodl,** Giordano Bruno footnote
March 12: **J. Breuer,** Fechner's Ideas on the Developmental History of Organisms 23
May 7: M. Straszewski, Ideas for a History of the History of Philosophy
May 9: **F. Wickhoff,** Who is Ugly?
May 21: **[F. Wickhoff],** Discussing the Lecture of May 9th on Who is Ugly
June 7: E. Schweidler, The Significance of Aether for the problem of Absolute Motion
June 18: [E. Schweidler], Discussing the Lecture of June 7th on Aether
June 30: A. Riegl, Late Roman Art and Saint Augustine's Theory of Beauty
Nov. 9: **A. Höfler,** Remembering **Nietzsche**'s First Five Publications (plus Discussion)
Nov. 22: **(J.K. Kreibig), on Nietzsche**'s Master vs. Herd Morality
Dec. 4: **W. Jerusalem,** Typical Ways of Looking at Things
Dec. 13: **(J.K. Kreibig)**, Is **Nietzsche** Right that Evaluation Must Go Beyond Feeling?

1901

Jan. 18: **[W. Jerusalem],** Discussing the Lecture of December 4th on Typical Views
Jan. 24: **(J.K. Kreibig)**, Discussing the Lecture of December 13th
Feb. 4: (R. Sterneck), Zeno's Aporia about Movement
March 14: E. Lamezan, Attempt to Base Ethical Formula on an Aesthetic Foundation
April 24: **A. Höfler,** Words in Memory of **Gustav Fechner**'s 100th Birthday
May 13: **[W. Jerusalem],** Further Discussion of the December 4th Lecture
May 29: E. Schweidler, On the Concept of Entropy
June 7: O. Mikes, The Epistemological Content of **James**' *Principles of Psychology*
Nov. 9: **C. Ehrenfels,** The Socio-Ethical Significance of Leisure
Nov. 25: M. Kassowitz, The Crisis of Darwinism (A Discussion Follows.)
Dec. 14: [M. Kassowitz], Discussing the Lecture of November 25th on Darwinism

1902

Jan. 20: R. Wettstein, Modern Botany and Darwinism
Feb. 1: [M. Kassowitz], Further Discussion of the Lecture of November 25th
Feb. 10: **R. Eisler,** On the Epistemology of Aesthetic Criticism
Feb. 22: [M. Kassowitz], Still further Discussion of the Lecture of November 25th
March 8: **[(R. Eisler)],** Discussing the Lecture of February 10th
May 2: **J. Breuer,** The Crisis of Darwinism and Teleology

May 9: **[(J. Breuer)],** Discussing the Lecture of May 2nd. More on Darwinism
May 30: **A. Lampa,** On Continuity
June 12: **[(A. Lampa)],** Discussing the Lecture of May 30th on Continuity
June 26: **[(A. Lampa)],** Further Discussion of Continuity
Nov. 18: K. Twardowski, On Conceptual Representations
Nov. 25: (K. Mumelter), On the Arithmetician's Psychological Aids
Dec. 9: R. Kralik, Philosophy of Concept and Conceptsystem

1903

Jan. 7: **(J.K. Kreibig),** What is a Concept? (Connected with the Lectures of November 18th and December 9th by K. Twardowski and R. Kralik)
Jan. 22: V. Urbantschitsch, On the Influence of Subjective Facial Perceptions
Feb. 5: **(J.K. Kreibig),** Continuing the Lecture of January 7th on Concepts
Feb. 27: A. Menzel, On Natural and Cultural Science
March 10: [A. Menzel], Discussing the lecture of February 27
May 18: R. Sterneck, On the Elements of Consciousness (plus Discussion)
June 9: [R. Sterneck], Further Discussion of the Lecture of May 18th
June 22: A. Gerstel, On the Axioms of Geometry
Nov. 2: **(J.K. Kreibig),** The Philosophical Concepts of Goodness and Value
Nov. 28: **F. Jodl,** On the Problem of Materialism in Psychology
Dec. 11: **[F. Jodl],** Discussing the Lecture of November 28th (Introduced by Some Remarks by L. Linsmayer on Neurons)

1904

Jan. 9: **R. Eisler:** The Will to Suffering, a Psychological Paradox (plus Discussion)
Feb. 12: **W. Jerusalem,** Memorial Address for the 100th Anniversary of Kant's Death
Feb. 22: **[R. Eisler],** Discussing the Lecture of January 9th on the Will to Suffering
April 14: **E. Müller,** On Multidimensional Spaces
April 23: **R. Goldscheid,** Epistemological vs. Unavoidably Volitional Thinking (with Particular Reference to Philosophy of History)
May 6: **[R. Goldscheid],** Discussing the Lecture of April 23rd
May 13: **S. Exner,** On the Central Process of Seeing (with Demonstrations)
June 6: W. Schultz, Speech Theory and Introjection
Oct. 14: H. Arnim, The Stoic Doctrine of Fate and Freedom of the Will
Nov. 19: J.J. Hollitscher, Some Preliminary Questions about Sociology. footnote
Nov. 25: **W. Ostwald,** The Energetic Theory of Happiness (plus Discussion) [24]
Dec. 1: **G. Jäger:** On Stereoscopic Vision (with demonstrations)
Dec. 17: **T. Gomperz,** H. Spencer's Education and What it Teaches Us (& Discussion)

1905

Jan. 9: L. Linsmayer, On the Representation of Space (with Discussion)
Jan. 14: E. Ehrlich, Sociology and Jurisprudence. footnote
Jan 21: **L. Boltzmann,** On a Thesis of Schopenhauer (with Discussion). [25]
Feb. 11: K. Siegel, Spatial Representation and Spatial Concept (with Discussion)
March 2: **A. Stöhr,** Theory of Space (plus discussion of the January 9th and February 11th Lectures of L. Linsmayer and K. Siegel)
March 11: [L. Linsmayer, K. Siegel, & **A. Stöhr]** Continuation of the Discussion of January 9th, February 11th, and March 2nd
March 17th: M. Benedikt, Human and Animal Brains (with Pictures & Discussion)
May 17: **W. Jerusalem,** Psychological Logic & Newest Opponent (& Discussion). [26]
May 24: **[W. Jerusalem],** Discussing the Lecture of May 17 on Psychological Logic

June 5: W. Schultz, Concepts and Perceptions of Time (plus Discussion)
October 14: **F. Klein,** Questions on the Border between Mathematics and Philosophy
Oct. 28: **L. Boltzmann,** Explaining Love and the Entropy Law by Means of Principles of Statistical Probability
Nov. 14: W. Schultz, The Problem of Time with Special Attention to the Question: Are there Sensations of Time? (plus Discussion). footnote 27
Nov. 21: G. Seidler, On the Socio-Psychological Foundation of National States
Dec. 2: (J. Hollitscher), On the Theory and Practice of Philosophy of History
Dec. 15: R. Sterneck, Attempt at a Psychological Theory of Apparent Distances. 28

1906

Jan. 11: **V. Stern,** Arguments Against Materialism (plus Discussion)
Jan 19: **(A. Höfler),** Discussing **F. Klein**'s Lecture of October 14, 1905 and his Theses about Non-Evident Geometry and Perception of Shape
Jan. 30: **R. Wallascheck,** Artistic Illusion
Feb. 10: M. Benedikt, Measuring the Cranium (with Demonstrations)
Feb. 24: **(W. Jerusalem),** On Genetic and Biological Aesthetics
March 8: **J. Ofner,** Schiller as Predecessor of Scientific Socialism
March 16: **(W. Jerusalem),** Discussing the Lecture of February 24 on Aesthetics
April 4: G. Simmel, Of the Essence of Truth
May 4: (O. Friedländer), Main Trends in Current Psychology
May 26: **H. Swoboda,** On the Laws of the Life of the Soul and the Way to Find Them
June 20: **F. Jodl,** Psychic Lawfulness and the Methods of Psychology
July 5: (J. Herz), Attempts at Constructing Types in Philosophy
October 22: V. Urbantschitsch, Over Optical Memory Pictures
Nov. 7: K. Siegel, The Table of Categories in Herder's Metacritique
Nov. 23: ***(H. Hahn),*** Does Geometry Ultimately Rest on Facts?
Dec. 5: ***(H. Hahn),*** Continuing the Discussion of November 23 on Geometry
Dec. 20: E. Schweidler, On the Objective Validity of Statistical Probability

1907

Jan. 16: **R. Goldscheid,** The Concept of Direction and Its Significance for Philosophy
Jan. 31: (W. Schultz), The Historical Origin of Concepts of the Knowing Subject
Feb. 11: **R. Eisler,** On the Problem of the Origin of Art
Feb 26: **[R. Goldscheid]** Discussing the Lecture of January 16 on Direction
March 7: R. Sterneck, On the Theory of Just Noticable Differences
April 18: **E. Lucka:** Memory and Fantasy
May 11: **A.E. Haas:** The Most General Laws of an Event and their Significance for the Cosmological Problem
May 27: **M. Adler,** Toward an Epistemology of the Social Sciences
June 13: **[M. Adler],** Discussing the Lecture of May 27
June 21: **H. Swoboda,** Thinking and Suffering
Oct. 28: R. Reininger, The Basic Question of Epistemology and Its Typical Solutions
Nov. 11: **(A. Höfler),** The Answer to Dr. Reininger's Question. footnote 29
Dec. 4: ***(P. Frank),*** Hans Driesch's Philosophy of Nature: Its Relation to Vitalism
Dec. 18: **E.P. Schwiedland,** Economy and Justice

1908

Jan. 14: **A. Stöhr, Fechner**'s Hypothesis of the Cosmo-Organic Condition of Matter
Jan. 24: **W. Jerusalem,** Pragmatism, a New Philosophical Method. footnote 30
Feb. 4: F. Hamburger, Typicality, Assimilation, and Inheritance

Feb. 18: S. Mekler, The Question of Platonic Letters
March 5: ***O. Neurath:*** War and Moral Principles
March 26: A. Adler: Theory of Inferiority about One's Organ and its Significance for Philosophy and Psychology
May 9: J. Kyri, The Creative Experience of the World
June 1: (K. Roretz), On **Roberto Ardigo** and His Significance for Modern Philosophy
Oct. 31: K. Siegel, Natural Lawfulness and Vitalism; a Question of Principle
Nov. 9: **(A. Höfler),** Questions which Physics has about Philosophy
Dec. 9: O. Ewald, Epistemological Criticism and Epistemological Theory
Dec. 16: [O. Ewald], Discussing the Lecture of December 9th on Epistemology

1909

Jan. 14: **A. Stöhr,** The Problem of Time
Jan. 22: Does the Evidence Favor an Internal or External Criterion of Truth or None?
Jan. 28: H. Przibram, Living Power or Living Matter?
Feb. 3: H. Przibram, Lecture of January 28th Continued
Feb. 10: B. Hatschek, Celebrating the 100th Birthday of Charles Darwin
Feb. 17: **A. Stöhr,** The Exciting Part of the Vitalism Question
March 3: W. Schmied-Kowarzik, Space and Tme, a Psychological Consideration
March 17: **H. Gomperz,** On Evaluating Personality
May 5: **H. Swoboda,** Spirit and Soul
June 9: K. Roretz: Concept of Purpose in Psychological & Epistemological Thinking
June 23: H. Pichler, Christian Wolff's Ontology
Nov. 8: E. Schweidler, The Theory of Radioactivity and its General Significance
Nov. 25: K.K. Schneider, Four Objective Criteria of the Psychical
Dec. 4: ***P. Frank,*** Is There an Absolute Motion? footnote 31
Dec. 10: ***[P. Frank],*** Discussing the Lecture of Dec. 4th about Absolute Motion

1910

Jan. 3: H. Schmidkunz, Academic Pedagogy
Jan. 11: **H. Swoboda,** On Soul-full and Soul-less Humanity
Jan 17: ***(O. Neurath),*** The Concept and Range of Validity of the A Priori
Jan. 28: W. Schultz, How World Views Originated from Myth
Feb. 12: R. Sterneck, On Apparent Monocular Directions of Vision
Feb. 14: ***(O. Neurath),*** Discussing the Lecture of January 17th on the A Priori
Feb. 28: [H. Schmidkunz], Discussing the Lecture of January 3 on Academic Pedagogy
March 8: A. Trebitsch, The Mental Tendency Toward Unity
May 10: **A. Stöhr:** On Monocular Graphics
May 20: W. Schmied-Kowarzik, Intuition, a Contribution to the Psychology of Aesthetic Experience
June 4: **V. Stern,** On the Philosophy of My Father
June 13: (A. Schwadron), *De naturae saltibus*
Oct. 31: R. Reininger, Kant's Critical Idealism in Its Epistemological Significance
Nov. 17: W. Schultz, The Gnostic Basileïdes
Dec. 5: **J.K. Kreibig**, On Theory of Perception

1911

Jan. 16: **[J.K. Kreibig]**, Discussion of the Lecture of December 5th on Perception
Jan. 31: K. Schrötter, On the Psychology and Logic of Lying
Feb. 14: **H. Swoboda,** Plan of a Metapsychology
March 3: [W. Schmied-Kowarzik], Shape and Conscious Immediacy [*Gestalt und*

Anschauung], Discussion Provoked by the May 20, 1910 Lecture on Intuition
March 13: [W. Schmied-Kowarzik], Further Discussion of March 3rd Topics
May 5: **A. Stöhr,** The Brain and What Stimulates Imagination
May 16: O. Ewald, Toward an Analysis of the Concept of Immortality
May 22: [O. Ewald], Discussing the Lecture of May 16th on Immortality
June 2: **A. Haas,** On Physical Objections Against the Idea of the Infinity of the World
June 23: [O. Ewald], Further Discussion of the May 16th Lecture on Immortality
June 26: E. Hanslick, Lawfulness in the Evolution of Culture
Nov. 6: ***V. Kraft,*** Toward a Critique of Epistemological Positivism. footnote **32**
Nov. 20: **[*V. Kraft*],** Discussing the Lecture of November 6th on Positivism
Nov. 29: H. Pichler, Possibility and Freedom from Contradiction
Dec. 4: **[*V. Kraft*],** Further Discussion of the November 6th lecture on Positivism
Dec. 20: A. Trebitsch, Knowledge [*Erkenntnis*] and Logic

1912

Jan. 15: ***[V. Kraft],*** Still Further Discussion of Epistemological Positivism
Jan. 26: R. Reininger, On **Vaihinger**'s Philosophy of "As If"
Feb. 3: **V. Stern,** On the Ontological Proof of God
Feb. 12: [A. Trebitsch], Discussing the Lecture of December 20th
Feb. 24: K. Schrötter, The Roots of Fantasy
Feb. 26: **(J.K. Kreibig)**, Discussion on Theodor Lessing's Axioms of Value
March 16: S. Kornfeld, On the Feeling of Justice
May 7: **C. Ehrenfels,** New Proofs for the Existence of God. footnote **33**
May 9: **C. Ehrenfels,** Continuation of May 7th Lecture on Proofs that God is Real
May 11: **C. Ehrenfels,** Further Continuation of the Lectures of May 7th & 9th
May 20: **[C. Ehrenfels]** Discussing the Three Lectures on New Proofs about God
June 1: ***O. Neurath,*** The Problem of Maximum Pleasure
Nov. 11: W. Schultz, The Social Ethics [*Sittenlehre*] of Zarathustra
Nov. 18: Discussion Evening on: What is Philosophy? Also: What is *Weltanschauung*?
Dec. 2: **A. Höfler,** Showing Models and Experiments of What is Psychological in Colored Bodies
Dec. 9: More Discussion on What is Philosophy and also What is *Weltanschauung*

1913

Jan. 20: Still more Discussion on what is Philosophy and What is *Weltanschauung*
Jan. 27: ***O. Neurath,*** Those Confused by Descartes and his Auxiliary Motive (Toward a Psychology of Decisionmaking)
Feb. 8: Discussing ***Neurath's Talk*** of January 27th on Descartes & Decisionmaking
Feb. 10: E. Mally, On the Concept of Object in **Meinong**'s Theory of Objects
Feb. 21: **H. Swoboda,** Experience [*Erlebnis*] in Its Significance for Psychology and Theory of knowledge
Feb. 24: (O. Pommer), Discussion Evening on the Lectures of February 10 & 21
Feb. 28: K. Schrötter, On Homesickness
April 29: J. Herz, On the Problem of the Human Constitution
May 20: (L. Linsmayer): Discussion Evening on Our Constitution & Disposition
May 26: A. Trebitsch, Law-Based Causality in Light of the Search for Unity
June 11: H. Eibl, Speculation from the First Centuries A.D. on Philosophy of History
Nov. 13: **A. Stöhr,** Is Metaphysics Possible? (Celebration of the 25th Anniversary of the Society)
Nov. 19: Discussion Evening on All of the Things which are Called "Metaphysics"

Nov. 24: G.S. Fullerton, What is New in New Realism
Dec. 1: Another Discussion Evening on the Meaning of "Metaphysics"
Dec. 15: R. Suppantschitsch, What is Epistemological in Mathematics

1914

Jan. 19: H. v. Arnim, Plato's Earliest Series of Writings in their Planned Connection
Feb. 5: Meeting to Mourn the Death of our Honored and Longtime Leader **Friedrich Jodl.**
Feb. 9: Discussion Evening: What Are Plato's Ideas?
Feb. 16: More Discussion on Plato's Ideas
Feb. 21: A. Seibt, Means and Ends
March 2: ***O. Neurath,*** Classification of Systems of Hypotheses (Concerning Optics)
March 6: **A. E. Haas,** The Significance of Greek Philosophy of Nature for Physics
March 9: A. Vetter, On the Philosophy of Applied Art
March 16: (O. Pommer), Discussion on the Concepts of Disposition, Power, & Energy
May 25: A. Menzel, Social Criteria for Standards of Value
June 8: Discussion on Ethical and Social Criteria for Standards of Value
June 22: More Discussion on Ethical and Social Criteria for Standards of Value
Oct. 24: K. Twardowski, Functions and Structure
Nov. 14: K. Twardowski, Functions and Structure II
Dec. 18: [K. Twardowski], Discussing the Lectures of October 24 & November 14

1915

Jan. 12: **L. Hartmann,** Historical Tendencies
Jan. 25: [K. Twardowski], More Discussion on Functions and Structure. footnote
Feb. 23: E. Bernheimer, Are There Scientific Laws of Complex Phenomena? 34
March 1: Discussion Evening on the Lectures of January 12 & February 23. footnote
March 22: (***P. Frank*** **& A. Höfler**), Discussing the Problem of Physical Relativity. 35
April 22: (***P. Frank*** **& A. Höfler**), More Discussion on Physical Relativity
May 26: **H. Thirring,** The Problem of Physical Relativity. footnote
June 30: Discussion Evening, Is Ethics Also Valid for Peoples & National States? 36
Nov. 16: Continuing the Discussion on the Validity of Ethics for Peoples & States
Nov. 29: More Discussion about the Validity of Ethics for Peoples & National States
Dec. 13: **C. Ehrenfels,** Report on His New Work: *Kosmogonie*
Dec. 14: **C. Ehrenfels,** Continuing His lecture of December 13

1916

Jan: 31: L. Linsmayer, On Evolution
Feb. 21: Discussion Evening on the Concepts of Evolution, Stagnation [*Stillstand*], Degeneration, and Destruction (On the Ideas of **Julius Wiesner)**
Feb. 23: S. Oppenheim, On Determining the Color of Stars
March 8: More Discussion on Evolution, Stagnation, & Degeneration (**J. Wiesner**)
March 13: **A. Stöhr,** Heraclitus
March 15: **A. Stöhr,** Continuing the Lecture on Heraclitus
March 18: R. Neumann, Toward a Descriptive Psychology of Vowels
March 22: K. Joël: Classical Idealism and Its Significance for the Present
May 9: F. Löwy-Cleve, From Book: *Die Philosophie des Anaxagoras*, reconstructed
May 11: F. Löwy-Cleve, More from the Book: *Die Philosophie des Anaxagoras*
May 12: F. Löwy-Cleve, Still More from: *Die Philosophie des Anaxagoras*
June 5: More Discussion on the Concepts of Evolution, Stagnation, Degeneration, and Destruction (On the Ideas of **Julius Wiesner**)

Nov. 6: ***V. Kraft,* Mach as Philosopher** footnote 37
Nov. 14: W. Schmied-Kowarzik, Leibniz - For His 200th Birthday
Dec. 1: Discussion Evening on Monads-Atoms-Quanta (Connected to Leibniz Lecture)
Dec. 15: ***O. Neurath,* Mach's Position on the Problems of Gravitation and Inertia**

1917

Jan. 8: **Discussion of the Mach Lectures of *Kraft and Neurath*** (Nov. 6 & Dec. 14)
Jan 22: **More Discussion of Mach's Ideas and the Lectures of *Kraft & Neurath***
Feb. 5: J. Obermann, Al Ghazzali, the Critic [*Kritizist*] of Arabic Philosophy
March 17: S. Meyer & **F. Ehrenhaft** On Atoms, Electrons, & Quanta (with Demonstrations by Spinthariscope and Photographic Depiction of Particle Paths). note 38
May 5: Discussion Evening on **Ehrenhaft,** Monads, Atoms, and Quanta
May 7: (W. Ambros, **H. Molisch**) Discussion Evening on Egoism & Altruism in Biology and Politics (Partly on **Becher**'s Theory of the External Utility of Gall Phenomena in Plants and the Hypothesis of Beyond-the-Individual Souls); J.K. Kreibig who would die on November 8, 1917 also helped lead Discussion.)
May 22: More on **Becher,** Egoism, & Altruism in Biology and Politics
June 20: Still More on **Becher,** Egoism, & Altruism in Biology and Politics
Nov. 20: **A. Höfler,** Words in Remembrance: On the 30th Anniversary of **Gustav Fechner**'s Death

1918

Jan 10: W. Stern, The Basic Ideas of Personalistic Philosophy
Jan. 21: Discussion Evening on the Concepts of Personality, Law-Based Human Causality, Human Determination, and Human Purpose (on W. Stern's lecture)
Feb. 11: R. Reininger, Theories of Individuation and Holism in Kant & Schopenhauer
Feb. 20: S. Oppenheim, The Impersonal Micrometer
March 12: R. Klumak, Macrocosmic Individual Shapes [*Gestaltindividuen*]
March 19: M.E. Gans, Spinoza's Philosophizing
April 5: ***O. Neurath,*** Schelling and Faraday (with Experiments)
May 6: Discussion Evening on Both Principles: Natural Science is Completely Independent of Philosophy and Philosophy is Repeatedly Dependent on Natural Science (Caused by **Neurath**'s Concluding Remarks to His April 5th Lecture)
May 27: Discussing the Question: Can & Should there be "Philosophy of Nature"?
June 11: More Discussion on Can and Should there be "Philosophy of Nature"
June 25: W. Schmied-Kowarzik, The Aesthetic Character of Arts which are United by the Same Basic Purposes
Nov. 18: C.K. Schneider, The Manysidedness of the Teleology Problem
Nov. 25: F. Röder, Law-based Causalism or Conditional Causation? footnote 39
Dec. 9: Discussing the Lectures of November 18th & 25th

1919

Jan. 24: More Discussion of Law-Based Causalism or Conditional Causation
Feb. 25: A. Seibt, The Problem of Goaldirectedness in the Psychological Field Plus Further Discussion of Law-Based Causalism or Conditional Causation
Feb. 28: M. Sternberg, Judging Causes of Illness under the Influence of Philosophical Currents of the Tme
March 17: K. Ettmayer, Discussing the Problem of Linguistic Understanding
March 24: M. Graf, The Philosophical Content of German Music
April 5: [K. Ettmayer], Discussing the Lecture of March 17 on Linguistics
April 7: ***E. Zilsel,*** Toward a Theory of Ideal Personality

May 5: **A. Haas,** Physics as Geometrical Necessity
May 7: M. Benedikt, Introducing Philosophical Instruction into All Universities, Colleges [*Fakultäten*]
May 12: R. Lach, The Psychology of Primitive and Archaic Music
May 31: **A. Höfler,** Can Feeling Give Knowledge? An Emotional Presentation Plus Discussion of the Question: What can be True and False about the Judgements of Great Personalities? (For Example, Carlyle's Hero Worship?)
June 2: More Discussion on Great Personalities and Hero Worship
June 23: ***E. Zilsel:*** Toward a Theory of Ideal Personality. Part II the Value Problem
Nov. 3: R. Neumann, Bird Calls and Music

1920
March 8: A. Wesselsky, On the Self-Governing Criteria and Goals of Mankind (Some Ongoing Thoughts for a Philosophy of Action)
March 10: Discussing the Question: Is Schopenhauer's Will without Logical and Psychological Intellect Possible?
May 11: **A. Haas,** Atomic Harmony of the Spheres
May 14: ***O. Neurath,*** Spengler's Concept of Culture
May 15: T. Weindl, Pure (Radical) Dualism in its Historical Development and its Current Significance
June 11: Discussion of Culture and Civilization in **Neurath**'s Lecture on Spengler
June 18: M. Sternberg, The Significance of Scholastic Philosophy for Current Medical Thinking
June 25: **E. Müller,** The Concept of Infinity in Recent Mathematics
Nov. 4: A. Wesselsky, The Primacy of Willing in Kant and Schopenhauer
Nov. 8: (A. Seibt), Continuing the Discussion of March 10 on whether Schopenhauer's Will is Possible without Logical and Psychological Intellect
Nov. 29: (A. Seibt), Continuing the Discussion Further on Schopenhauer's "Blind Will"
Dec. 6: Discussion Evening on Culture, Civilization, & Spengler's Ideas
Dec. 10: More Discussion on "Decline of the West", & "Prussiandom & Socialism" (on Spengler's Ideas)

1921
Feb. 11: R. Stigler, Instinctivism and Intellectualism
Feb. 14: F. Schulhof, Aristotle's Theory of the Senses footnote
Feb. 25: Remembrance Evening for **Alexius Meinong** (Died November 27, 1920). 40
1. O. Redlich, **Meinong's** Beginnings as a Historian
2. E. Martinak, **Meinong** as Man and Teacher
3. R. Meister, From **Meinong's** Philosophy

Feb. 28: J. Herz, On the Problem of Periodicity
March 7: H.T. Horwitz, Laws in the History of the Development of Technology
May 13: Remembrance Evening for **Adolf Stöhr** (Died February 10, 1921)
1. R. Reininger, **Stöhr** as the Personification of a Thinker
2. S. Kornfeld, **Stöhr** as Psychologist and Moralist

May 20: R. Urbantschitsch, Inner Secretion and its Determining Influence on our Bodily and Spiritual Life
June 6: F. Schulhof, Campanella as Psychologist
June 20: (W. Wrchovsky), On T.K. Österreich's Book: *Der Okkultismus in modernen Weltbild*

Oct. 31: H. Torzyner, Number [in a Grammatical Sense] in the Problem of the Origin of Language
Nov. 4: **A. Höfler,** Delving into and Discussing the Philosophy of **Alexius Meinong**
Nov. 25: **W. Jerusalem,** Knowledge as Sociological Problem
Nov. 30: Discussing **Meinong**'s Ethical and Value Theories
Dec. 2: ***H. Hahn,*** Bolzano's "Paradoxes of Infinity"
Dec. 9: A. Wesselsky, The Philosophical Significance of Infinitesimal Methods
Dec. 20: **A.E. Haas,** The Problem of Ultimate Matter and Modern Physics

1922 footnote

Jan. 21: ***O. Neurath, Ernst Mach's History of Optics and Theory of Science.*** 41
Feb. 3: Discussion Evening on the Lectures of June 25, 1920 and December 2 and 9, 1921 on the Notion of Infinity in Mathematics
Feb. 17: J. Herz, Does the Succession of Philosophical Systems Follow a Law?
March 3: **E. Müller, Epistemology and Mysticism**
March 4: Wilhelm Wrchovsky, Inspiration, Intuition, and Ecstacy
May 22: Remembrance Evening for our Honorary President and Active Leader [*Obmann*]: Hofrat Universitäts Professor **Dr. Alois Höfler** (who died on February 26, 1922)

1. R. Reininger, **Höfler**'s Activity for the Philosophical Society
2. O. Pommer, **Höfler** as Philosopher
3. R. Meister, **Höfler** as Teacher

June 7: R. Pick, Antique Ethics in its Relation to religious and Cultural Movements
June 19: H. Eibl, Aristotle's Problem with Time
Nov. 20: H. v. Arnim, The Way to the Historical Socrates.
Dec. 4: C.K. Schneider, What is German in Spengler's "Decline of the West".

1923

Jan. 29: Discussion on Professor Schneider's Lecture of December 4, 1922.
Feb. 5: J. Herz, On the Different Kinds of Knowledge [Erkenntnis]
Feb. 19: ***H. Hahn,*** On **Bertrand Russell** 's *Introduction to Mathematical Philosophy*
Feb. 26: ***O. Neurath,*** The Conceptual Building of Economic Theory. Logical Focus.
March 5: H. Cysarz, The Concept of Life in Light of Psychology [*Geisteswissenschaft*]
May 18: K. Reichl, The Problem of **Stöhr**'s Theory of Metaphysical Matter.
May 28: F. Kaufmann, The Idea of Logical Foundation Science
June 4: ***H. Hahn,*** A Discussion of the Lecture on **Bertrand Russell** of February 19
June 11: R. Neumann, A Report on Charles A. Strong's *The Origin of Consciousness*
June 19: ***V. Kraft,*** On Müller-Freienfels: Irrationalism, Sketch of an Epistemology
Nov. 22: ***M. Schlick,*** The Question of Vitalism as an Epistemological Problem
Dec. 6: G. Roffenstein, On the problem of Unconsciousness

1924

Jan. 18: K. Bühler: On the Biological Origin of Language
January 31: E. Oberhuber: On the Philosophical Significance of the Physical Phenomenon of Mediumism
Feb. 14: M. Jasek, Bernhard Bolzano: Results of New Investigations in Vienna and in Czechoslovakia
Feb. 21: C.K. Schneider, The "Phaidon" of Plato and Mendelsohn
March 6: H. Przibram, The Apogenetic Theory of Development
March 13: ***O. Neurath,*** The Philosophy of History Problem about the Development of Architecture

May 10: R. Reininger, 200th Year Anniversary Celebration of Birth of Immanuel Kant
May 19: F. Koch, Plotinus's Concept of Beauty and Goethe's Artistic Creation
May 26: J. Herz, Questions in the Philosophy of History
June 2: O. Pommer, **Meinong**'s *Zur Grubdlegung der allgemeinen Werttheorie* (1923)
Nov. 20: **H. Gomperz,** Pro and Anti Plato from Atticus to Voltaire
Dec. 4: R. Prandau, Discussion on Epistemological Questions in Ethics and Religious Philosophy with Reference to Kant.

1925

Jan. 15: E. Bormann, Categorical Deduction from One and Two Premises
Feb. 19: Dr. Hochfeld (leader of the Johannes Rehmke Society), Basic Scientific Philosophy
Feb. 26: F. Kainz, Structure and Typology of Artistic World Views.
May 18: H. Cysarz, Space and Time in Mental Science [*Geisteswissenschaft*]
May 25: E. Fröschels, Freedom inspite of Natural Laws
June 6: Discussion of Dr. Fröschels' Lecture of May 25
June 19: R. Stanka, Sociology of the Present
Nov. 13: R. Meister, The Pansophy of Comenius
Dec. 4: ***M. Schlick,*** The Concept and Possibility of Metaphysics

1926

Jan. 22: H. v. Arnim, Ethics of Life Lived According to Nature
Jan. 29: W. Schmied-Kowarzik, Objective Spirit [*Geist*] and Its Forms
March 12: F. Eckstein, Contribution to the Philosophy of the Straight Line
May 21: ***V. Kraft,*** The Basis for the Validity of the Principles of Knowledge
May 28: O. Brechler, The Teaching of Mani and Its Impact on European Thought
June 11: J. Sauter, Baader's Position in German Idealism, Especially with Kant
June 18: **E. Müller,** Epistemology of the Problem of Space
Nov. 19: **A.E. Haas,** Objective and Human Physics
Nov. 26: **G. Rabel,** Goethe's Conception of Kant
Dec. 10: J. Sauter, Baader and Phänomenologie

1927

Jan. 28: ***H. Hahn,*** On the Newer Logical Theories
Feb. 11: ***R. Carnap,*** The Essence [*Wesen*] of Concepts
Feb. 18: R. Meister: Pestalozzi's Relation to the Philosophy of His Time
Feb. 25: F. Redl: Setting and Criticizing Values
March 4: M. Haubfleisch, Ways to Solve the Mind-Body Problem
March 11: O. Ewald, The Ethical Impulse in the Philosophy of the Present
May 13: C. Bühler, Basic Probelms of Social Psychology
May 20: R. Meister, Work and Play as Done by Children and Adults
May 27: F. Eckstein, Leibniz's Concept of System and Immortality
June 3: M. Haubfleisch, On **Julius Schultz**: Body and Soul.
June 10: A. Molitor, Bolzano as Philosopher of Religion
Nov. 18: R. Meister, History of Philosophical Studies at the University of Vienna
Dec. 2: R. Lach, History of the Development of Musical Thinking
Dec. 16: F. Redl, The "good" Paul Häberlin

1928

Jan. 13: **H. Gomperz,** Plato's Autobiography
Jan. 27: F. Kaufmann, Guilt and Responsibility
Feb. 17: ***H. Feigl,*** Accident and Law (an Examination in Philosophy of Nature)
Feb. 24: F. Eckstein, Modern Science and The Doctrine of Eternal Recurrence

March 2: ***V. Kraft,*** Ethics as Science
June 15: F. Redl, A Lecture on Bühler's "Crisis of Psychology"
Nov. 23: R. Links, Philosophical Fundamentals of the Conceptual Structure of Psychoanalysis
Nov. 29: L. Bertalanffy, The Current State of Problems in Evolution Theory
Dec. 7: F. Redl, Kolbenheyer as Philosopher

1929

Jan. 7: O. Höfler (Uppsala), Historicism and Relativism. footnote 42
Jan. 11: L. Bertalanffy, Discussion of Bertalanffy's Lecture of November 29, 1928
Jan. 25: O. Ewald, Systematic Philosophy or Philosophy of Life?
Feb. 15: F. Garbeis, C.G. Bardili as intermediary between Leibniz and Hegel
March 8: A. Handlirsch, On Orthogenesis, Its Suitability and Practicality
March 15: P. Kuranda, On Beginnings of German Political Education and Philosophy
March 16: E. Utitz, On the Essence *[Wesen]* of Charakterologie
May 3: E. Schweidler, On the Natural Rational Unity in Physics
May 10: M. Haubfleisch, On E. Rothacker: Logic and Systematics of Psychology
May 17: O. Rossi, Max Scheler's Theory of Value
June 7: J. Herz, **Adolf Stöhr** and the Philosophy of Matter
Nov. 22: **F. Strunz,** A Critique of the Religious Writings of Paracelsus
Dec. 6: F. Kröner, On the Idea of a Systemology

1930

Jan. 17: A. Molitor, **Bergson's** Teaching of Instinct and the Psychobiology of Animals without Backbones
Jan. 31: R. Meister, The Systematic Connection Between Locke's Pedagogical and Philosophical Views
Feb. 21: ***V. Kraft,*** Critical Lecture on **Husserl's** "Formale und transcendentale Logik"
March 7: K. v. Ettmayer, Problems in the Psychology of Speech
March 21: **H. Gomperz,** Can Deduction Lead to "New" Results?
May 23: M. Haubfleisch, Lecture on M. Heidegger's *Sein und Zeit*
June 13: F. Redl, The Ethics of Nicolai Hartmann
Nov. 14: ***M. Schlick,*** Is There a Material A Priori?
Nov. 28: F. Kainz, Report of the 4th Congress of Aesthetics and the Science of Art in Hamburg from October 7 to 9, 1930
Dec. 12: H. Piesch, On the Mental Relationship Between Kant's Ethics and that of Meister Eckhart

1931

Jan. 16: J. Petrovici (Romanian Minister of Education), German Philosophy in Romania
Feb. 13: H. Eibl, The Effect of Augustine's Thoughts on Modern Philosophy
Feb. 27: F. Redl, The Psychology of Adolescence
May 15: R. Meister, Campanella's Dialogue about the [Political] State of the Sun
May 29: F. Trojan, Report on the 5th International Congress for Moral Education (Paris, 1930)
Nov. 27: R. Reininger, On the 100th Anniversary of Hegel's Death
Dec. 4: K. Roretz, What is Reality?

1932

Jan. 22: F. Koch, Goethe and German Idealism
Feb. 19: E.K. Winter, Platonism as Psychological Problem

March 4: F. Kröner, On R. Reininger's "Metaphysik der Wirklichkeit"
April 22, F. Borkenau-Pollak, Descartes' Rationalism and the Religious Crisis of His Life
May 13: F. Trojan, Writings of the American Moralist, F.C. Sharp
May 20: ***B. Juhos,*** Stages of Causality
June 10: H. Latzin, Homologous Space in Unperceivables [*Unanschaulichen*]
Nov. 18: K. Siegel, Dualism in Spinoza's System (300th Anniversary of His Birth
Dec. 2: F. Kröner, Jasper's "Philosophy"

1933:

Jan. 20: O. Hartmann, Man, Speech, and the World
Feb. 3: K. Siebert, Theory of Psychoanalytic Dreams and Mistakes from the Standpoint of Modern Psychology
Feb. 17: F. Redl, "Constructive Understanding" ["*emporbildende Verstehen*"] of Spranger and the Psychoanalytic Method
March 3: L. Bolterauer, The question about the Meaning of Life as a Philosophical Problem
March 17: A. Molitor, On the Essence [*Wesen*] and Definability of Truth
May 26: D. Petrasovics, Thoughts on Indian Social Philosophy
June 13: R. Meister, Philosophical Content and Poetic Form of **Nietzsche**'s *Also Sprach Zarathustra*
June 20: E. Liljequist (Lund), Swedish Personal Idealism and Kant
Nov. 10: K. Bühler, On the Relations of Theory of Language to Philosophy
Nov. 17: H. Lindemann, The Sensations of Speech and the Problem of Other Minds
Dec. 1: M. Haubfleisch, The Absolute and the Ego in the Metaphysics of the Present

1934

Jan. 12: **H. Gomperz,** Science and Action
Jan. 26: A. Molitor, On Gustav Wolff, Life and Knowledge - A Philosophy of Biology
Feb. 9: R. Meister, On F. Schleiermacher on the 100th Anniversary of His Death
Feb. 16: K. Siebert, On the Construction of Value
June 15: K. Wolf, Karl Bühler's Theory of Expression
Nov. 9: A. Kastil, **Franz Brentano**'s Concept of Truth. With an Introduction into the Publication of His *Nachlass*
Nov. 30: A. Molitor, Neo-Positivism and Metaphysics
Dec. 7: ***V. Kraft,*** The Empirical Foundations of Value Theory

1935

Jan. 11: H. Hartmann (Berlin), The Gestalt Philosophy of Aristotle
January 18: ***M. Schlick,*** On the Concept of Wholeness [*Ganzheit*]
Feb. 8: R. Meister, The *scientia experimentalis* of Roger Bacon
March 1: A. Seibt, Locke on Free and Unfree Human Behavior
May 17: P. Merlan, Plato's Form of Philosophical Communication
May 24: K. Wolf: Language in a System of Two Fields [*Zweifeldersystem*]
Nov. 8: R. Meister, The Origin of **Nietzsche**'s *Also Sprach Zarathustra* from
Dec. 6: E. Fröschels, On the Problem of Free Will

1936

Jan. 17: ***M. Schlick,*** On Unsolvable Problems
March 6: F. Kainz, On the Essence [*Wesen*] and Accomplishments of Language
May 8: F. Garbeis, On the Background of [Kant's] Critical Theory
June 5: M. Haubfleisch, Nicolai Hartmann: On the Foundation of Ontology
Oct. 23: ***F. Waismann, Schlick*** as Philosopher - In Memory of ***Moritz Schlick***

Nov. 13: A. Kastil: **Franz Brentano**'s Criticism of Anti-Metaphysicians
Dec. 4: A. Molitor: New Realism and New Scholasticism

1937

Jan. 15: K. Mumelter, The Fourth Dimension of Mathematics and Occultism

March 5: A. Kastil, Is the difference Between Wholeness and Totality a Factual One? (Remarks on ***M. Schlick***'s Lecture of January 18, 1935)
April 23: M. Haubfleisch, Lecture from Karl Neisser's Nachlass; F. Kröner on Karl Neisser's Book *Ptolomäus oder Kopernikus*. footnote 43
June 4: ***F. Waismann,*** What is Logical Analysis?
Oct. 5: R. Meister, 50th year Celebration of the Philosophical Society
Dec. 10, E. Frauwallner, On Different Periods of Indian Philosophy

1938

Jan. 14: A. Kastil, The World View of the Philosopher **Franz Brentano** (Celebrating the 100th Anniversary of His Birth)
May 13: R. Reininger, Arthur Schopenhauer. (Celebrating the 150th Anniversary of His Birth). footnote 44
June 10: F. Weinhandl (Kiel), The Intellectual Line [*Gestaltgedanke*] in the Philosophy of the New Germany

SECTION 4. **MACHIST MODE TWO**

MACH'S CONTACT WITH OR INFLUENCE ON MANY VIENNESE INTELLECTUALS

A. = Influence alleged by Person or Biographer:
B. = Wrote Book or Article on Ernst Mach:
C. = The Person Corresponded with Mach:
D. = Discussions or Lectures Presented in the Philosophical Society (1888-1938):

Friedrich Adler[45] (1879-1962) [INFLUENCE: PHILOSOPHY AND PHYSICS]

A. "Ernst Machs Überwindung des mechanischen Materialismus", *Die Neue Zeit,* **35** (1916) 111.
B. *Ernst Machs Überwindung des mechainschen Materialismus,* Wien, 1918, plus several articles.
C. 42 letters from Mach and 15 to Mach (1904-13).

Guido Adler (1855-1941) [INFLUENCE: CHAIR IN MUSICOLOGY]

A. Kurt Blaukopf, *Pioniere Empiristischer Musikforschung*, Wien, 1995.
B. Review of Ernst Mach's " Beitrag zur Geschichte der Akustik" in the journal: *Vierteljahrschrift Für Musikwissenschaft,* **8** (1892) 533-536.
C. 3 letters and 3 postcards from Mach (1884-1890); 1 letter to Mach (1896).

Max Adler (1873-1937) [INFLUENCE: PHILOSOPHY]

B. "Mach und Marx. Ein Beitrag zur Kritik des modernen Positivismus." *Archiv für Sozialwissenschaft und Sozialpolitik,* **33** (1911) 348-400.

D. May 27, 1907: "Toward an Epistemology of the Social Sciences"
June 13, 1907: Discussion of the lecture of May 27.

Ursula Baatz [INFLUENCE: MACH'S INTEREST IN BUDDHISM]
A. "Ernst Mach und der Buddhismus", *Conceptus - Zeitschrift für Philosophie*, Nr. 56, **22** (1988) 19-33.
B. "Ernst Mach - The Scientist as a Buddhist?" in *Ernst Mach - A Deeper Look*, Dordrecht, 1992, pp. 183-199.

David Josef Bach (1874-1947) [INFLUENCE: PHILOSOPHY AND MUSIC]
A. Kurt Blaukopf, *Pioniere Empiristischer Musikforschung*, Wien, 1995, pp. 164-167.
B. "Ernst Mach", *Frankfurter Zeitung*", no. 49, Erstes Morgenblatt, Feb. 18, 1908.
C. 1 letter to Mach (1901)

Theodor Beer [INFLUENCE: PHYSIOLOGY, PSYCHOLOGY, & BUDDHISM]
A. T. Beer, A. Bethe, & J. v. Uexküll, "Vorschläge zu einer objektivierenden Nomenklatur in der Physiologie des Nervensystems", *Zentralblatt für Physiologie*, **13** (1899) No. 6.
B. *Die Weltanschauung eines modernen Naturforschers. Ein nicht-kritisches Referat über Machs "Analyse der Empfindungen"*, Dresden/Leipzig, 1903.
C. 22 letters to Mach. (1902-1913)

Kurt Blaukopf (1914- [INFLUENCE: PHILOSOPHY AND MUSIC]
A. "Friedrich Jodl und Ernst Mach über Bachs Dissertation", *Pioniere Empiristischer Musikforschung* (Wissenschaftliche Weltauffassung und Kunst, Band 1), Institut Wiener Kreis (Hrsg.), 1995, pp. 162-163.
B. Kurt Blaukopf, *Philosophisches Denken in Österreich*, Wien, 1946.

Ludwig Boltzmann (1844-1906) [INFLUENCE: PHILOSOPHY AND PHYSICS]
A. *Populäre Schriften*, Leipzig, (1905) 1925, pp. 338-339.
B. "Models", *Encyclopaedia Britannica*, 10th edition, 1902. Reprinted in 11th edition, Volume 10, Cambridge, England, 1911, pp. 638-640.
C. 7 letters to Mach (1892-1905) in Joachim Thiele, *Wissenschaftliche Kommunikation*, Kastellaun, 1978, pp. 45-52.
D. January 17, 1895: "On a Physical Application of Probability Statistics"
November 22, 1899: Contributed to Discussion on: "Are There Strongly Proven Grounds Against Solipsism?" (Unpublished)
November 25, 1904: Contributed to Discussion after W. Ostwald's Lecture "The Energetic Theory of Happiness" (Published in *Beilage zum XVIII Jahresbericht*, 1905)
January 21, 1905: "On a Thesis of Schopenhauer" (Published in *Beilage zum XVIII Jahresbericht*, 1905.)
October 28, 1905: "An Explanation of Love and the Entropy Law by means of Statistical Probability"

Franz Brentano (1838-1917) [INFLUENCE: PSYCHOLOGY]
A. "Zur Würdigung des Positivismus von Ernst Mach", a manuscript from 1900.
B. *Über Ernst Machs "Erkenntnis und Irrtum"*, aus dem Nachlass herausgegeben und eingeleitet von R.M. Chisholm & J.C. Marek, Wien, 1988.

C. 6 letters from Mach & 7 to Mach (1895-1910) in *Über Ernst Machs...* pp. 203-27)
D. April 27, 1888: "Methods of Historical Investigation in Philosophical Fields"
March 27, 1889: "On the Concept of Truth"
December 17, 1889: "Schelling's Philosophy"
December 16, 1890: "Modern Errors in Our Understanding of the Laws of Deduction"
March 22, 1892: "On the Future of Philosophy"
January 29, 1893: "The Phenomenal Green"
[November 9, 1934: A. Kastil on Brentano's concept of truth and on his *Nachlass*]
[November 13, 1936: A. Kastil on Brentano's criticism of anti-metaphysicians]
[January 14, 1938: A. Kastil on Brentano's world view in celebration of the 100th anniversary of his birth]

Josef Breuer (1842-1925) [INFLUENCE: PHYSIOLOGY]
A. Joachim Thiele, *Wissenschaftliche Kommunikation,* Kastellaun, 1978, pp. 62-63.
B. "Über die Funktion der Otolithen-Apparate". *Arch. Physiol.* **48** (1891) 195-306.
C. 3 letters to Mach (1882-1908) in Thiele, op. cit., pp. 59-62.
D. February 26, 1894: "The Static Mind"
March 5, 1894: Continuing the discussion of February 26.
January 15, 1900: Introductory lecture to the discussion of M. Kassowitz's theory of "The Old and New Vitalism".
January 29, 1900: More discussion on vitalism.
March 12, 1900: "Fechner's Ideas on the Developmental History of Organisms"
May 2, 1902: "The Crisis of Darwinism and Teleology".
May 8, 1902: Discussion on the lecture of May 2. (Parts of the discussion on Darwinism by Breuer and others included in *Beilage zum XV. Jahresbericht,* 1902.)

Rudolf Carnap (1891-1970) [INFLUENCE: EPISTEMOLOGY & SEMANTICS]
A. "Preface to the Second Edition", *The Logical Structure of the World,* Berkeley, 1969, p. vii.
B. *Der Logische Aufbau der Welt,* (originally written 1922-1925) 1928.
D. February 11, 1927. "The Essence [*Wesen*] of Concepts".

Christian Freiherr von Ehrenfels (1859-1932) [INFLUENCE: GESTALT THEORY]
A. "Über Gestaltqualitäten", *Vierteljahrschrift für wissenschaftliche Philosophie,* **14** (1890) 242-292.
B. Barry Smith, "The Theory of Gestalt Qualities", in *Austrian Philosophy,* La Salle, 1994, pp. 243-250.
D. January 30, 1892: "About Unprovable Knowledge".
February 9, 1892: "Richet's Book on Thought-Transfer & so-called 'Bright-Sight'".
March 8, 1892: Discussion on the lecture of February 9.
November 16, 1892: Discussion on some theses about moral assessment.
December 7, 1892: More discussion.
February 22, 1893: "Semper as Aesthetician".
April 27, 1893: "Probability Grounds Against Monism".
May 10, 1893: Discussion of April 27th lecture.
May 10, 1894: "On the Influence of Metaphysical Views on Ethical Education".
May 22, 1894: Discussion of May 10th lecture.

June 10, 1894: More discussion.
June 23, 1896: "The Problem of Psychological Egoism".
November 9, 1901: "The Socio-Ethical Significance of Leisure". Published in *Beilage zum XV. Jahresbericht* (1902).
[Leipzig, 1902: Ehrenfels contributes to "Die Krises des Darwinismus" which is published as part of *Beilage zum XV. Jahresbericht* by the Society.
May 7, 1912: "New Proofs for the Existence of God".
May 9, 1912: Continuation of lecture on 'New proofs for the Existence of God".
May 11, 1912: Further Continuation of lecture.
May 20, 1912: Discussion on the 3 lectures on New proofs about God.
December 13, 1915: Report on his new work *Kosmogonie*.
December 14, 1915: Continuation of his lecture on *Kosmogonie*.

Felix Ehrenhaft (1879-1952) [INFLUENCE: ECONOMY OF THOUGHT]
A. Josef Mayerhöfer, "Der Kampf um die Atomistik und die Relativitätstheorie in Wien", *Mitteilungen der Österreichischen Gesellschaft für Geschichte der Naturwissenschaften*, **12** (1992) 39.

D. March 17, 1917: "On Atoms, Electrons, & Quanta with Stefan Meyer (with demonstrations by spinthariscope and photographs of particle paths)".
May 5, 1917: Discussion on the March 17th lecture.

Sigmund Exner (1846-1926) [INFLUENCE: PERCEPTION & SENSATIONS]
A. Ernst Mach, *Grundlinien der Lehre von den Bewegungsempfindungen*, Leipzig, 1875.
C. 1 letter to Mach (1894).
D. May 29, 1888: "Some New Studies on Sensations of Movement (with Examples)".
October 30, 1895: "On Sensations of Direction".
January 16, 1899: "Studies on the Periphery of Localized Vision".
May 13, 1904: "On the Central Process of Seeing".

Paul Feyerabend (1924-1994) [INFLUENCE: PHYSICS & PHILOSOPHY]
A. *Zeitverschwendung*, Frankfurt am Main, 1995, p. 47.
B. "Mach's Theory of Research and its Relation to Einstein", *Farewell to Reason*, London, 1995, pp. 192-218.

Philipp Frank (1884-1966) [INFLUENCE: PHILOSOPHY & PHYSICS]
A. "Die Bedeutung der physikalischen Erkenntnistheorie Machs für das Geistesleben der Gegenwart", *Naturwissenschaften*, **5** (1917) 65-72.
B. "Einstein, Mach, and Logical Positivism" in *Albert Einstein: Philosopher-Scientist*, Volume One, edited by P.A. Schilpp, La Salle, (1949) 1982, pp. 269-286.
C. 2 letters to Mach (1910); one letter is in J. Blackmore & K. Hentschel (eds.), *Ernst Mach als Aussenseiter*, Wien, 1985, pp. 82-83.
D. December 4, 1907: Leads discussion on Driesch's philosophy and vitalism.
December 4, 1909: "Is There an Absolute Motion?"
December 10, 1909: Discussion on the December 4, 1909 lecture.
March 22, 1915: With A. Höfler he leads discussion on physical relativity.
April 22, 1915: More discussion on physical relativity.

Rudolf Goldscheid (1870-1931) [INFLUENCE: ECONOMY OF THOUGHT]
A. Wilhelm Jerusalem, *Gedanken und Denker - Neue Folge*, Wien, 1925, p. 196.
B. *Höherentwicklung und Menschenökonomie*, Leipzig, 1911.
C. 12 letters to Ernst Mach (1907-1911)
D. April 23, 1904: "On the Necessity of Volitional next to Epistemological Thinking (with Particular Reference to Philosophy of History)". This lecture was included in the *Beilage zum XVII Jahresbericht*, 1904.
May 6, 1904: Discussion on the lecture of April 23.
January 16, 1907: "The Concept of Direction and its Significance for Philosophy"
February 26, 1907: Discussion on the lecture of January 16.

Heinrich Gomperz (1873-1942) [INFLUENCE: NEUTRAL MONISM]
A. Philip Merlan, "Foreword", in Heinrich Gomperz, *Philosophical Studies*, edited by Daniel S. Robinson, Boston, 1953, pp. 9-11.
B. "Ernst Mach", *Archiv für Geschichte der Philosophie*, **29** (1916) 321-328.
C. 13 letters from Mach and 6 to Mach (1902-1910) in R. Haller & F. Stadler (eds.), *Ernst Mach - Werk und Wirkung*, Wien, 1988, pp. 244-257.
D. March 17, 1909: "On Evaluating Personality"
November 10, 1924: "For and Against Plato from Atticus to Voltaire"
January 13, 1928: "Plato's Autobiography"
March 21, 1930: "Can Deduction Lead to 'New' Results?"
January 12, 1934: "Science and Action"

Theodor Gomperz (1832-1912) [INFLUENCE: PHILOSOPHY]
A. Heinrich Gomperz, "Ernst Mach", *Archiv für Geschichte der Philosophie*, **29** (1916) 325-326.
C. 6 letters from Mach and 11 to Mach (1882-1909) in R. Haller & F. Stadler (eds.), *Ernst Mach - Werk und Wirkung*, Wien, 1988, pp. 212-228.
D. May 29, 1888: "Report on the New Archive for the History of Philosophy"
February 17, 1898: "Philosophy of History"
December 17, 1904: "Spencer's Education and What it Teaches us (& discussion).

Arthur E. Haas (1884-1941) [INFLUENCE: HISTORY OF SCIENCE]
A. *Antike Lichttheorien*, PHD Thesis, Universität Wien.
B. "Historische Analyse des Energieprinzips", *Gesellschaft für Geschichte der Medizin und Naturwissenschaften*, Köln, 1908.
D. May 11, 1907: "The Most General Laws of an Event and their Significance for the Cosmological Problem".
June 2, 1911: "On Physical Objections Against the Idea of Infinity of the World".
March 6, 1914: "The Significance of greek Philosophy of Nature for Physics".
May 5, 1919: "Physics as Geometrical Necessity"
May 11, 1920: "Atomic Harmony of the Sheres"
December 20, 1921: "The Problem of Ultimate Matter and Modern Physics".
November 19, 1926: "Objective and Human Physics"

Gottlieb Haberlandt (1854-1945) [INFLUENCE: BIOLOGY & PHYSIOLOGY]
B. *Sinnesorgane im Pflanzenreich*, Leipzig, 1901. [Both men had been influenced by Gustav Fechner's psycho-physics and attribution of minds to plants.]
C. 9 letters to Mach (1900-1914).

Hans Hahn (1879-1934) [INFLUENCE: PHENOMENALISM]

A. Rudolf Carnap, "The Development of My Thinking", *The Philosophy of Rudolf Carnap, La Salle*, 1963, p. 20.

B. Hans Hahn, "Superfluous Entities or Occam's Razor", *Hans Hahn - Empiricism, Logic, and Mathematics*, edited by Brian McGuinness, Dordrecht, 1980, p. 17.

D. November 23, 1906: "Does Geometry Ultimately Rest on Facts?"
December 5, 1906: Discussion of the November 23rd lecture.
December 2, 1921: "Bolzano's Paradoxes of Infinity".
February 3, 1922: Discussion on infinity.
February 19, 1923: "On Bertrand Russell's *Introduction to Mathematical Philosophy*".
June 4, 1923: Discussion of the lecture on Russell.
January 28, 1927: "On the Newer Logical Theories"

Ludo Hartmann (1865-1924) [INFLUENCE: MONISM, POPULAR EDUCATION]

A. Friedrich Stadler, *Vom Positivismus zur "Wissenschaftliche Auffassung"*, Wien, 1982, p. 108-109.

B. *Arbeit über historische Entwicklung*, 1905. (Dedicated to Ernst Mach.)

C. 9 letters to Mach (1906-1914), and 4 more if "Ludwig Hartmann" is the same as "Ludo Hartmann" (1913-1916).

D. January 12, 1915: "Historical Tendencies"
March 1, 1915: Discussion on Jauary 12 lecture and whether there are laws of complex phenomena.

Friedrich von Hayek (1899-1993) [INFLUENCE: COLOR THEORY]

A. *The Sensory Order*, London and Chicago, 1952.

B. "Diskussionsbemerkungen über Ernst Mach und das sozialwissenschaftliche Denken in Wien" in *Symposium aus Anlass des 50. Todestages von Ernst Mach*, Freiburg im Breisgau, 1967, pp. 41-44.

Alois Höfler (1853-1922) [INFLUENCE: PSYCHOLOGY & PEDAGOGY]

A. "Ernst Mach", *Zeitschrift für den physikalischen und chemischen Unterricht*, Zweites Heft, **29** (1916) p. 60.

B. "Zur Geschichte und Wurzel der Machschen Philosophie", *Zeitschrift für den physikalischen und chemischen Unterricht*, Erstes Heft, **23** (1910), 1-16.

C. 36 letters from Mach and 3 to Mach ; the majority of letters on both sides seem to be lost. 35 extant letters from Mach are only from 1887 to 1891.

D. June 22, 1888: "On Paradoxes in Teaching How to Determine Distance and Size"
June 11, 1891: "Seven Theses in A. Frick's 'The Steady Filling of Space by Mass'"
January 9, 1892: "Relating Political to Philosophical Education"
June 10, 1892: "Remembering Theodor Meynert and His Relations to the Society"
December 13, 1894: "Mach's Lecture on the Principle of Comparison in Physics"
January 24, 1895: Continuing the discussion of December 13.
November 9, 1900: "Remembering Nietzsche's First Five Publications"
April 24, 1901: "Words in Memory of Gustav Fechner's 100th Birthday"
January 19, 1906: A discussion of F. Klein's Thesis about Non-evident Geometry

and Perception of Space. ("Grenzfragen der Mathematik und Philosophie" by Klein and Höfler published in *Beilage zum XIX. Jahresbericht*, 1906.)
November 11, 1907: "The Answer to Dr. Reininger's Question about Epistemology and 'Its Typical Conclusions'".
November 9, 1908: "Questions which Physics has about Philosophy"
December 2, 1912: "Showing Models and Experiments of what is Psychological in Colored Bodies".
March 22, 1915: Discussion with P. Frank on the question of physical relativity
April 22, 1915: More discussion with P. Frank on physical relativity
November 20, 1917: "Words Honoring the 30th Anniversary of Fechner's Death".
May 31, 1919: "Can Feeling Give Knowledge? An Emotional Presentation".
November 4, 1921: "Discussing the Philosophy of Alexius Meinong".
[May 22, 1922: Honoring our departed leader and honorary President A. Höfler:
1. R. Reininger, "Höfler's Activity for the Philosophical Society".
2. O. Pommer, "Höfler as Philosopher".
3. R. Meister, "Höfler as Teacher".]

Richard Hönigswald (1875-1947)
A. William M. Johnston, *The Austrian Mind*, Berkeley, 1972, p. 185.
B. *Zur Kritik der Machschen Philosophie. Eine erkenntnistheoretische Studie*, 1903
C. 2 letters to Mach (1905-1906)

Edmund Husserl (1859-1938) [INFLUENCE: PSYCHOLOGISM]
A. "Amsterdam Vorträge", *Husserliana*, Bd. IX, Den Haag, 1968, p. 302.
B. A review of E. Mach's "Über das Princip der Vergleichung in der Physik", *Verhandlungen der Gesellschaft Deutscher Naturforscher und Ärzte*, Theil I (1894) 44-56 in *Archiv für systematische Philosophie*, **3** (1897) 241-244.
C. 1 letter from Mach and one letter (a long one) to Mach (1901) in Joachim Thiele, *Wissenschaftliche Kommunikation, Kastellaun*, 1978, pp. 153-157.
D. [February 21, 1930: V. Kraft, "Critical Lecture on Husserl's 'Formale und transcendentale Logik'".]

Wilhelm Jerusalem (1854-1923) [INFLUENCE: ECONOMY AND METHODOLOGY]
A. "Ernst Mach", *Die Zukunft*, **95** (1916) p. 327.
B. *Gedanken und Denker: Neue Folge*, Vienna and Leipzig, 1925, pp. 23-30.
C. 98 letters and postcards from Mach and 11 to Mach (1895-1915); 15 letters can be found in J. Blackmore & K. Hentschel, *Ernst Mach als Aussenseiter*, Wien, 1985.
D. June 26, 1889: "On the Education of Laura Bridgman".
February 24, 1892: "Wundt's Institute for Experimental Psychology in Leipzig"
March 28, 1895: "On the History of the Problem of Judgment"
December 4, 1900: "Typical Ways of Looking at Things"
January 18, 1901: Discussion on the lecture of December 4.
May 13, 1901: Further Discussion.
February 12, 1904: "Memorial Address on the 100th Anniversary of Kant's Death".
May 17, 1905: "Psychological Logic & Its Newest Opponent (with Discussion)".
May 24, 1905: Discussion of the lecture of May 17.
February 24, 1906: "On Genetic and Biological Aesthetics"
March 16, 1906: Discussion of the lecture of February 24.

January 24, 1908: "Pragmatism, a new Philosophical Method"
November 25, 1921: "Knowledge as a Sociological Problem"

Friedrich Jodl (1859-1914)

A. Georg Gimpl, "Fronde gegen die philosophierenden Physiker. Friedrich Jodl und seine Auseinanderung mit Ernst Mach und Ludwig Boltzmann", *Ego und Alter-ego. Wilhelm Bolin und Friedrich Jodl um die Aufklärung*, Georg Gimpl (Hrsg.), Festschrift für Juha Manninen, Verlag Peter Lang: Frankfurt a.M., pp. 287-330.
B. "Ernst Mach und seine Arbeit *Erkanntnis und Irrtum*" in Ernst Mach, *Erkenntnis und Irrtum*, 3. Auflage, 1917, pp. 464-470.
C. 3 letters from Mach and 1 letter to Mach (1909)
D. November 25, 1896: "Some Thoughts on the Causal Problem".
February 18, 1899: "Natural Law and Chance".
February 17, 1900: "Giordano Bruno".
November 28, 1903: "On the Problem of Materialism in Psychology".
December 11, 1903: Discussion of last lecture as stimulated by some remarks of L. Linsmayer on Neurons.
(January 21, 1905: Discussion with Boltzmann on Schopenhauer after B's lecture, included in *Beilage zum XVIII Jahresbericht*, Leipzig, 1905, pp. 60-62.)
June 20, 1906: "Psychic Lawfulness and the Methods of Psychology".
[[February 5, 1914: Meeting to Mourn the death of our Leader Friedrich Jodl.]]

Bela Juhos (1901-1971) [INFLUENCE: OPPOSITION TO "METAPHYSICS"]

A. "Wie stellt sich die neuere Erkenntniskritik zur Philosophie Schopenhauers?" in C.A. Emge und O. v. Schweinichen (Hrsg.), *Gedächtnisschrift für Arthur Schopenhauer zur 150. Wiederkehr seines Geburtstages*. (Philosophische Untersuchungen, Bd. 6), Berlin, 1938, p. 138.
B. *Erkenntnisformen in Natur- und Geisteswissenschaften*, Leipzig, 1940.
D. May 20. 1932: "Stages of Causality".

Jürgen Kaizik (1950-

B. *Versuch einer Begegnung - Mach gegen Boltzmann*, Franz Deuticke, Wien, 1986. [a fictional dialogue plus interpretation and evaluation]

Victor Kraft (1880-1975) [INFLUENCE: PHILOSOPHY OF SCIENCE]

A. Friedrich Kainz, "Viktor Kraft", *Almanach der Österreichischen Akademie der Wissenschaften*, **125** (1976), p. 527.
B. "Ernst Mach als Philosoph", *Almanach der Österreichischen Akademie der Wissenschaften*, **116** (1966) 373-387.
D. November 6, 1911: "Toward a Critique of Epistemological Positivism".
November 20, 1911: Discussion of the lecture on November 16.
December 4, 1911: More discussion.
November 6, 1916: "Mach as Philosopher".
January 8, 1917: Discussion on the Mach lecture and on one by O. Neurath.
January 22, 1917: More discussion.
June 19, 1923: "On Müller-Freienfels: Irrationalism & a Sketch of Epistemology".
May 21, 1926: "The Basis for the Validity of the Principles of Knowledge".
March 2, 1928: "Ethics as Science".
February 21, 1930: "Criticism of Husserl's 'Formale und transcendentale Logik'".

December 7, 1934: "The Empirical Foundations of Value Theory".

Johann Clemens Kreibig (1863-1917)

B. Review of Mach's Erkenntnis und Irrtum in *Deutsche Literaturzeitung* (Berlin), **27** (1906) 2071-2075.

C. 2 letters to Mach (1907)

D. **March 14, 1895: Continuing discussions of 13/12/94 & 24/1/95 on Ernst Mach's lecture "On the Principle of Comparison in Physics".**

April 25, 1895: "How Can We Refute Popular Scepticism about Ethics?"

May 9, 1895: Discussion on lecture of April 25.

May 30, 1895: More Discussion.

November 22, 1900: "Lecture on Nietzsche's Understanding of Herd-Morality and Human-Morality" (with Discussion).

December 13, 1900: "Is Nietzsche's Summons Justified? Must one Seek out Values which Transcend Feelings?

January 24, 1901: Discussion of lecture of December 13.

January 7, 1903: "What is a Concept?"

February 5, 1903: Discussion of lecture of January 7.

December 11, 1903: Discussion of Jodl's lecture of November 28 on materialism in psychology.

December 5, 1910: "On the Theory of Perception".

January 16, 1911: Discussion of December 5th lecture.

February 26, 1912: Discussion on Theodor Lessing's *Wertaxiomatik*.

November 16, 1915: Discussion on the question: "Is Ethics Also Valid for Peoples and Nations?".

November 29, 1915: Discussion of lecture of November 16.

May 7, 1917: Discussion evening on egoism and altruism in biology and politics.

May 22, 1917: More Discussion.

June 20, 1917: Still more discussion.

Eduard Krischek

C. 18 letters and 2 postcards to Mach (1863-1897).

Eduard Kulke (1831-1897) [INFLUENCE: SURVIVAL OF THE FITTEST IN MUSIC]

A. *Über die Umbilding der Melodie*, Prag, 1884, p. 2.

C. 104 letters from Mach and 26 to Mach (1865-1896)

D. January 18, 1890: "On Both Basic Problems of Beauty"

November 26, 1895: "On Winckelmann's Theory of Beauty"

Anton Lampa (1868-1938) [INFLUENCE: PHILOSOPHY AND PHYSICS]

A. Lise Meitner, "Looking Back", *Bulletin of the Atomic Scientists*, Volume **20**, November 1964, pp. 2-7.

B. *Ernst Mach*, Prag, 1918.

C. 17 letters and postcards to Mach (1900-1915). One letter is in *Ernst Mach als Aussenseiter*, edited by J. Blackmore and K. Hentschel, Wien, 1985, pp. 80-81. See also A. Kleinert, *Gesnerus*, **35** (1975) 285-292.

Emil Lucka (1877-1941)

B. "Das Erkenntnisproblem und Machs *Analyse der Empfindungen*. Eine kritische

Studie", *Kantstudien*, **8** (1903) 396-447.

D. April 18, 1907: "Memory and Fantasy". Published as "Gedächtnis und Phantasie" in *Beilage zum XX. Jahresbericht*, 1907.

E. Lucka was a member of the *Wiener Dichterkreis* with Wildgans, Zweig, & Werfel.

Josef Mayerhöfer (1914- [INFLUENCE: HISTORY OF SCIENCE]

A. "Ernst Mach in Wien", *Mitteilungen der Österreichischen Gesellschaft für Geschichte der Naturwissenschaften*, **9** (1989) 19-42.

B. "Der Kampf um die Atomistik und die Relativitätstheorie in Wien", *Mitteilungen der Österreichischen Gesellschaft für Geschichte der Naturwissenschaften*, **12** (1992) 23-50.

Alexius Meinong (1853-1920)

C. 17 letters from Mach (many published in R. Kindinger (ed.), *Philosophenbriefe*, Graz, 1965) and 6 letters to Mach (1890-1909).

D. [April 25, 1896: "Professor Meinong's Theory of Evident Suppositions", (by K. Zindler).]

[February 10, 1913: "On the Concept of Object in Meinong's Theory of Objects", (by E. Mally).]

[[February 25, 1921: "Meinong as Teacher and Human Being", (by E. Martinak).
"Meinong's Beginnings as a Historian", (by O. Redlich)
"From Meinong's Philosophy", (by R. Meister)]]

[November 4, 1921: "Out of and into the Philosophy of Alexius Meinong", (by A. Höfler).]

[June 2, 1924: "On Meinong's *Zur Grundlegung der allgemeinen Werttheorie*", (by O. Pommer).]

Carl Menger (1840-1921) [INFLUENCE: ECONOMY OF THOUGHT]

A. Friedrich Stadler, *Vom Positivismus zur "Wissenschaftlichen Weltauffassung"*, Wien, 1982, p. 105.

C. 1 letter to Mach (1906)

D. Member of the University of Vienna Philosophical Society (1913).

Karl Menger (1902- 1988) [INFLUENCE: FUNCTIONAL RELATIONS]

A. "The Mathematical Implications of Mach's Ideas: Positivistic Geometry, the Clarification of Functional Connections" in *Ernst Mach - Physicist and Philosopher*, edited by R.S. Cohen and R.J. Seeger (Boston Studies in the Philosophy of Science, Volume 6), Dordrecht, 1970, pp. 107-125.

B. "Introduction to the Sixth American Edition 1960", in Ernst Mach, *The Science of Mechanics*, La Salle, 1960, pp. v-xxi.

Richard von Mises (1883-1953) [INFLUENCE: PHILOSOPHY]

A. *Kleines Lehrbuch des Positivismus*, Berlin, 1951.

B. *Ernst Mach und die empirische Weltauffassung*, (Verein Ernst Mach, Heft 7) Wien, 1938.

Robert Musil (1880-1942) [INFLUENCE: NO EGO, SUBJECT, OR SELF]

A. *Der Mann Ohne Eigenschaften*, Berlin & Lausanne, 3 volumes, 1930-1943.

B *Beitrag zur Beurteilung der Lehren Machs*, PHD Diss, University of Berlin, 1908.

Otto Neurath (1882-1945) [INFLUENCE: PHILOSOPHY, HISTORY & OPTICS]

A. "Principielles zur Geschichte der Optik", *Archiv für die Geschichte der Naturwissenschaften und Technik,* **5** (1915) 371-389.

B. "Verein Ernst Mach", *Erkenntnis,* **1** (1930) 76.

C. 2 letters to Mach (1915) in Joachim Thiele, *Wissenschaftliche Kommunikation,* Kastellaun, 1978, pp. 99-101.

D. March 5, 1908: "War and Moral Principles".

January 17, 1910: Led discussion on the concept and range of the validity of the a priori.

February 14, 1910: More discussion on the a priori.

June 1, 1912: "The Problem of Maximum Pleasure".

January 27, 1913: "Toward a Psychology of Decisionmaking".

February 8, 1913: Discussion on the lecture of January 27.

March 2, 1914: "Classification of Systems of Hypotheses (Concerning Optics)"

December 15, 1916: "Mach's Position on the Problem of Gravitation and Inertia".

January 8, 1917: Discussion of Kraft's and Neurath's lectures on Mach.

January 22, 1917: More discussion.

April 5, 1918: "Schelling and Faraday (with Experiments".

May 6, 1918: Discussion on whether natural science is independent of philosophy as suggested by Neurath's April 5th lecture.

May 27, 1918: Discussion on whether 'philosophy of nature' can and should exist.

June 11, 1918: More discussion.

May 14, 1920: "Spengler's Concept of Culture".

June 11, 1920: Discussion on Neurath's May 14th lecture.

January 21, 1922: "Ernst Mach's History of Optics and Theory of Science".

February 26, 1923: "The Conceptual Construction of Economic Theory".

March 13, 1924: "The Philosophy of History Problem about the Development of Architecture.

Julius Ofner (1845-1924)

C. 3 letters to Mach (1895-1914).

D. March 8, 1906: "Schiller as Predecessor of Scientific Socialism" as "Schiller als Vorgänger des wissenschaftlichen Sozialismus" was published in *Beilage zum XIX. Jahresbericht,* 1906.

E. Member (1913) of the Austrian *Herrenhaus* (Upper House)

Wolfgang [Pascheles] Pauli Sr. (1869-1955) [INFLUENCE: PHYSIOLOGY]

A. Ernst Mach to Wolfgang Pauli, January 12, 1914.

B. Pauli was proofreader for Mach's *Analyse der Empfindungen,* 1902. See Darmstadt edition, 1987, p. XXXI

C. 1 letter from Mach and 4 letters to Mach (1913-1914)

Wolfgang Pauli Jr. (1900-1958) [INFLUENCE: POSITIVISM]

A. Pauli to Born, Zürich, April 15, 1954 in Max Born, *The Born-Einstein Letters,* New York, 1971, p. 223. (See also March 31, 1954 letter from Pauli to Born.)

E. Mach was Pauli's Godfather. See Mach to W. Pauli Sr., January 12, 1914 for thoughts on how to raise Pauli Jr.

Josef Popper-Lynkeus (1838-1921) [INFLUENCE: LOVE OF HUMAN PROGRESS]
A. Heinrich Löwy, "Die Erkenntnistheorie von Popper-Lynkeus und ihre Beziehung zur Machschen Philosophie", *Naturwissenschaften*, **20** (1932) 770-772.
B. "Ernst Mach" in *Sonntagsbeilage der Vossischen Zeitung* (Berlin), 26. März & 2. April.
C. 33 from Mach and 46 to Mach (1872-1915). Ten letters by Mach are published in *Ernst Mach als Aussenseiter*, editors J. Blackmore & K. Hentschel, Wien, 1985

Gabriele Rabel [INFLUENCE: PHENOMENALISM]
A. "Mach und die Realität der Aussenwelt", *Physikalische Zeitschrift*, **21** (1920) 433-437.
C. 3 letters from Mach and 1 to Mach (1906-1908). One letter by Mach is published in *Ernst Mach als Aussenseiter*, edited by J. Blackmore & K. Hentschel, Wien, 1985, pp. 47-48.
D. November 26, 1926: "Goethe's Conception of Kant".

Erwin Schrödinger (1887-1961) [INFLUENCE: EPISTEMOLOGY & SENSATIONS]
A. *My View of the World*, Cambridge and New York, 1964, p. viii.
B. "Grundlinien einer Theorie der Farbenmetrik im Tagessehen", *Annalen der Physik*, **63** (397-426, 427-456, and 481-520.
C. 1 letter to Mach (1910); it has been commented on by J. Blackmore, *Ernst Mach*, Berkeley, 1972, p. 262-263 and by W. Moore, *Schrödinger - Life and Thought*, Cambridge, England, 1992, p. 44.

Josef Schumpeter (1883-1950) [INFLUENCE: EPISTEMOLOGY]
A. Friedrich Stadler, *Vom Positivismus zur "Wissenschaftliche Weltauffassung"*, Wien, 1982, pp. 105-109.
D. Corresponding member of the University of Vienna Philosophical Society (1913).

Friedrich Stadler (1951- [INFLUENCE: IMPACT ON THE VIENNA CIRCLE]
A. *Vom Positivismus zur "Wissenschaftlichen Weltauffassung" - Am Beispiel der Wirkungsgeschichte von Ernst Mach in Österreich von 1895 bis 1934*, Wien & München, 1982, 168-208.
B. Rudolf Haller & Friedrich Stadler (Hrsg.), *Ernst Mach - Werk und Wirkung*, 1988.

Viktor Stern
B. "Die logischen Mängel der Machschen Antimetaphysik und die realistische Ergänzung seines Positivismus", *Vierteljahrschrift für wissenschaftliche Philosophie und Soziologie*, **38** (1914) 372-415.
C. 1 letter to Mach (1905).
D. January 11, 1906: "Arguments Against Materialism (plus Discussion)"
June 4, 1910: "On the Philosophy of My Father" as "Die Philosophie meinen Vaters" was included in *Beilage zum XXIII Jahresbericht*, 1910.
February 3, 1912: "On the Ontological Proof of God"

Adolf Stöhr (1855-1921) [INFLUENCE: PHYSIOLOGY & PSYCHOLOGY]
A. William M. Johnston, *The Austrian Mind*, Berkeley, 1972, pp. 199-201.
B. "Ernst Mach" [Nekrolog] *Die Feierliche Inauguration des Rektors der Wiener Uni-*

versität für das Studienjahr 1916/17 am 6. Nov. 1916, Wien, 1916, pp. 37-46.

C. 5 letters to Mach (1905-1915) including 2 in Joachim Thiele, *Wissenschaftliche Kommunikation*, Kastellaun, pp. 222-223.

D. May 24, 1892: "Wiesner's Theory of Plasma".
March 2, 1905: "Theory of Space".
March 11, 1905: Discussion of Stöhr's lecture of March 2 and others.
January 14, 1908: "Fechner's Hypothesis of the Cosmo-Organic State of Matter".
January 14, 1909: "The Problem of Time" as "Das Zeitproblem" was included in *Beilage zum XXII. Jahresbericht*, 1909.
February 17, 1909: "The Exciting Part of the Vitalism Question".
May 10, 1910: "On Monocular Graphics" as "Monokulare Plastik" was included in *Beilage zum XXIII. Jahresbericht*, 1910.
May 5, 1911: "The Brain and What Stimulates Imagination".
November 13, 1913: "Is Metaphysics Possible?"
November 19, 1913: Discussion on everything which is called "metaphysics".
December 1, 1913: More discussion on the meaning of "metaphysics".
March 13, 1916: "Heraclitus".
March 15, 1916: Continuing the lecture on Heraclitus.
[[May 13, 1921: Remembrance evening for Adolf Stöhr:
1. Robert Reininger: "Stöhr as the Personification of a Thinker".
2. Siegmund Kornfeld: "Stöhr as Psychologist and Moralist".]]
[June 7, 1929: Josef Herz: "Adolf Stöhr and the Philosophy of Matter".]

Hermann Swoboda (1873-1963)

B. Hannelore Rodlauer, *Otto Weininger Eros und Psyche - Studien und Briefe 1899-1902*, Wien, 1990. (See especially the letters between Swoboda and Weininger on Mach)

C. 3 letters to Mach (1908).

D. November 22, 1899: "Are There Strongly Proven Grounds Against Solipsism?"
November 29, 1899: Discussion on lecture of November 22.
June 21, 1907: "Thinking and Suffering".
May 5, 1909: "Spirit and Soul".
January 11, 1910: "On Soul-Full and Soul-Less Humanity".
February 14, 1911: "Plan of a Metapsychology".
February 21, 1913: "The Significance of Experience [*Erlebnis*] for Psychology and Theory of Knowledge".
February 24, 1913: Discussion on lecture of February 21 (O. Pommer).

Hans Thirring (1888-1976) [INFLUENCE: "MACH'S PRINCIPLE"]

A. "Über die Wirkung rotierender ferner Massen in der Einsteinschen Gravitationstheorie", *Physikalische Zeitschrift*, **19** (1918) Nr. 3, v. 1. Feb. S. 33-39.

B. "Ernst Mach als Physiker", *Almanach für das Jahr 1966/Österreichische Akademie der Wissenschaften*, Wien, **116** (1967) 361-372.

D. May 26, 1915: "The Problem of Physical Relativity".

Richard Wahle (1857-1935) [INFLUENCE: PHILOSOPHY & PSYCHOLOGY]

A. William M. Johnston, *The Austrian Mind*, Berkeley, 1972, pp. 201-203.

B. *Gehirn und Bewusstsein*, 1884.

D. June 27, 1893: "The Discovery of the Godhead".

Richard Wallaschek (1860-1917) [INFLUENCE: MUSIC BOOK PUBLICATION]
A. Kurt Blaukopf. *Pioniere Empiristischer Musikforschung*, Wien, 1995, pp. 146-149.
B. *Anfänge der Tonkunst*, Leipzig, 1903.
C. 7 letters to Mach (1902-1913).
D. January 30, 1906: "Artistic Illusion"

Otto Weininger (1880-1903) [INFLUENCE: PHILOSOPHY]
A. Hannelore Rodlauer, *Otto Weininger Eros und Psyche - Studien und Briefe 1899-1902*, Wien, 1990, p. 24.
B. Otto Weininger, *Geschlekt und Charakter*, München, (1903), but reprint undated.
C. See Weininger's correspondence with Hermann Swoboda for many references to Ernst Mach and his ideas.

Julius Wiesner (1838-1916)
C. 2 letters to Mach, undated.
D. [May 24, 1892: "Wiesner's Theory of Plasma", (A. Stöhr)
May 18, 1898: "On the Law of Unity in the Inner Structure of Organisms"
[February 21, 1916: Discussion evening on Wiesner's ideas of evolution, stagnation and degeneration.
[March 8, 1916: More discussion.
[June 5, 1916: Still more discussion.
E. Member of the *Herrenhaus* [Upper House in the Austrian parliament].

Ludwig Wittgenstein (1889-1951) [INFLUENCE: THOUGHT EXPERIMENTS AND POSSIBLY PHENOMENALISM]
A. *Tractatus logico-philosophicus*, London, (1922), 7th impression 1958, 6.36, 6.3631, 6.36311, 6.5, [See H. Visser, "Wittgenstei's Debt to Mach's *Popular Scientific Lectures*" in *Mind*, **91** (1982) 102-105 and his article in this book,]
B. John W. Cook, "Neutral Monism", *Wittgenstein's Metaphysics*, Cambridge, 1994, pp. 14-30. See also Professor Henk Visser's article in this book.
C. A letter from Wittgenstein to Bertrand Russell (January 16, 1913) gives an opinion about Mach's style.

Rudolf Wlassak (1865-1930) [COMPARING VIEWS OF MACH AND AVENARIUS]
A. Ernst Mach, "Meine Verhältnis zu R. Avenarius und andern Forschern", *Analyse der Empfindungen*, siebente Auflage, Jena, 1918, pp. 38-46.
B. Rudolf Wlassak, *Ernst Mach. Gedächtnisrede*, Leipzig, 1917, 47 Seiten.
C. 7 letters to Mach (1899-1909).
D. May 31, 1897: "Richard Avenarius's Empiriocriticism". footnote 46
June 15, 1897: "On Richard Avenarius's Empirical Concept of the World".
April 25, 1898: "Physiological and Mentalistic Views about Theory of Feeling".
May 2, 1898: Discussion on the lecture of April 25.

Edgar Zilsel (1891-1944) [INFLUENCE: MONISM, UNITY OF SCIENCE]
A. Friedrich Stadler, *Vom Positivismus zur "Wissenschaftlichen Weltaffassung"*, Wien, 1982, pp. 92-97.
B. "Philosophischen Bemerkungen", *Der Kampf*, **22** (1929) 178-186.

D. April 7, 1919: "Toward a Theory of Ideal Personality"
June 23, 1919: "Toward a Theory of Ideal Personality. Part II. The Value Problem".

NOTES

* See also the "Late Appendix" [1888-1914], in *Ludwig Boltzmann - His Later Life and Influence, Book Two* (Boston Studies in the Philosophy of Science Volume 174), Kluwer: Dordrecht, 1995, pp. 267-315. Many of the documents used in this book and the above work on Boltzmann were originally located by Henk Visser and Hannelore Rodlauer.

1 For many years the preferred German was . . . *Universität zu Wien.*

2 Ernst Mach was still registered as a member in 1913, the year he left for Vaterstetten, Bavaria. He resigned from the Austrian Academy of Sciences during that year, but no evidence has surfaced yet that he also resigned from the Philosophical Society. See John Blackmore & Klaus Hentschel (Hrsg.), *Ernst Mach als Aussenseiter,* Braumüller: Wien, 1985, pp. 113-116.

3 (Alois Höfler), *Rückblick auf die ersten fünfundzwanzig Vereinsjahre der Philosophischen Gesellschaft an der Universität zu Wien mit dem 26. Bericht für das Jahr 1912/13,* im Selbstverlag der Philosophischen Gesellschaft: Wien, 1913, p. 38-39. For a much more accessible source see Black-more, op. cit. (1995), p. 300.

4 Wilhelm Ostwald, "Energetische Theorie des Glücks - Anhang" in *Wissenschaftliche Beilage zum achtzehnten Jahresbericht (1905) der Philosophischen Gesellschaft an der Universität zu Wien.* Verlag von Johann Ambrosius Barth: Leipzig, 1905, p. 4. See also Ibid. (Blackmore, 1995), p. 305.

5 *Rückblick,* op. cit. (1913), p. 3. See in addition Richard. Meister, "Die Geschichte der Philosophischen Gesellschaft an der Universität Wien 1888-1938", in Robert Reininger, *50 Jahre Philosophische Gesellschaft an der Universität Wien 1888-1938,* Verlag der Philosophischen Gesellschaft: Wien, undated but presumably 1938 or 1939, pp. 3-19.

6 Boltzmann had taken eight courses under Zimmermann as a student during the 1860's.

* Boltzmann wrote poetry, played piano well, and had a box in the Vienna Opera house.

7 Each year's *Bericht* m contained from fifteen to thirty pages with basic information about members, officers, finances, and brief descriptions of their activities.

8 The yearbook was originally titled in the following fashion: *Wissenschaftliche Beilage zum fünfzehnten Jahresbericht (1902) der Philosophischen Gesellschaft an der Universiät zu Wien, Vorträge und Besprechungen über . . .* But in the year 1912 the title was altered to *Jahrbuch der Philosophischen Gesellschaft an der Universität zu Wien . . .*

9 The major non-members were Kant and Bolzano.

10 (Höfler) 1913, op. cit. (See footnote #3.)

11 The University of Vienna Philosophy Library has the *Berichte* until 1916-17.

12 This was part of Jodl's substitute for holding separate discussion sessions.

13 These are not easy to locate, except sometimes by author of each article as a booklet.

14 Please see *50 Jahre Philosophische Gesellschaft an der Universität Wien 1888-1938,* Herausgegeben vom kommissarischen Leiter der Gesellschaft Univ.-Prof. Dr. Robert Reininger, Verlag der Philosophischen Gesellschaft an der Universität Wien, Druck E. Kainz vorm. J.B. Wallishauser, Wien VIII, undated (1938-39?).

15 These were titled *Wissenschaftliche Jahresberichte,* but unlike earlier *Berichte* there was no mention of the activities, officers, or number of members of the Society.

16 Schlick's circle usually met in the mathematics seminar in the *Institut für Physik* on Boltzmanngasse in Vienna every Thursday evening. Gomperz's circle met at his home at Grünbergstrasse 25

on Saturdays. See Friedrich Stadler, "Heinrich Gomperz und Karl Popper im Kontext des Logischen Empirismus" in *Heinrich Gomperz, Karl Popper und Die Österreichische Philosophie*, herausgegeben von Martin Seiler und Friedrich Stadler, (Studien zur Österreichischen Philosophie, Band XXII), Rodopi: Amsterdam, 1994, p. 6.

17 Sir Karl Popper, *Unended Quest - An Intellectual Autobiography*, Fontana/Collins: Glasgow, 1980, p. 84.

18 *Rückblick*, op. cit (1913), p. 45. (See footnote # 14.)

19 As told to the editor by Karl Nawratil, Robert Reininger's friend and biographer.

20 R. Reininger, "Verzeichnis der Vorträge und Aussprachen", *50 Jahre Philosophische Gesellschaft an der Universität Wien 1888-1938*, Verlag Philosophischen Gesellschaft an der Universität Wien, Vienna, 1938, pp. 21-37.

21 It is possible that this discussion and those following on the same topic may well have helped persuade the philosophy faculty of the University of Vienna to recommend Mach to a position in philosophy.

22 There is a report Ludwig Boltzmann intervened rather emotionally in the discussion on solipsism. Hannelore Rodlauer, *Otto Weininger - Eros und Psyche, Studien und Briefe 1899-1902*. Verlag der Österreichischen Akademie der Wissenschaften: Vienna, 1990, p. 24.

23 Sigmund Freud was originally scheduled to lecture during this period, but had influenza. (See Rodlauer, Ibid., pp. 67-68.)

24 The Philosophical Society privately published a version of Wilhelm Ostwald's address which includes some dialogue at the end between Ostwald and Boltzmann.

25 Mach at the urging of Heinrich Gomperz sent a letter to Boltzmann that his intended title to his lecture was unsatisfactory, so Boltzmann changed it. But it was Friedrich Jodl who claimed credit for forcing Boltzmann to moderate the title to the one listed above.

26 Jerusalem defended Mach's psychological understanding of logic. Husserl had insisted that logic was idealized or consisted of universals and was not "psychologistic". While Husserl remained a corresponding member of the Society for many years it is not certain how often his position in Göttingen allowed him to attend its sessions. What is known is that he helped set up his own philosophical society in Göttingen in 1907.

27 Mach had argued in his *Beiträge zur Analyse der Empfindungen* in 1886 that space and time were sensations, but he later backtracked, admitted doubt and sometimes considered them to be relations while often distinguishing between what he called "physiological", "physical", and "metric" space and time.

28 Ernst Mach's book *Erkenntnis und Irrtum* had just appeared with several chapters on psychological and physiological approaches to space, time, geometry, and distance, and which seem to have stimulated several lecturers in the Society. See also the titles of lectures for January 19 and 30, 1906 and later ones as well. For his own point of view, see Ernst Mach, *Knowledge and Error*, with an introduction by Erwin N. Hiebert, Reidel: Dordrecht, 1976, chapters XX to XXIV.

29 Neither Reininger's lecture of October 28 nor Höfler's reply of November 11 were published in the Society's *Beilage* for that year, but it would be interesting to see if the discussion led by Alois Höfler was as critical of Reininger's lecture as one is inclined to imagine.

30 Wilhelm Jerusalem, one of Ernst Mach's closest friends and philosophical ally on most questions, is known as the man who introduced pragmatism into Austria, though one could argue that Ludwig Boltzmann had already done so in a philosophy lecture in 1903. Jerusalem is also known as a leading figure in the attempt to organize a sociology society in Austria.

31 Frank seems to have accepted both Ernst Mach's and Einstein's objections to "absolute motion". But it is not clear whether he distinguished between Mach's phenomenalistic objections and Einstein's specifically physical objections.

32 Perhaps because Viktor Kraft wrote the first major history of the Vienna Circle and because he

attended many of its meetings he has generally been considered a Logical Positivist himself, but in fact he had long been opposed to phenomenalism, and it is hard to believe that he became sympathetic to its later physicalistic facade and obsession with language as if epistemology & ontology were "external questions" or "merely linguistic".

33 Lectures and discussions on religion were rather rare in Alois Höfler's science-focused Society, but became increasingly common after 1922, that is, during Reininger's period as leader. Ehrenfels as a co-founder of the Society and a fellow alumnus of Brentano could naturally talk on virtually any subject he wanted.

34 This is still a very live issue, since it raises the question whether scientific laws are part of nature or are merely idealizations to help describe appearances as Mach thought or in inadequate fashion are attempts to describe or explain complex historical phenomena as many humanists think.

35 Frank almost certainly would have discussed Mach's position in close connection to that of Einstein. Höfler on the other hand, probably opposed Mach's position or at least distinguished it from that of Einstein.

36 It might seem that the Society was bold to discuss this issue in the middle of The Great War, but even by the second year some Austrians and Germans wanted peace, especially many socialists. Nor would there be dynamic civilian leadership like Lloyd George, Clemenceau, or Wilson would offer in the West to spur on the Central Powers to victory. Loss of morale would grow. On the other hand, while the topic of the discussion above could be considered subversive in wartime, the Society wisely listed no one as discussion leader, which could reduce the liklihood of being arrested for "defeatism". Mach was known as a pacifist as well as several of his followers. One of his closest followers, Friedrich Adler, would shoot and kill the Austrian Prime Minister several months later in an effort to force an end to the war.

37 Ernst Mach had died in February 1916. The lectures of Kraft and Neurath would help commemorate his life and work.

38 Ehrenhaft had already become notorious for his theory of fractional electron charges in his dispute with Millikan.

39 At least for most historians this is still a live issue of some importance. If History is a social science, then presumably causes can be identified with laws, but if it is a humanity, then causes as agents or forces would seem more reasonable. Another way of looking at it would be to distinguish between idealized and non-idealized science.

40 The years 1921 and 1922 would see the death of many members of the "old guard" in the Society, Meinong, Stöhr, and of course Höfler. Unfortunately, the Society was not yet in a position to resume publishing any of the many spoken obituaries on these men.

41 Mach's book *Die Prinzipien der physikalischen Optik* was posthumously published under the editorship of his son Ludwig in 1921. See Michael Stöltzner's article in this volume. Neurath published "Die Verirrten des Cartesius und das Auxiliarmotiv" in the Society's Yearbook for 1913 and "Zur Klassifikation von Hypothesensystemen (mit besonderer Berücksichtigung der Optik)" in the Yearbook for 1914/15. pp. 39-63. Outside the Society he published: "Prinzipielles zur Geschichte der Optik", *Archiv für Geschichte der Naturwissenschaften und Technik*, **5** (1915) 371-389.

42 Otto Höfler became a Professor of German. He was Alois Höfler's son.

43 These two talks Commemorated the death of Karl Neisser who had been one of the original founders of the Society during the late 1880's.

44 The Nazis invade Austria on March 12, 1938.

45 Adler split his life between Vienna and Zürich and may be discussed again in our volume which includes Mach's Legacy in Zürich.

46 Rudolf Wlassak, like Friedrich Adler, will probably be discussed in the forthcoming volume *Re-Discovering Ernst Mach's Influence*, which will include a section on Zürich. Avenarius who didn't like Zürich, but taught there from 1876 to 1896, will be discussed with Petzoldt elsewhere.

Appendix

COLLATERAL PHILOSOPHERS

by themselves

1. THE QUEEN OF THE SCIENCES[1]

by Wilhelm Dilthey[2]

Philosophy is growing steadily more important in its primary function of grounding and connecting the thinking which has developed in our particular fields of culture. The positivistic philosophy of d'Alembert, Comte, Mill, and Mach is significant in that it grows precisely out of an occupation with the particular sciences from within, follows their method, and always applies the confirmed standard of their universally valid knowledge. In another field the philosophical thinking of Carlyle or Nietzsche is positive in its very effort to generalize and to confirm the policies contained in the experience of life and which has been worked out by the poets and prose-authors on the conduct of life. It is natural therefore that in this informal way philosophy is influencing more and more the whole mental life of modern times. The spirit of method, generalization, and inter-scientific connection, which was dominant in the natural science of Galileo, Kepler, and Newton, was then supported by the positivist tendencies of d'Alembert and Lagrange in furthering French study of nature, and has continued to influence us in Germany through Kantian Criticism and investigation into nature, especially in the work of Ernst von Baer, Robert Mayer, Helmholtz, and Hertz. And this same philosophic spirit has come to the fore by means of the recent socialist theories which focus on the particular sciences of society and history. So it is characteristic of the present situation in philosophy that its strongest influences emanate not from the systems, but from this free philosophical thinking, which pervades the sciences and all literature. For even in the field of fine letters authors like Tolstoy and Maeterlinck also exert an important philosophical influence. The drama, the novel, and now also lyric poetry have become vehicles of the most vital philosophical activity.

The philosophical spirit is present wherever a bold thinker, free from the fashion of a philosophical system, will examine what appears to be odd and obscure in humans, like

J. Blackmore, R. Itagaki, and S. Tanaka (eds.). Ernst Mach's Vienna 1895-1930. 315-336.

instinct, authority, or faith. It has been present wherever investigators with a strong consciousness of method retrace their steps from their science to its ultimate grounds of justification or push concern forward to generalizations which connect and confirm several sciences. It is also present wherever the values of life and our ideals are re-examined. What appears anywhere in chaotic or hostile struggle within an age or the heart of a man is to be reconciled through thought; what is obscure is to be clarified; and what stands there in bare juxtaposition is to be mediated and connected. This spirit leaves no feeling of value and no striving in its immediacy, no precept and no knowledge in its isolation. For every valid thing it seeks the ground of its validity. In this sense, the eighteenth century chose to rightly call itself the philosophical century in virtue of its effective rule of reason over the dark, unconscious, instinctive forces within us . . . .

Philosophy has proved itself to be a group of very diverse functions, which, through insight into their regular connection, are collectively regarded by many as the essence of philosophy. A function always refers to a purposive system, and consists of a group of related activities which occur within this systematic whole. The concept is not derived from the analogy of organic life, nor does it refer to some faculty or original ability. The functions of philosophy are related to the purposive structure both of the philosopher and society. They are activities in which the person turns in upon himself and at the same time produces outward effects. In this, the activities are also akin to the deeds of religion and poetry. So philosophy is an activity which springs from the need of the individual mind for reflection on its behavior, for inner pattern and security of action, and for a fixed form in its relation to the whole of human society. And philosophy is at the same time a function rooted in the structure of society and necessary for the completeness of social life. Accordingly, philosophy is a function which occurs uniformly in many persons and unites them in a social and historical continuum. In this latter sense, it is a cultural system. For the characteristics of such a system are above all: uniformity of the activity in each individual member of the system, and solidarity of the individuals in whom this activity takes place. If this solidarity assumes fixed forms, then organizations arise in the cultural system. Among all purposive systems those of art and philosophy bind the individuals together least, for the function which the artist or philosopher performs is dependent on no institutional forms of life. In this realm, there is the highest freedom of the mind. The membership of philosophers in such organizations as a university or an academy may increase his contribution to society. Nevertheless, his permanent principle of life is the freedom of his thought, which must never be infringed, and which has proven itself indispensable not only for his own philosophical character but also for the confidence in his absolute sincerity, and thus for his influence.

The most general characteristic belonging to all functions of philosophy is rooted in the nature of objective awareness and conceptual thought. So regarded, philosophy appears to us as only the most consistent, vigorous, and comprehensive form of thinking, while such is separated from the empirical consciousness by no fixed boundary. From the form of conceptual thinking it follows that judging will advance to higher generalizations, like the forming and classifying of concepts into a conceptual architectonic, and this to the very highest apex, as this relating proceeds to an all-inclusive system, and becomes grounded in ultimate principle. In this activity, thinking refers to the common goal of all of the thinking acts of various human beings such as the systematic unity of sense perception is the world. To form this world the plurality of things is ordered in space, and the variety of their changes and movements in time. All feelings and volitional acts of this vast world are arranged through the definite location of bodies belonging to them and the perceptual elements woven into them. All values, purposes, and goods are posited in these feelings or volitional acts and fitted into them. Human life is embraced by them. And now thinking has arrived to express and to unite the whole content of our perceptions, lived experiences, values, and purposes, as they are truly lived and given in our empirical consciousness (experience and the sciences of experience). In so doing, thinking moves from the concatenation of things and of changes in the historical world toward a world-concept. For a ground for itself it focuses on a world principle, a world-cause. It seeks to determine the value, sense, and meaning of the world, and it asks us to help develop a sound world purpose. Whenever this procedure of generalizing, integrating, and of logical grounding is borne forward by the cognitive drive and frees itself from the particular need and the limited interest, it passes over into philosophy. And wherever the subject, who relates himself to this world in his activity, then rises in the same way to reflection on this activity of his, the reflection becomes philosophical. Accordingly, the fundamental characteristic in all functions of philosophy is this drive of the mind which transcends mere attachment to the determinate, finite, limited interest and strives to fit itself into an inclusive, definitive idea far above every theory which has arisen from a restricted need. This drive of thought springs from its rationality and attempts to meet the needs of human nature which in their complexity well nigh defy reliable analysis, needs which would find joy in knowledge, ultimate security in relation to the world, and which endeavors to release life from its restricted conditions. In fact, every mental attitude seeks a fixed point, free from relativity.

2. THE LOVE OF WISDOM[3]

by W.P. Montague[4]

The first question of a philosophic kind that I can remember considering concerned the nature of the soul and its relation to the body. Having been informed by my mother that the soul was that which made you laugh and cry and think and move, I asked if you could get it out by boring very carefully up through the foot and leg until you reached it somewhere in the chest. To this question my mother was providing a hesitant and somewhat puzzled negative, when my father broke in impatiently with the warning that I must never think of the soul in that way, that it was not at all the kind of thing that had a place inside the body from which it could be fished out.

I mention this little incident of my early childhood because the question I asked then is still the kind of question I have been asking ever since, and the reproof for it which was administered by my father is the same reproof that I have received many times from my teachers and colleagues in philosophy.

I feel myself to be a thing in a world of things. And the thing which I am does not seem to me to be the thing my body is. The two are alike in being substantive rather than adjectival. They are alike also in being agents and patients in space and time, but they are contrasted with respect to the laws and processes pertaining to them. The intimate union in existence of these entities so disparate in essence has become to me ever more mysterious. All my thinking has been oriented with regard to the psycho-physical problem. And blessed or cursed by this "animistic complex", and goaded on by it, I have sought continuously for an intellectual theory that will bridge a dualism imposed upon me by feeling or intuition and increasingly confirmed by the evidence of experience.

Next to the metaphysical question of the mind and body there came, and continued, the religious problem. Church services in the early morning with my mother gave me a poignant sense of the beauty of the Christian religion. Sunday-school and the atmosphere of a small New England community gave me an equally poignant sense of the falsity and incredible ugliness of the authoritarian and ascetic aspects of the same doctrine. Cool, condescending approval or equally cool and tolerant contempt, which are usual alternative attitudes toward Christianity that are prescribed by the "genteel tradition" in American philosophy, have never done justice to its baffling mixture of what is best with what is worst. Love and enthusiasm for one half of the Church, righteous hate and contempt for the other, are sternly called for.

Somewhere between twelve and fifteen I read Edward Bellamy's *Looking Backward*, Abbott's *Flatland*, and articles on theosophy by Blavatsky and others. Bellamy, in addition to making me a socialist, made me realize that the business of the good life was also institutional as well as an individual affair. *Flatland* and the theosophy stimulated and made more explicit what might be called my own latent metaphysical interests. The characteristically theosophical attempt to explain the mind and its doings in terms of material or pseudo-material categories appealed to my animistic attitude and has thus seemed to me then, as indeed it still does now, to be the most delightful of pastimes.

It was not really until the middle of my sophomore year that my formal education took on significance. For I suddenly found myself suspended from college, and very miserably I made visits to my various professors to get instructions for the enforced period of my homework. Josiah Royce, whose course in the History of Philosophy I had been taking with some genuine interest, received me kindly, and inquired what I planned to fit myself for. I replied that I didn't know, but supposed that I should have to be a lawyer. Whereupon he asked me encouragingly if I had ever thought of the academic career. I never had, and I hardly knew the meaning of the phrase, but the fact that here was a great man showing me sympathy and faith in my dark hour stirred me unforgettably. By hard and eager study I passed well the final examination in his course. It had opened a new world to me. Everything was changed, and I was happy. In my Junior and Senior years in college, and in the three years in the graduate school that followed, I gave my whole time, so far as I was permitted, to philosophy. The classes suddenly ceased to be tedious interludes between parties, and became themselves the most exciting of parties.

Everyone knows the extraordinary company who dispensed philosophy at Harvard at the end of the 'nineties: Palmer, James, Royce, Münsterberg, and Santayana. Each of them had a distinctive philosophy, and each of them preached it with the force of conviction. Their methods of teaching were almost as different as their viewpoints. Palmer's lectures were incomparably the most finished, both as to content and as to form, of all that I have ever heard; and whether because of it, or in spite of it, their literary perfection, their pedagogical effectiveness was extraordinary. I had the good luck to be his assistant or reader, and so had an opportunity to observe the progress in philosophic comprehension made by the large group of undergraduates in his course, each of whom was required to submit four papers during the year. Quite apart from this ideal of pedagogical technique, and in addition to many specific illuminations, I owed to Palmer the realization that a naturalistic relativism, as regards the varying *content* of the good, is quite compatible with a Kantian rigorism as regards the invariant *form* of right or duty.

The classroom lectures of James were in striking contrast to those of Palmer, and, indeed, to the polished nature of his own formal papers. He would utter his thoughts spontaneously, just as they came. As a result were most uneven in quality. The roughness and irregularity of them were, however, more than balanced by the simplicity and directness of his conversational manner. It was an inspiration to say and hear a great man work out his own thoughts in the presence of his students. The informal and colloquial language coming from a scholar and a genius possessed a peculiar piquancy. I have seen him stop in the middle of a sentence with some such remark as: "By George, Mr. Smith, perhaps you were right after all in what you said a few minutes ago, and I had never thought of that." Mr. Smith thus honored, would, of course, become the envy of all of us and the devoted slave of his master. Nor would he realize that James with characteristic and unconscious generosity had probably read into his comment a richness and meaning of which Smith himself had been quite innocent.

The parts of James's philosophy that concerned his Pragmatism and his Radical Empiricism left me uninterested or actually repelled, but to the more mystical and less explicitly developed phases of his thought I owe much. His conception of the subconscious and hidden energy of men, his Transmission hypothesis, his defence of indeterminism and of a finite God, I greedily accepted. When taken in connection with the Tychism of his friend, Charles Peirce, they furnished me with a large part of my philosophical faith. For Peirce himself I had a kind of worship. And while his intellect was cold and clear, his metaphysical imagination was capricious, scintillating, and most unbridled, and his whole personality was so rich and mysterious that he even seemed a being apart, a super-man. I would rather have been like him than like anyone else I ever met. And on the two or three occasions in which he carefully criticized and then praised some of my student writing I was truly transported with happiness.

It is to Santayana especially that I owe my realization of Plato's truth. I took his course at about the same time that I read Huxley's *Evolution and Ethics* and Stevenson's *Pulvis et Umbra*. In all three the teaching was to the effect that value did not depend on existence, nor right on might, nor ethics on religion. An atheistic nature red with tooth and claw could in no sense absolve man from his obligation to actualize ideals of beauty and goodness; nor could it deprive him of the consolation of knowing that these ideals were always and eternally there; be the world what it might. The fact that I did not share the pessimism of Huxley and Santayana as to the existing world in no way lessened my debt of gratitude.

There was still another thing I got from George Santayana's courses: support and confirmation of my temperamental belief in a world of substances. Georg Hegel's hideous slogan, "Not sub*stance* but sub*ject*", so fertile in steriilities of every kind, had been pretty generally adopted by professional philosophers, and it was a relief to hear Santayana, who had gone through that epistemological hell with his common sense unscathed, talk delightfully and quietly about the various substances of which the world might conceivably be composed.*

It was, however, to Josiah Royce[5] that my debt was greatest, though in another sense it was also least. He taught me almost all that I know about the world's great philosophers, and yet in his own philosophy there was hardly anything that seemed to me true. His lectures on the history of metaphysics, however, were glorious affairs. He not only made the various systems clear in themselves he portrayed their inner-relations so luminously and with such originality and depth of insight that the whole course of speculative thought was presented in a series of magnificent vistas and then given the unity of a single great picture of the human mind objectified. He had an unexampled power of making abstract examples concrete and almost sensuously vivid. It was the ideas themselves rather than the biographical details or even the identities of their authors that he emphasized; and I think all of his students were impressed with the fact, that however it may be in other history, in the history of philosophy at least, it is not the *who's who* but the *what's what* that really matters. I not only got from Royce my knowledge and appreciation of philosophy, but I got from him the kindest and most painstaking assistance in working out my own philosophical problems. He gave me this technical help through my five years of advanced study, and he accompanied it with continuous personal interest and affectionate counsel. He had, moreover, started me going, and trusted me when I had been down and out. I owed everything to him, and it seemed mean and disloyal for me not to become his disciple. He was my dear teacher, and I longed to call him my master, but I couldn't because his idealistic premises seemed to me false. Despite this I felt the tug of the will to believe wherever I could find elements within my teacher's philosophy that did not seem definitely false. And in a system so rich there were, of course, many such elements. I remember particularly the hypothesis about the varying time-span in nature put forward in the second volume of *The World and the Individual*, and how I jumped at it with almost tearful gratitude as a clear good thought that might even be true. For it was a new and interesting contribution to the great pan-psychist tradition. It could be studied and appraised on its own merits in the light of what was known and presumed about the world. And it did not in the least depend upon the monstrous premise on which the rest of the book had been based, that great nature, in which we are such recent and humble participants, is itself the product of our social consciousness and of the funny little techniques of our communication which we have developed.

My doctor's thesis, which was rather formidably entitled *An Introduction to the Ontological Implicates of Practical Reason*, had been suggested by Kant's *Critique* and by Hegel's *Phenomenology*. Sheldon[6] and I worked a great deal together, and we each felt that the principle of *Negativität* was inadequate to the great enterprise of categorical deduction, which translated into modern terms, meant the search for a principle of a hierarchical organization in the domain of subsistence or of essence. We were cheerfully confident of being able to remedy the defect. So Sheldon took the principle of the *Self-Repeater* and I took the principle of growth or increase, which was perhaps suggested by Bergson's *durée* to which James had introduced us. Royce christened my principle the *Pleon*, and helped me with his customary generous kindness to lick the thing into shape. I then supplemented my revision of the Hegelian deduction of the categories with an adaptation of Kant's treatment of the sense of duty as revealing a reality more ultimate than anything contained in the world of sense-perception, and attempted to show that in the experience of moral obligation that the Pleon revealed itself in its true colours as the principle of reality itself. The thesis was lengthy, and must have been pretty awful as judged by contemporary standards; but it was fun to write, and it got me the degree.

During the following year I secured my first regular job under Professor Howison[7] in the University of California. Howison, like Royce, held a kind of variant of Post-Kantian idealism, but it was more original than Royce's though not nearly so well organized and supported. It was a sort of Fichtean monadism in which God was no Absolute but merely *primus inter pares*, and the world was the phenomenal manifestation of the society of eternal persons who, by timelessly recognizing one another and one another's recognitions of one another, etc., `a la Leibniz, created the time process and generally kept things going on. This pluralism was in refreshing contrast to the anthropopphagous absolutes of Royce and Bradley; and to its author at any rate, it had come as a genuine *aperçu* . He preached to us it with fiery and unflagging earnestness, and so great was the force of his personality that people all over the Pacific Coast began and have even continued to study the philosophy of German Idealism.

The spirit at Berkeley was very different from Harvard. Under Howison, philosophy was not a fencing-match in which friendly gentlemen exchanged playful thrusts and courteously applauded their adversaries. It was a grim thing of life and death for the soul, a veritable religion in which either you were orthodox (Howisonian) or you were damned. I with my realism was mostly damned, and would have been completely so had it not been that Howison was really two persons - a grand inquisitor who would burn your body to save your soul and also one of the kindest men alive. I was lucky enough to win his affection and not only his professional condemnation, so for four years he kept me on, scolding me angrily for my rotten views and even warning his students against my courses, and then, together with Mrs. Howison, showering me and my family with every sort of friendly kindness and material help.

It was for me a strange, hard-working time, and in striking contrast to the dreaming and brooding years at Cambridge. I learned to teach, and found I could do it all right, which was a great relief, because both I and my Harvard teachers had had grave doubts about my being able to. I formed a friendship with Overstreet,[8] who was at that time devoted to the study of Plotinus and in high favor with Howison, and that helped me as it has ever since. I took two interesting courses, one with professor Slate in physics and one with Professor Stringham in mathematics. The course with Slate was a splendid one, and it started me reading hard in such books on science as I could understand. Stringham's course had an even stronger effect. He was a man much like Peirce. he had a Pythagorian sense for the things that really count, and a Lewis Carroll humour. I had no ear for music, but I think I got from his functions and series the kind of experience that musicians must get from their music. Numbers had always for me an almost pathological fascination, and an almost tangible objectivity. To regard the number 37, for example, as the product of a human mind, a result of the counting activity by which we reached it, seemed not only false but idiotic. The numbers stand in vast and infinite array with all their still more multitudinous inter-relations full of unending and delightful surprises. They are what they are, and must through all time be what they have been, more steadfast than the stars and more clear and beautiful than existing things can ever hope to be. It would be easier and less absurd to suppose that Baedeker had by his descriptions created the Jungfrau peak at which I am now looking than to suppose that the ephemeral mathematicians of this planet had created by their technique of procedure the timeless truths which they discover. Stringham's course revived all of my old interest in mathematics, and the result was disastrous, for I became an addict and began to neglect my philosophy in order to play amateurishly with the new problems. Girolamo Cardano's solution of the Cubic seemed intolerably complicated, and I got the feeling that the ease with which equational knots of any degree could be tied must mean that an equally general formula for untieing them must exist. On that presumably hopeless quest I would spend days and nights quite unpersuaded by my mathematical colleagues that the general equations of even the fifth degree had actually been proved to be insoluble by the ordinary methods which I was using. Then also there were the several simple-seeming series for the summation of which no formula had yet been found; and always and best of all, there was the ancient lure of the primes. I suppose it was vanity or laziness or an unacknowledged distrust of my own competency to master the real stuff I needed that prevented me from fitting myself by serious study for this new game. I preferred instead to wait until the house was quiet, and then like a secret drinker unlock my cupboard and take out with guilty joy, not the good little black bottle, but the nice blank book in which I kept my scribblings and hold high revel with myself through the night, exploring the lovely grottos in which the roots of my equations lurked, bound by mysterious threads to the combinations of the known and visible coefficients on the surface up above. In the more than twenty-five years in which I have indulged this vice I have discovered - at least I think I have discovered - a few old and pretty things, though certainly not enough as yet to justify the hours filched from philosophy.

In spite of this disastrous by-product of my work with Stringham, and yet perhaps partly because of it, but still more because of the courses with Slate and the reading which had grown out of it I began to take [epistemological] realism with a new seriousness. Science was discovering most exciting things. With the atoms not only being arranged in families; but actually caught and counted, it seemed a piece of pedantic insolence for philosophers who knew nothing of the new work, and cared nothing for it, to label the atoms and their electronic constituents as "vicious abstractions from the organic unity of experience", or similar barren nonsense. If philosophy was to play its historic role it must acquaint itself with the new scientific truths and exploit whatever speculative possibilities they contained. Idealism began to seem to me, not just falsity to be neglected, but a positive menace debouching the minds of the youths who studied it. There was need for a definite campaign to help deliver philosophy from the fog of confusion that threatened to obliterate it. This feeling as to the increasing importance of getting the epistemological situation properly cleared up, not as an end in itself, but as an indispensable prerequisite to a worthwhile philosophy, made me very glad to accept the call to Columbia in 1903.

The new place was quite different in atmosphere from either Harvard or Berkeley, and wonderfully stimulating. Sheldon was there and we could now resume both our golf and our duets about an ultimate romantic metaphysics. Woodbridge[9] was as realistic as I was myself - perhaps even more realistic - at least as concerned the world of concrete existence. Indeed, it was almost disconcerting to find that what had so long been my daring and dangerous heresy was now taken by the head of my department to be a mere matter of course. And yet there were moments even in those early days when I had misgivings that the new realistic accord was itself not so close as it appeared and as I wanted it to be. For Woodbridge seemed to accept nothing of that inner veil of sensation through which we all must pass to reach the outer world. When a distant star would hit him in the eye with its light, his body would bow or gesture in the direction from which the light had come; and he would say in effect: "*That* is my perception of the star - *voilá tout*!" What could it mean to identify the publically observable bodily antecedents and consequents of an experience with the experience itself, the latter being obviously private and not observable by others?

It was the first case of acute behaviorism that I had seen, and the first, I believe, that existed. To believe in the outer world was indeed very good, but to purchase that belief at the cost of denying the inner world was too high a price even for realism. The baby had been emptied out with the bath. But even though the theory was queer it was original and never put forward before. And in philosophy, at least, we should do everything once. So I was not very much troubled if my friend for some unaccountable reason, chose to make-believe that his own sensations were non-existent. Moreover, René Descartes had tried to dump the theory on animals [as if they had no mind or consciousness], and it was a pretty sporting thing for a man to be willing to try it out on himself.

Under the stimulus of this new and to me most delightful association, I changed from the representative or dualistic theory of human perception to a presentative or monistic theory. Not that I ever doubted the existence of my own mental states and their location inside my skull but I got to believe that the objects immediately present in perception were not those states but their meaning or implicates, which as such could coincide (though they need not), not only in essence, but in position and date with the things and events in the extra-cranial world. The states that do this revealing are thus not themselves revealed. They are not motions, but they are describable in physical terms.

What kind of event it is that can reveal events at other places and times than its own, is but one phase of the psycho-physical problem. That intra-cerebral states can and do reveal without creating such events is sufficient for the field of epistemology.

Beginning about 1910 there came the association with Perry, Holt,[10] and the others which resulted in the publication of *The New Realism*. We set out with high hope of success, confident in one another and in the sympathy of our big brothers in Europe, Russell, Moore, and Meinong. We wanted, first of all, to introduce into philosophy the two methods which had been so profitably employed in science: the method of co-operative work and the method of isolating problems and tackling them one by one. And in addition to these methodological policies we also had (at least so I thought) several epistemological theses all of us believed to be true and which we intended to establish.

To me, at least, these theses committed us to no decision as to the more properly metaphysical issues concerning the ultimate nature of the world, or even of the nature of the mind and its functions. On these latter questions I was pretty certain that I was not in agreement with my confrères, except possibly with Pitkin.[11] Our realism was thus not a philosophy; it was rather a prolegomenon to philosophy as a project and a declaration of independence that would make it possible to investigate the nature of things on their own merits and without dragging in the tedious and usually irrelevant fact that they could be experienced by us.

But even within the domain of epistemology, or closely connected with it, the cognitive function was interpreted by Perry and Holt; as well as by Woodbridge and Watson;[12] to be nothing more than a "specific response", i.e. a motion of the organism, or some part of it, elicited by a stimulus. For reasons already stated this seemed to me preposterously false. Magnify the importance of bodily movements on earth as much as we please, they can never be other than north or south; east or west; up or down, or in directions intermediate to them. But the peculiar *rapport* between an individual and the objects of which he is aware extends to past and future and to the realm of the abstract. And it is impossible for the body to move in the "direction" of such entities. But what is perhaps an even more serious difference between my neo-realist colleagues and myself has developed in recent years. The objects which appear in distorted illusory and hallucinatory experience, and which had always been for me, not existential, but merely subsistential complexes of essences that could not on reflection be believed to be in real space, however vividly they might appear there at the moment of perception, have been given an existential status by most of those calling themselves neo-realists.[13] The result is to make of the space in which we live and move a dumping ground for all the contents of dreams and illusions, actual and even possible. Any space would crack under such a strain, and of course "neo-realist space" has cracked and broken up into a series of "private spaces" and a "public space" as a construct of each individual. Are there as many public spaces as there are individuals who construct them? Do they interpenetrate completely or partly? Or are they mutually external and side by side in a super-public space which is not a construct? These last questions must also be asked about these private spaces. And where were the individuals who construct public space before they did their constructing, i.e. when they were babies or embryos? I would rather be an idealist, at least a Kantian idealist, than swallow any such mess. For Kantian space, even when reduced to an apriori form of human perception could still keep up, though in straightened circumstances, a semblance of Euclidean respectability and enjoy a dignified priority in its own contents.

It has seemed to me to be a certainty that the things of which the existing world consists must at each instant have ultimately univocal [geographical] positions, regardless of all of the conflicting perspectives in which they may appear. And if we make temporal cross-sections of these things as we always can and thus treat them as histories or continuous series of events, and accept in addition the Special Theory of Relativity, the situation is not essentially altered; for though in that case [historical] space and time will have become interdependent aspects of the four-dimensional continuum "space-time", yet even so, each event will enjoy a univocal "date-locus" or absolute "position" in the new continuum. How otherwise could events be unambiguously inter-related by Eddington "intervals",[14] which are to remain invariant, regardless of perspectives which vary with the motions of the systems from which they are taken? Nor do spaces and times of even these varying perspectives have places for hallucinatory contents. Relativity to the photographic plates on moving systems is not the same relativity to the apperception-masses of the men suffering from nightmares or delirium.

If neo-realism is to mean for us an ontological egalitarianism in which existential status is to be accorded to every content of perceptual experience, whether veridical or illusory, then such a theory is not *realism* at all. We may give new names, such as "sense data" or "sensa" or "sensibilia", to the rabble of experiential contents, and insist piously on calling them "physical" in every sentence; but they are not properly physical; for they are properly not things at all. They are not agents and patients; they cannot go under their own power; and perhaps worst of all, they are totally incapable of an orderly arrangement in a single milieu; whether a space or a time or the space-time of relativity. They are in fact , nothing but the well-known adjectival sense-impressions of Hume and Mill and Mach and Pearson masquerading under fancy appellations. And in treating them as the sole constituents of the world of existence, New Realism has surrendered to the old phenomenalism.

These sad trends of behaviorism and positivism, which have taken place in the last ten years, and have attempted to describe, spoiled my interest in the movement from which I had hoped so much good would result; and once more as in the old days at Harvard and Berkeley, I am left without a party....

3. THE STUDY OF EXPERIENCE[15]

by De Witt H. Parker[16]

My interest in philosophy can be dated exactly from the moment when in the little public library of Summit, New Jersey, a mere boy of eleven years, I chanced upon White's *History of the Warfare of Science with Theology*[17] and read on through the afternoon until closing hour; then hastened home to announce to my astonished sisters that Adam and Eve never existed, Noah never gathered the animals "two by two" into the Ark, and that Jonah was never swallowed by a whale. Years of fervid quest followed, during which, without teachers and without fellow-inquirers, I tried to discover whether "the faith of my fathers" was true. Small wonder that with a brain seething with such thoughts, I lost interest in the more usual pursuits of boyhood, so unexciting by comparison! I read all the books relating to the subject I could lay my hands on, making later use, on the family's removal to Boston, of the ample resources of the Public Library there; first in order: books of "higher criticism" beginning with Renan[18] and Strauss;[19] then books of popular philosophy, like those of Fiske,[20] Haeckel,[21] and Carus,[22] dominated by the concept of evolution, and even weightier works, such as Martineau's[23] *Seat of Authority in Religion* and [Herbert] Spencer's *First Principles*. Under the influence of this reading I passed from orthodoxy to a Unitarian "theism" (fortified incidentally by Felix Adler's *Creed and Deed*,[24] a book that I know has left a permanent impress on my mind); thence to a mystical pantheism and nature worship; and finally, after reading Mathew Arnold's *God and the Bible*, [Charles] Darwin's *Descent of Man*, and more books of Spencerian and Comtean positivism, to entire rejection, as I thought, of all theological beliefs.

When I entered Harvard in 1902 I had, therefore, lived through six years of lonely and unguided reflection, with a purely negative result. Yet I came with a tremendous eagerness to discover, if it were possible, the real nature of things, and a profound conviction that philosophy, and philosophy alone, could inform me. My life's purpose was henceforth fixed. In my Freshman year I elected History of Philosophy, ancient philosophy being given by Santayana and modern philosophy by Professor R.B. Perry. Few of my later experiences will stand comparison in vividness and freshness of joy with my first readings in genuine philosophy in connection with this course, namely, the reading of Plato's *Republic*, Aristotle's *Ethics*, Descartes' *Meditations*, Spinoza's *Ethics*, Hume's *Treatise*, and Berkeley's *Dialogues*. How trivial by contrast all my previous reading seemed, as the shadow to the substance! Later on I was to listen to Josiah Royce lecture on Kant and Symbolic Logic, Santayana on Plato, Palmer on Goodness, Münsterberg on Eternal Values, William James on Tychism[25] and Radical Empiricism. To me and to my fellow-students, philosophy, as represented by these men, whom we deemed immortal, was still in the making, and "to be alive was very heaven" in their company. Yet I was not able to count myself a disciple of any of them. A regrettable shyness, and a too great reverence for my professors, prevented me from seeking them out and getting personally acquainted with them; and besides, there were always, in the case of each one of them, some important matters of theory, with regard to which I could not follow them.

To my thoroughly impractical self, William James's pragmatism was meaningless as philosophy; in his proofs for the existence of the Absolute, Royce seemed to be always "working for an answer", like the small boy who has peeked into the last part of the exercise book and already knows the solution of the problem; Santayana's materialism could not stand up in my mind against Berkeley and Hume; the new realism of Holt and Perry I thought incapable of accounting for subjectivity, for knowledge, error, and illusion; Palmer's ethical pietism seemed inconsistent with the fact of evil, and Münsterberg's contrast between science and life too sharply drawn. Yet from each of my Harvard teachers I was able to learn things invaluable for my philosophical development: from Royce my interest in Symbolic Logic; from James the standpoint he called radical empiricism with a bias towards pluralism and indeterminism; and from Santayana an awakening to the problems of value (I still think Santayana's whole metaphysics, with its artifices of matter and essence, however original and beautiful in expression, a sad mistake, while his interpretations of culture remain incomparable in our generation; from Perry and Holt a discontent with traditional solutions. Aside from my debt to my teachers, I owed most, during my student days, to Berkeley and Hume; to Hegel and Nietzsche; to Bradley, Mach, and Avenarius, and outside of the field of philosophy, to my study of Greek and mathematics.

Since leaving Harvard I have taught at the University of Michigan for nearly twenty years, with the exception of two years spent at the University of California. During these years I have made a special study of aesthetics, both historical and theoretical, falling successively under the influence of Santayana, Lipps, and Croce, but eventually being able to work my way to an independent standpoint, as represented in my books, *The Principles of Aesthetics*, 1920, and *The Analysis of Art*, 1926. Plato and Aristotle, whom I teach, have persistently occupied my thought; my knowledge of the German idealists and of Leibniz and Spinoza has been intermittently renewed and extended; Bergson and Peirce, both strangely neglected during my student days, have influenced me powerfully. Like most of my contemporaries, I fancy, I have been tantalized and stimulated by the writings of Bertrand Russell, so fresh and mercurial; and through his work I have maintained my interest in symbolic logic, a subject which I teach occasionally to minute classes of advanced students. My book, *The Self and Nature*, 1917, would represent the impact of all of these influences for the first ten years of my professional career. The book is merely an impressionistic sketch, which is full of inconsistencies and other shortcomings; but I still adhere to its more essential positions. Very recently, as a result doubtless of ethical problems raised by the Great War, I have become interested in the problems connected with the general nature, classification, and criticism of Values. With respect to these values I have derived the greatest help from the ideas of Hegel and Nietzsche, Freud and the Austrians, and more recently from the protagonists of the *Gestalt* concept. Throughout my career at Michigan I have profited exceedingly from the friendship and the intellectual companionship of my colleagues. Professors Wenley,[26] LLoyd,[27] and Sellars.[28] From Sellars I am sure I have borrowed some points in theory of knowledge, while dissenting vigorously from his disguised materialism.

With the above as an autobiographical introduction, I shall now attempt to give a sketch of my philosophical position, confining myself mostly to metaphysics.

For me the proper standpoint of philosophy is what William James called "radical empiricism", which is as I understand it, the demand that philosophers shall use only concepts representing concrete experience. This point of view was already occupied by Hume when for every idea he sought the corresponding impression. Hence, I should say, Hume's method remains essentially sound, while his results are often defective owing to faulty analysis of experience. Thus, the key to the understanding of reality lies within experience, and to find the metaphysical meaning and validity of such concepts as cause, law, stuff, substance and the like, we must always look to their experience-equivalents. Moreover, since experience itself is a part of the system of the universe, it is reasonable to believe that it provides us with a fair sample of the whole. I see no reason for believing that there is any other stuff or substance except experience; hence, to my thinking, materialism, naturalism - evolutionary or not - epiphenomenalism, and the like are the *proles horrenda* of philosophic perversion . . . .

4. ORIGINAL THINKING[29]

by William James[30]

When Clerk-Maxwell was a child it is written that he had a mania for having everything explained to him, and that when people put him off with vague verbal accounts of any phenomena he would then interrupt them impatiently by saying, "Yes, but I want you to tell me the *particular* go of it!" Had his question been about truth, only a pragmatist could have told him the particular go of it. I believe that our contemporary pragmatists, especially Messrs. Schiller and Dewey, have given the only tenable account of this subject. It is a very ticklish subject, sending subtle rootlets into all kinds of crannies, and hard to handle in the sketchy way that alone befits a public lecture. But the Schiller-Dewey conception of truth has been so ferociously attacked by rationalistic philosophers, and so abominably misunderstood that here, if anywhere, is the point where a clear and simple statement should be made.

I fully expect to see the pragmatic view of truth run through the classic stages of a theory's career. First, you know, a new theory is attacked as being absurd; then it is admitted to be true, but obvious and insignificant; finally it is seen to be so important that its adversaries claim that they themselves discovered it. Our doctrine of truth is at present in the first of these three stages, with symptoms of the second stage having begun in certain quarters. I wish that this lecture might help it beyond the first stage in the eyes of many of you.

Truth, as any dictionary will tell you, is a property of certain of our ideas. It means their "agreement", just as falsity means their disagreement with "reality". Pragmatists and intellectualists both accept this definition as a matter of course. They begin to quarrel only after the question is raised as to what may precisely be meant by the term "agreement", and what by the term "reality", when reality is taken as something for our ideas to agree with.

In answering these questions the pragmatists are more analytic and painstaking, the intellectualists more offhand and irreflective. The popular notion is that a true idea must copy its reality. Like other popular views, this one follows the analogy of the most usual experience. Our true ideas of sensible things do indeed copy them. Shut your eyes and think of yonder clock on the wall, and you get just such a true picture or copy of its dial. But your idea of its [internal] "works" (unless you are a clockmaker) is much less of a copy, yet it passes muster, for it in no way clashes with the reality. Even though it should shrink to the mere word "works", that word still serves you truly; when you speak of the "time-keeping function" of the clock, or its spring's "elasticity", it is hard to see exactly what your ideas can copy.

You perceive that there is a problem here. Where our ideas cannot copy definitely their object, what does agreement with that object mean? Some idealists seem to say that they are true whenever they are what God means that we ought to think about that subject. Others hold the copy view all through, and speak as if our ideas possessed truth just in proportion as they approach to being copies of the Absolute's eternal way of thinking.

These views, as you can see, invite pragmatic discussion. But the great assumption of the intellectualists is that truth means essentially an inert static relation. When you've got your true idea of anything, there's an end of the matter. You're in possession; you know; you have fulfilled your thinking destiny. You are where you ought to be mentally; you have obeyed your categorical imperative; and nothing more need follow on that climax of your rational destiny. Epistemologically you are in stable equilibrium.

Pragmatism on the other hand, asks its usual question. "Grant an idea or belief to be true," it says, "What concrete difference will its being true make in any one's actual life? How will the truth be realized? What experiences will be different from those which would obtain if the belief were false? What, in short, is the truth's cash-value in terms of experience?"

The moment pragmatism asks this question, it sees that the answer: *True ideas are those that we can assimilate, validate, corroborate and verify. False ideas are those that we cannot.* That is the practical difference it has made for us to have true ideas; that, therefore, is the meaning of truth, for it is all that truth is known-as.

This thesis is what I have to defend. The truth of an idea is not a stagnant property inherent in it. Truth *happens* to an idea. It *becomes* true, is *made* true by events . . . .

Here rationalism is instantaneously up in arms against us. I can imagine a rationalist to talk as follows:

"Truth is not made", he will say; "it absolutely obtains, being a unique relation that does not wait upon any process, but shoots straight over the head of experience, and hits its reality everytime. Our belief that yon thing on the wall is a clock is true already, although no one in the whole history of the world should verify it. The bare quality of standing in that transcendent relation is what makes any thought true that possesses it, whether or not there be verification. You pragmatists put the cart before the horse in making truth's being reside in verification-processes. These are merely signs of its being, merely our lame way of ascertaining after the fact, which of our ideas already has possessed the wondrous quality. The quality itself is timeless, like all essences and natures. Thoughts partake of it directly, as they partake of falsity or of irrelevancy. It cannot be analyzed away into your pragmatic consequences."

The whole plausibility of this rationalist tirade is due to the fact to which we have already paid so much attention. In our world, namely, abounding as it does in things of similar kinds and similarly associated, one verification serves for others of its kind, and one great use of knowing things is to be led not so much to them as to their associates, especially to human talk about them. The quality of truth, obtaining *ante rem*, pragmatically means, then, the fact that in such a world innumerable ideas work better by their indirect or possible than by their direct and actual verification. Truth *ante rem* means only verifiability, then; or else it is just a case of the stock rationalist trick of treating the *name* of a concrete phenomenal reality as an independent prior entity, and placing it behind the reality as its explanation. Professor Mach quotes somewhere[31] an epigram of Lessing's:

Sagt Hänschen schlau zu Vetter Fritz
"Wie kommt es, Vetter Fritzen,
Dass grad' die Reichsten in der Welt,
Das meiste Geld besitzen?"

Hänschen [little Hans] here treats the principle of "wealth" as something distinct from the facts denoted by the man's being rich. It antedates them; the facts become only a sort of secondary coincidence with the rich man's essential nature.

In the case of "wealth" we all see the fallacy. We know that wealth is but a name for concrete processes that certain men's lives play a part in, and not a natural excellence found in Messrs. Rockefeller and Carnegie, but not in the rest of us.

Like wealth, health also lives *in rebus*. it is a name for processes, like digestion, circulation, sleep, etc., that go happily, though in this instance we are more inclined to think of it as a principle and to say the man digests and sleeps so well *because* he is so healthy. With "strength", we are, I do think, more rationalistic still, and decidedly inclined to treat it as an excellence pre-existing in the man and explanatory of the Herculean performances of his muscles.

With the word "truth" most people go over the border entirely, and then treat the rationalistic account as self-evident. But really all these words for *things* are exactly similar. Truth exists *ante rem* just as much and as little as the other things do . . . .

"The true", to put it briefly, is only the expedient in the ways of our thinking, just as "the right" is only the expedient in the way of our behaving.

5. RESPECT FOR CONCRETE DISTINCTIONS[32]

by Arthur Lovejoy[33]

My approach to realism was thus through a criticism of those idealistic philosophies prevalent when my reflection on these matters began; but this criticism, when its implications were examined, turned out to require, as its positive corollary, a conception of knowledge as indirect and substitutional; it involved, in short, a theory of representative ideas, at least with respect to cognition of past and future events and of the experiences of other selves. And with this the epistemological basis of every sort of idealism collapsed. If knowledge was, in fact, mediate in certain cases, it just might conceivably be mediate in some others - for example, in the case of sense-perception . . . .

In the usual epistemological sense of the term, a temporalistic or a pluralistic "idealism" would obviously not be idealism at all; for example, any consistent monadology is clearly a form of realism, since it implies that there are, between two ["windowless"] monads, real relations which do not fall within the experience of either of them or of any third monad. If, then, a *particular* knower at a particular time can truly know realities which do not have their existence within, or in dependence upon, his consciousness at *that time*; and if there may subsist any relations, even though they be relations between consciousnesses, which are not in any consciousness - then there is no decisive presumption to be drawn from what are purely epistemological considerations against even physical realism. Meanwhile, a vigorous revolt against idealism had been taking place in other quarters, motivated by other reasons. Some of these seemed to me valid, and therefore reinforced the argument based upon the temporalistic and pluralistic considerations I have mentioned. Especially sound and salutary seemed to me the attacks made by several writers upon the doctrine of the "internality" of all relations, which had been so important among the logical motives of post-Kantian idealism. This I had already come to reject as incompatible with both temporalism and pluralism, but I had not subjected it to the separate and direct analysis which was now brought to bear upon it by others.

Nevertheless, the theses that were most distinctive, and in some sense "new", in the neo-realistic movements which became so very conspicuous in American and British philosophy during the first decades of the century, I could by no means accept, and the primary reason for this was the opinion about the nature of knowledge that was implicit on the grounds upon which I had rejected idealism. The new types of realistic theory were profoundly influenced by William James's supposed discovery that "consciousness" as distinct from "content" or "objects", does not exist at all, or by G.E. Moore's supposed discovery, about the same time, that "consciousness" *does* exist, but is entirely distinct from its content and is incapable of generating or modifying its immediate data. If either of these premises were to be adopted, then it would follow that all knowing is immediate, that the "content"(or the "content" of consciousness) which is present in the cognitive experience is existentially identical with the object known . . . .

Now it appeared to me, for the reasons I have previously indicated, to be certainly false to say that when I recall an event of yesterday, or foreknow an event of to-morrow, or apprehend my neighbor's thought or his feelings, that these objects of my knowledge also literally "become themselves contents of (my) consciousness", in the same way in which my own present sense-datum or imagery, or feeling, is now a content of my consciousness. If yesterday's experience is in the past, it cannot be "numerically identical" either with the present event of my experiencing or with the content which is now existent simultaneously with my present experiencing. Attempts have sometimes indeed been made to meet this objection, either by suggesting that a present remembrance need have no compresent content - which seems to me to be an absurdity - or by conceiving of the past event as somehow persisting into the present. With respect to the latter method of evading the difficulty I am in agreement with the remark of Mr. Bertrand Russell:

> This is mere mythology. The event which takes place when I remember is quite different from the event remembered. People who are starving can recollect the last meal they ate, but the recollection does not appease their hunger. There is no mystic survival of the past when we remember merely a new event having a certain relation to the old one.[34]

In short, a consideration of memory alone would suffice, even if there were no other reasons, to lead me as a "temporalist" to reject epistemological monism, i.e. the generalized hypothesis of a direct knowledge without any intermediating "ideas", which is the essence of the "new" realism.

But in fact there are many other reasons. The most important of these is the utter inability of any theory which would regard knowledge as always being an immediate apprehension of its intended object to be able to account for error and illusion. The fact of error is the acid test for any epistemology; and the implication, in this regard, of neo-realism was all too truly indicated by the early American champions of that philosophy when they tried to identify it, "broadly speaking" with the theory - so "naive" that no one, I suppose, had ever previously asserted it - which "makes no distinction between seeming and being", but regards objects as "being precisely what they appear to be".

It is a further implication of the [neo-realist or behaviorist] theory that knowing is always a direct apprehension of objects which would just be the same if no cognitive event had occurred at all, that all data of perception, whether those commonly regarded as veridical or those called illusory, are genuine parts of the physical world. To this conclusion the neo-realist is led no doubt, by more than one logical motive. Indeed, the desire to eliminate the duality of "mental" and "physical", at least with respect to content of consciousness (in distinction from a possible function of merely "being conscious"), is now widely manifest in contemporary philosophy. It is of course a natural desire, since, if it could be satisfied, then the result would be a great unification of our conception of the nature of things - a unification in realistic instead of idealistic terms. It is manifested, therefore, in other contemporary realists such as Mr. Bertrand Russell, who reject epistemological monism. But here as well, I have been unable to follow the road which so many philosophical minds of my generation have found inviting and apparently satisfying. To explain the considerations which have deterred me from doing so would manifestly require, first, that I attempt to define the nature of the "physical world", and second, examine in detail each of the classes of experienced content which seem to me to resist inclusion in that world, when so defined. It would also require from me, in order to be relevant to the latest phase of the controversy, a discussion of Mr. Russell's recent ingenious hypothesis that our concepts and apparently all other contents [normally thought of as in consciousness] are "in our heads", and that, e.g., the bit of what I call the "blue sky" which is now present in my visual field is really part of my brain - a task made the more difficult because I find it hard to see how Mr. Russell's general theory [reducing everything physical to "sensibilia"] permits us to suppose that we have any members [of our bodies] at all which are properly describable in terms of sensibilia as "heads". These are, of course, too vast and complicated undertakings to be attempted here. But if I may assume that membership in the physical world means presence in a single, public, spatial, or spatio-temporal order, in which the laws of physics hold good, then [like Wm. P. Montague] I cannot see how the content of dreams, hallucinations, illusions, or even the contents of veridical perception as the epistemological dualist must conceive them, can be given a place in the world. This does not imply that the existence of these contents is not dependent upon processes themselves physical and conforming to the generalizations of the physicists. But physical science itself would be impossible if such entities as dream-monsters, the dipsomaniac's pink rats, or my present memory of vanished objects - all with the qualities and behavior which they are experienced as exhibiting - were placed upon the same footing as "scientific objects" - The distinction between the "mental", in the sense of that which *merely* "appears" in an individuated private field of content and the "physical", in the sense of an independent, orderly, perduring system of entities or "happenings" having dynamic or causal relations *inter se* of the kind which physical science often hypothetically describes, was perhaps one of the very earliest, and is still, I am disposed to think, one of the most indispensable of all the philosophical acquisitions of mankind. The "bifurcation" of experience, if not of "nature" (the use of the term in this connection is question-begging) is the beginning of wisdom in metaphysics.

This account will, perhaps, suffice to convey a very general notion of the primary considerations (though these are by no means all) which have brought me to that position mentioned at the outset, intermediate between the idealism accepted by most good philosophers when my acquaintance with philosophy began, and the type of realism enthusiastically adopted by so many good philosophers during the past quarter-century . . . .

6. RESPECT FOR ABSTRACT DISTINCTIONS[35]

by Morris R. Cohen[36]

If philosophy is viewed as love of wisdom or general knowledge, I may say that such a love was awakened in me between my seventh and tenth year by my grandfather, a poor tailor in the Russian town of Nesviesh. Though he never learned to write and had only the most moderate reading knowledge of Hebrew, he nevertheless had become the master of an extraordinary amount of knowledge and wisdom. Walks and talks with him first stimulated my imagination about the world at large and its long history. From him, also, I acquired a certain ineradicable admiration for the ascetic virtues and a scorn for the life of wealth, ease, creature comforts, and all that goes under the old name of worldliness.

For the rest, I had a varied education. I began like other orthodox boys, with the Bible, and then went on to the Talmud. The first non-religious book which I read was a Hebrew copy of Josephus. When in 1890, I was sent back to live with my mother in the larger city of Minsk, I began to read Yiddish books on Hebrew history and soon drifted into very secular romances. In 1892 I was brought to New York, and in 1895 I entered the scientific course of its City College, where, besides mathematics, physics, chemistry, and biology, I learned to appreciate French literature.

Possibly the circumstances of my youth in Russia, which prevented me from participating in the usual boyish games, emphasized my inclination to indulge in idle disinterested speculation. But in any case, my own limited experience is in accord with the Aristotelian view that philosophy grows out of our naive curiosity or wonder about the world at large. To philosophize has always seemed to me to be as natural and desirable in itself as to sing, to dance, to paint or mold, or to commune with those we love . . . .In the spring of 1899 I gained the friendship of Thomas Davidson.[37] His personal affection and his own touchingly unrestrained faith in my abilities increased my zest in life and in philosophic study. he aroused in me the great dream of a group of congenial spirits co-operating to create a philosophical encyclopedia that should do for the culture of the twentieth century what the Brothers of Sincerity did for the Saracen culture of the tenth or what Bayle, D'Alembert, and Diderot did for the culture of the eighteenth century. my heart was thus set on the systematic and comprehensive aspect of philosophy and I was led to read generally along diverse lines . . . .

It was the study of Bertrand Russell's *Principles of Mathematics*, some months after I was appointed to teach mathematics at the City College, that finally liberated me from the feeling of helpless philosophical bewilderment and enabled me to undertake an independent journey. The demonstration that pure mathematics asserts only logical implications and that logical relations or implications cannot be identified with either psychological or physical events, but are involved as determinants of both, seemed to me to offer a well-grounded and fruitful starting point for philosophy. For whatever the opinions of philosophers, they must rely on the validity of logical reasoning to establish their position; and at no time could I take seriously any attempt to question the fruitfulness of mathematical method in building up scientific knowledge . . . .

The doctrine that abstract logical or mathematical relations are real has justified for me the hypothetico-deductive procedure of science in which we follow the implications or effects of one single aspect or factor of a situation. For not all things that happen together are relevant to each other. Scientific search is difficult just because of the exuberant multiplicity of existential coincidences which are irrelevant to our inquiry as to the order of meaning. But if scientific inquiry is successful it discovers abstract relations which do characterize the world of phenomena. Thus instead of the two alternatives of either swallowing the whole universe or starving intellectually, the method of logical realities showed me how to bite into it.

In subsequent reflection this metaphor of "biting into the world" proved itself peculiarly apt in suggesting that opposing considerations must be taken into account in explaining anything, and indeed, that nothing is definite apart from numerous pairs of polar opposites such as form and matter, identity and difference, mediacy and immediacy, and so forth. The genuine existence of mathematical relations, and the principle of polarity, opened for me the path to systematic philosophy.

It was the discovery of the logical or hypothetical character of pure mathematics which would show me in a new and clear light the relation of the scientific method to induction, to skepticism, and to *a priori* rationalism. This would lead me to a much better understanding of the Medieval and still fundamental issue of conceptualism (or subjectivism) versus [ontological] realism.

Mill's contention that the truths of mathematics were approximations or inductions from experience became impossible after the distinction had been drawn between physics as a science of existence and pure mathematics as the development of the logical implications of *all* kinds of assumptions. That there are no two rational numbers whose ratio is the square root of two or π is not and cannot be an induction from experience. Experience alone cannot prove the absolute impossibility of things that have not occurred.

This disposed of the type of absolutist nihilism (which sometimes may call itself empiricism, relativism, or skepticism) which has denied the existence of all necessary truths and tries to maintain that even propositions like "2 plus 2 equals 4" are contingent as if on some other planet the result might well be different. One does not have to go to another planet to see that two pints of water and two pints of alcohol will not necessarily make four pints of the mixture. But the hypothetical truths of pure mathematics are not affected by empirical physical facts that do not conform to its postulated conditions. The truth of the proposition that "two and two equals four" (assuming the usual definitions and postulates of arithmetic) is not at all a question of physical observation, nor of psychological habit but of exact logical proof. It can be demonstrated that its denial involves self-contradiction, and thus such cannot maintain itself. It cannot, therefore, be the truth that all assertions are merely personal opinions; or if you call them opinions, it is certainly true that the opinion "2 plus 2 equals 5" is not as true as "2 plus 2 equals 4".

However, it was the same realization of the logical and hypothetical nature of pure mathematics which also destroyed classical *a priori* rationalism. It destroyed the ground of any assertion that we truly know a priori and with apodictic certainty that physical space must be Euclidean, just as the development of modern physics has rendered vain Kant's further argument that we know a priori that physical nature must follow the Newtonian laws of mechanics. Indeed, renewed faith in formal logic showed me that since all proof rests on assumption, it is vain for any philosophy to pretend to be able to prove all of its material propositions. It must make indemonstrable assumptions in regard to existence, value, or duty. This is particularly cogent against Kant's transcendental method, i.e. the attempt to prove certain propositions true because they explain how experience is possible. We cannot explain any experience or anything else without assuming something; and it is downright logical fallacy to assert that because our assumptions explain something that they are therefore demonstrably true. Obviously Kant does not and cannot offer any cogent proof that there may not be some other set of assumptions which will also explain the facts of experience.

That an assumption is not proved is, of course, not an argument against its truth. And so long also as assumptions will differ there will be ground for preferring the one that best explains the assumed facts. But the recognition of possible alternative assumptions bankrupts the pretension that philosophic assertions are all necessary and not merely probable truths. Doctrines as to the material nature of the world or of our duties in it can thus never be more than merely probable.

The principles or laws of every science are the rules or constant relations which hold amidst all the changes in its field; and so the fundamental laws or postulates of pure mathematics or logic are the invariant forms or relations which hold of all possible objects. Logical laws are thus neither physical nor mental but the laws of all possible significant being. No material fact can be deduced from purely formal considerations, but formal relations are assumed in inferring any fact from any others, and indeed no fact can be formulated except in terms of the forms or universal relations embodied in it. The world which is the object of science (since all science involves logic) is thus a union of form and matter. it is rational in the sense that its phenomena do conform to the laws of possibility that are objects of logic; but the element of chance or contingency can never be eliminated from it, since all proof must rest on unproved assumptions. The world might also be said to contain an irrational element in the sense that all form is the form of something which cannot be reduced just to form alone. The duality or polarity of terms and relations cannot be eliminated by reducing everything to terms alone, or to relations alone. We may say that everything which is intelligible can be expressed in logical form; but that which is expressible has no valid claim to absolute totality. Nor is there any contradiction in speaking [like Professor Wittgenstein] of the inexpressible, since it is of the essence of all expression to point to something beyond itself.

NOTES

1 Wilhelm Dilthey, *The Essence of Philosophy*, The University of North Carolina Press: Chapel Hill, 1956, pp. 72-75.

2 Wilhelm Dilthey (1833-1911) was born in Biebrich and was educated at Heidelberg and Berlin. After three other positions he became professor of philosophy at Berlin in 1882 where he advanced influential ideas on the understanding and philosophy of history in a neo-Kantian and Hegelian spirit. Our selection is from a rather late work *Das Wesen der Philosophie* which first appeared in 1907.

3 Wm. Pepperell Montague, "Confessions of an Animistic Materialist", in *Contemporary American Philosophy, Volume II*, ed. by George P. Adams and Wm. Pepperell Montague, Macmillan, circa 1930, [the first half only] pp. 135-147.

4 William Pepperell Montague (1873-1953) describes his own life and views before 1930 in the article above but we can add that he called his epistemology "subsistential realism", thought that the world was part of the body of God, and published the following books: *The New Realism* (1912), *The Ways of Knowing* (1925), *Belief Unbound* (1930), and *The Ways of Things* (1940). It is interesting that his much later namesake, Richard Montague, (1930-1971) was also quite interested in mathematics and symbolic logic, but apparently without the same feeling that he was wasting his time.

* My first philosophical article was, "A Plea for Soul Substance", *Psychological Review*, November 1899, was written with his kind advice and encouragement.

5 Josiah Royce(1855-1916) who was born in Grass Valley, California in frontier days and studied at the University of California and Johns Hopkins and taught at Harvard from the early 1880's until his death. His erudite and grandiloquent lectures on absolute idealism were long remembered, but his greatest influence on many students was probably his courses on the history of philosophy and his interest in symbolic logic. Royce and James in their conversations and debates appear to have stimulated each other in opposite but friendly ways.

6 Wilmon Henry Sheldon was born in 1875 and taught at several East Coast universities including Dartmouth and Yale. He attempted to overcome "the strife of systems" as he called it by developing a comprehensive synthetic worldview which is connected with his theory of polarity, a notion in logic which Morris Cohen also focused on. See Sheldon's *The Strife of Systems and Productive Duality* (1918). Much later publications include *Process and Polarity* (1944) and *Sex and Salvation* (1955).

7 George H. Howison (1834-1916) like Dilthey, Royce, and Bradley was part of the great idealist movement of the late 19th and early 20th century which had sprung up under the influence of Kant and Hegel and which attempted to defend many aspects of religion and ethics which they felt had been challenged by Charles Darwin's theory of evolution and the increasing "worldliness" partly inspired by the industrial revolution and the resulting rise in economic expectations. But Howison, who earlier had been at Harvard himself, strongly opposed Josiah Royce's emphasis on a unified absolute and preferred instead a league of independent minds as if becoming part of God were the goal.

8 Harry Allen Overstreet (1875-1970) began his career as a professional philosopher with works on *The Basal Principle of Truth Evaluation* (1904) and a book on Plotinus (1909) but in later years he became prolific writer in popular psychology, adult education, and philosophy of life. His best-known works are perhaps *A Guide to Civilized Living* (1934) and *The Mature Mind* (1950).

9 Frederick James Eugene Woodbridge (1867-1940) would become a leader of the Neo-Realists and one of the founders of the *Journal of Philosophy*.

10 Edwin B. Holt (1873-1946) like Woodbridge and even Montague who had temporarily abandoned epistemological dualism for a largely presentist monism, became prominent "New Realists", but in a sense their position and that of some of their "realist" colleagues was misleading since the

attempt to ignore consciousness or to deny it is real is equivalent in virtually all practical respects to alleging that consciousness is *all-inclusive*, which is very distinctly an idealist way of thinking. Genuine epistemological realists clearly accept consciousness as existent, at least in the sense of being mental and significant, but they restrict it in scope, and many also have begun to follow medical usage about its degree, stage, and type where dreaming, reflecting, alertness, shock, sedated conditions, opium-induced conditions, depression, exhaustion, concentration, excitement, worry, and coma, include different types and stages of consciousness, and which in turn can affect or be related to other activities in the brain or mind very differently.

11 Walter B. Pitkin (1878-1953) taught at Columbia from 1905 until his death. In addition to being one of the original six signers of the "Program and Platform of Six Realists" in 1910 from which the movement called "New Realism" arose he also published *The Psychology of Achievement* (1930) and *The Art of Learning* (1931)

12 Even though there was a Canadian philosopher of the time named John Watson (1847-1939), Montague is probably referring to John Broadus Watson (1878-1958), the famous psychologist and founder of behavioral psychology.

13 The distinction between "Scientific Realism" and "Metaphysical Realism" from 1980 to the present has many parallels with the older one between "New" or "Neo-Realism" circa 1910 and "Critical Realism" circa 1920.

14 Eddington's theory of quantized time allowed for fixed temporal intervals.

15 De Witt Parker, "Empirical Idealism", *Contemporary American Philosophy*, Volume II, edited by George P. Adams and Wm. Pepperell Montague, Macmillan, circa 1930, pp. 163-166.

16 De Witt H. Parker (1885-1949) taught many years at the University of Michigan. He was an idealist with a strong interest in aesthetics. See: *Principles of Aesthetics* (1920), *The Analysis of Art* (1926), *Human Values* (1931), and *Experience and substance* (1941).

17 This book by Andrew Dickson White (1832-1918) was originally published in 1896.

18 Ernst Renan (1823-1892) an ex-priest who became a leading French sceptic, historian, and literary stylist. See *La vie de Jesus* (1863), *Histoire des origines du christianisme* (7 vols. 1863-1883), and *L'Avenir de la Science* (written about 1850 but published in 1890).

19 David Friedrich Strauss (1805-1874) is best-known for *Leben Jesu* (1835-36) which described Christ in Hegelian style without miracles. He later criticized Schleiermacher's arguments in favor of Christianity and finally adopted a kind of scientific materialism.

20 John Fiske (1842-1901) became an influential philosopher and historian in his day. He was an avid supporter of Herbert Spencer's evolutionism which he combined with a vast, optimistic, providential theology. As a prolific historian, he strongly supported democratc government in its American form. His best-known book in philosophy was: *Outlines of Cosmic Philosophy* (1874).

21 Ernst Haeckel (1834-1919) was a leading zooligist and is usually acknowledged as Darwin's strongest and most influential supporter in Germany.

22 Paul Carus (1852-1919) is probably meant. He was the editor of *The Monist* and *Open Court* magazines which were dedicated to reconciling science and religion. Carus, who humself wrote several books on philosophy and religion was a good friend of Ernst Mach.

23 James Martineau (1805-1900) had this book published in 1890. He was English but of Huguenot ancestry and became a Unitarian minsister and hymn-writer. His sister was the somewhat better-known Harriet Martineau who became a Positivist and freely translated and condensed Auguste Comte's *Positive Philosophy* into English (1853).

24 Felix Adler (1851-1933) established the "Ethical Culture" movement in New York City with the aim of putting ethics at the center of religion. it is now an organization with societies in many different cities and even abroad.

25 This Greek expression meaning chance was apparently introduced into philosophy by Charles Sanders Peirce, a friend of William James. Peirce apparently believed for some time that chance

could be a literal feature of the universe, that is, as if there were areas with no causes or where causal explanation was not possible. This may have been partly suggested by Boltzmann's statistical interpretation of the second law of thermodynamics or by other statistical theories in science.

26 Robert Mark Wenley (1861-1929) attended the University of Glasgow and became a theology and biology focused philosopher. Much of his work shows an interesting literary style, especially in his biographical essays on Robert Flint and George Sylvester Morris.

27 Alfred H. LLoyd (1864-1927) was an out and out idealist who seemed more interested in the application of his philosophy than in spreading it. His books which included *Dynamic Idealism, an Elementary Course in the Metaphysics of Psychology* (1898) and his reply to William James titled *The Will to Doubt.* treated life as a kind of dialectical struggle. He also seems to have shared Hegel's view that human history is more and more the history of the development of human freedom.

28 Roy Wood Sellars (1880-1973) along with Arthur Lovejoy and George Santayana led the representist revolt against the presentist "realism" of Holt, Perry, Pitkin, and other philosophers who had signed or supported the "New Realist" manifesto of 1910. Sellar's book, *Critical Realism* (1916) even provided the title for the movement of 1920.

29 William James, "Pragmatism's Conception of Truth", *The Writings of William James,* edited by John J. McDermott, Modern Library/Random House: New York, 1968, pp. 429-430 & 436-438.

30 For Ernst Mach's correspondence with William James see Joachim Thiele, *Wissenschaftliche Kommunikation,* A Henn Verlag: Kastellaun, 1978, pages 168-176. Other letters by Mach are in the Houghton Library at Harvard University. See also Gerald Holton, *Science and Anti-Science,* Harvard University Press: Cambridge, 1993, pp. 1-55.

31 This editor also remembers the quotation but has dificulty finding it in Mach's works.

32 Arthur Lovejoy, "A Temporalistic Realism", in *Contemporary American Philosophy, Volume II,* ed. by George P. Adams and William. Pepperell Montague, Macmillan, circa 1930, pp. 98-100 & 101-103.

33 Reference to Mach and Petzoldt can be found in Lovejoy's most extensive contribution to epistemology and ontology, which first appeared in print at about the same time as the above essay. See his classic but difficult book: *The Revolt Against Dualism,* 2nd edition, Open Court: LaSalle, (1930) 1960, p. 99.

34 Bertrand Russell, *An Outline of Philosophy,* George Allen & Unwin: London, 1927, p. 198.

35 Morris R. Cohen, "The Faith of a Logician", *Studies in Philosophy and Science,* Henry Holt and Company: New York, 1949, pp. 3-4, 5, 7-8, & 9-11.

36 Morris R. Cohen (1880-1947) was once the best-known American Cohen in philosophy.

37 Thomas Davidson (1840-1900) was an enthusiastic philosopher and educator. On his coming to America in 1866 he was much influenced by William T. Harris in St. Louis and his Hegelianism. Later he helped organize the Aristotelian Society in London. Back in the United States he helped establish the Glenmore School for the Cultural Sciences (1889). In New York in 1898 he was associated with the People's Institute and also the Educational Alliance. He wrote books on Aristotle, Dante, and Goethe.

NAME INDEX
(includes text and footnotes)

J. Blackmore, R. Itagaki, and S. Tanaka (eds.). Ernst Mach's Vienna 1895-1930. 337-347.

Boston Studies in the Philosophy of Science

193. R.S. Cohen, M. Horne and J. Stachel (eds.): *Experimental Metaphysics*. Quantum Mechanical Studies for Abner Shimony, Volume One. 1997 ISBN 0-7923-4452-9
194. R.S. Cohen, M. Horne and J. Stachel (eds.): *Potentiality, Entanglement and Passion-at-a-Distance*. Quantum Mechanical Studies for Abner Shimony, Volume Two. 1997 ISBN 0-7923-4453-7; Set 0-7923-4454-5
195. R.S. Cohen and A.I. Tauber (eds.): *Philosophies of Nature: The Human Dimension*. 1997 ISBN 0-7923-4579-7
196. M. Otte and M. Panza (eds.): *Analysis and Synthesis in Mathematics*. History and Philosophy. 1997 ISBN 0-7923-4570-3
197. A. Denkel: *The Natural Background of Meaning*. 1999 ISBN 0-7923-5331-5
198. D. Baird, R.I.G. Hughes and A. Nordmann (eds.): *Heinrich Hertz: Classical Physicist, Modern Philosopher*. 1999 ISBN 0-7923-4653-X
199. A. Franklin: *Can That be Right?* Essays on Experiment, Evidence, and Science. 1999 ISBN 0-7923-5464-8
200. D. Raven, W. Krohn and R.S. Cohen (eds.): *The Social Origins of Modern Science*. 2000 ISBN 0-7923-6457-0
201. Reserved
202. Reserved
203. B. Babich and R.S. Cohen (eds.): *Nietzsche, Theories of Knowledge, and Critical Theory*. Nietzsche and the Sciences I. 1999 ISBN 0-7923-5742-6
204. B. Babich and R.S. Cohen (eds.): *Nietzsche, Epistemology, and Philosophy of Science*. Nietzsche and the Science II. 1999 ISBN 0-7923-5743-4
205. R. Hooykaas: *Fact, Faith and Fiction in the Development of Science*. The Gifford Lectures given in the University of St Andrews 1976. 1999 ISBN 0-7923-5774-4
206. M. Fehér, O. Kiss and L. Ropolyi (eds.): *Hermeneutics and Science*. 1999 ISBN 0-7923-5798-1
207. R.M. MacLeod (ed.): *Science and the Pacific War*. Science and Survival in the Pacific, 1939-1945. 1999 ISBN 0-7923-5851-1
208. I. Hanzel: *The Concept of Scientific Law in the Philosophy of Science and Epistemology*. A Study of Theoretical Reason. 1999 ISBN 0-7923-5852-X
209. G. Helm; R.J. Deltete (ed./transl.): *The Historical Development of Energetics*. 1999 ISBN 0-7923-5874-0
210. A. Orenstein and P. Kotatko (eds.): *Knowledge, Language and Logic*. Questions for Quine. 1999 ISBN 0-7923-5986-0
211. R.S. Cohen and H. Levine (eds.): *Maimonides and the Sciences*. 2000 ISBN 0-7923-6053-2
212. H. Gourko, D.I. Williamson and A.I. Tauber (eds.): *The Evolutionary Biology Papers of Elie Metchnikoff*. 2000 ISBN 0-7923-6067-2
213. S. D'Agostino: *A History of the Ideas of Theoretical Physics*. Essays on the Nineteenth and Twentieth Century Physics. 2000 ISBN 0-7923-6094-X
214. S. Lelas: *Science and Modernity*. Toward An Integral Theory of Science. 2000 ISBN 0-7923-6303-5
215. E. Agazzi and M. Pauri (eds.): *The Reality of the Unobservable*. Observability, Unobservability and Their Impact on the Issue of Scientific Realism. 2000 ISBN 0-7923-6311-6
216. P. Hoyningen-Huene and H. Sankey (eds.): *Incommensurability and Related Matters*. 2001 ISBN 0-7923-6989-0
217. A. Nieto-Galan: *Colouring Textiles*. A History of Natural Dyestuffs in Industrial Europe. 2001 ISBN 0-7923-7022-8

Boston Studies in the Philosophy of Science

218. J. Blackmore, R. Itagaki and S. Tanaka (eds.): *Ernst Mach's Vienna 1895–1930. Or Phenomenalism as Philosophy of Science*. 2001 ISBN 0-7923-7122-4
219. R. Vihalemm (ed.): *Estonian Studies in the History and Philosophy of Science*. 2001 ISBN 0-7923-7189-5

Also of interest:
R.S. Cohen and M.W. Wartofsky (eds.): *A Portrait of Twenty-Five Years Boston Colloquia for the Philosophy of Science, 1960-1985*. 1985 ISBN Pb 90-277-1971-3

Previous volumes are still available.

KLUWER ACADEMIC PUBLISHERS – DORDRECHT / BOSTON / LONDON

Printed in the United States
64430LVS00003B/4

9 780792 371229